ELECTRONS IN METALS

ELECTRONS IN METALS

AN INTRODUCTION TO MODERN TOPICS

C. M. HURD

National Research Council of Canada
Ottawa, Ontario

A WILEY-INTERSCIENCE PUBLICATION

JOHN WILEY & SONS, New York · London · Sydney · Toronto

Library of Congress Cataloging in Publication Data:
Hurd, Colin M
  Electrons in metals.

  "A Wiley-Interscience publication."
  Includes bibliographical references and index.
  1. Free electron theory of metals. 2. Metals.
I. Title.

QC176.8.E4H87   530.4'1   75-20241
ISBN 0-471-42220-7

Printed in the United States of America

10 9 8 7 6 5 4 3 2 1

PREFACE

As a field of science progresses, it becomes desirable periodically to introduce its new concepts and terminology to undergraduates and interested nonspecialists so that they are able at least to recognize the new ideas in the literature and the new directions in contemporary research. The object of this book is to do just this for that field known generally as the physics of metals.

My aim is to give a generally elementary introduction to the present understanding of the more familiar characteristics and properties that set metals apart from the other elements and that typify the metallic state. I have tried to produce an easily readable and descriptive book that I hope will give the average student a feeling for the direction of recent developments, as well as perhaps an alternative perspective of the fundamental concepts encountered by him in the more quantitative textbooks of his formal course. I am aiming primarily at advanced undergraduate and first-year graduate students planning to specialize in solid state physics, but I hope the book will also make useful supplementary reading for students following courses of engineering, metallurgy, and solid state chemistry—in short, those disciplines touching upon the study of metals.

The text assumes a knowledge of the Bohr theory of the atom, of the *aufbau* principle leading to the periodic table, and of the elementary principles of quantum mechanics; but it would be pretentious to say that it continues from any starting point, for no attempt is made at any systematic mathematical development. I have included only that mathematical development that seems to be essential for continuity in the text (to avoid, for example, the constant and irritating cross-reference to the standard textbooks), and that I consider a conscientious student can reasonably digest at a first reading. The price of this approach is that in some places standard material available elsewhere is repeated, and in other places the reader is asked to accept what is often an obscure result without the justification of theory. It seems that we lose on all counts—but this is the well-known difficulty of writing a descriptive view of what is essentially a highly mathematical subject.

It is a pleasure to acknowledge the continuous support of the National Research Council of Canada and the help I have received from various members of its staff; particularly Mr. J. Novak, who prepared most of the illustrations, and Mrs. H. S. Cuccaro and her staff of the Stenographic Services. A substantial part of the manuscript was written while I was a temporary member of the Département de Physique de la Matière Condensée of the University of Geneva. The helpfulness of particularly Profs. B. Giovannini, J. Sierro, and Mlle. M.-L. Zeier of that institution made my several months' stay there the more worthwhile. Conversations with Dr. A. Fert (Orsay) greatly improved my understanding of the subject of Sect. 3.4.6. Taylor and Francis Ltd. kindly gave permission to reproduce figures 5.3 and 5.6–5.11, which originally appeared in *Advances in Physics*.

No book of reasonable length can deal with all phenomena exhibited by metals, and I am aware that my choice of material cannot match the interests of all possible readers. The book was written from the rather selfish standpoint that I wanted to do it. Accordingly, the range of topics reflects a personal interest and does not pretend to be a balanced diet. In such circumstances one can only hope that the fraction of the readership disappointed by the selection will be the minority.

Each chapter is followed by a graded list of books and articles that I hope will be useful to the student interested in reading further afield. Standard textbooks are generally not included since I presume the student is familiar with them from his formal courses. These lists are intended rather to make known the various reviews—often nontechnical—that have been published in the more accessible journals. Throughout the text I have chosen to cite original research publications that mark particularly important advances or that illustrate the application of some particular concept. Not only should this practice extend the book's usefulness to a wider audience, but it could—and I hope it will—encourage the ambitious student to venture occasionally into the specialists' literature.

Suggestions, comments, and criticisms relating to any aspect of the following will always be welcome.

C. M. Hurd

Ottawa, Ontario, Canada
May 1975

CONTENTS

ELECTRONS IN METALS

1

PREAMBLE

Let us briefly set out the scales of the problems with which we will be concerned. First, there is the scale of dimensions. This refers to the rather well-known fact that modern science sees matter as organized into levels of structure. For example, if we study the mechanical properties of a lump of metal, we will observe the phenomenon called work hardening even on a macroscopic (everyday) scale using samples with dimensions of the order of centimeters. The modern understanding of this phenomenon is in terms of dislocations, which are regions of crystalline imperfection on the scale of $\sim 10^{-6}$ m. However, if we wish to understand the origin of a dislocation, we must examine its constituents and consider their behavior on the microscopic (atomic) scale of $\sim 10^{-10}$ m.

The description of work hardening, therefore, covers about eight decades on the dimension scale. These can be divided (although not with any well-defined separation) into two major domains: The macroscopic and the microscopic. This range of dimensions is typically that covered by solid state science in general and by metal physics in particular. However, if we include other important effects such as the nature of atomic structure, the origin of electron spin, or the electromagnetic force, then the range must be extended to the subatomic or even to the subnuclear level ($\sim 10^{-17}$ m).

Second, there is the scale of forces. This organization into levels of structure is seen not only in matter but also in the interactions* between its constituents at each level. There are four fundamental interactions in nature, and they account for all known natural phenomena (Fig. 1.1).

* In present-day physics the terms *force, interaction,* and *coupling* are used interchangeably.

1

FORCE	STRONG NUCLEAR	ELECTRO- MAGNETIC	WEAK NUCLEAR	GRAVITATIONAL
STRENGTH	1	$\sim 10^{-2}$	$\sim 10^{-14}$	$\sim 10^{-39}$
RANGE	SHORT $\sim 10^{-15}$m	∞ $1/r^2$ LAW	SHORT $\sim 10^{-15}$m	∞ $1/r^2$ LAW
EXAMPLE				

Fig. 1.1 Four fundamental interactions account for all known natural phenomena. The strong nuclear force, which binds the nucleons to form the nucleus, is the strongest; here it is represented by the coupling between a proton and a neutron. Next in order is the electromagnetic force, which acts on all particles possessing an electric charge; here it is shown as the attraction between a proton and an orbiting electron. Then follows the weak nuclear force, which governs the decay of certain radioactive nuclei but does not, as far as anyone knows, bind anything together; here it is shown as the decay of a nucleon by emission of a β particle. Last is the familiar gravitational force; here represented by an example on the cosmic scale. (After F. J. Dyson.)

The strong nuclear interaction is of very short range ($\sim 10^{-15}$ m, about the diameter of a strongly interacting particle) and binds the nucleons together in the atomic nucleus. The next force in order of strength is the electromagnetic force (or the electrostatic force if motion of the charged bodies is not involved). It acts on all particles that possess an electric charge, and its origin is considered in Sect. 4.3.1. Since it is one of the inverse-square-law forces of physics, its range is in principle unlimited. It binds electrons in space about nuclei to form atoms, it binds atoms to form molecules, and it binds atoms or molecules to form solids; it is thus essential in chemistry and biology. Next is the weak nuclear force, which (as far as is known at the moment) is of very short range and does not bind anything together, although it does control in some way the stability of elementary particles and thus the decay of certain radioactive nuclei. Finally, the fourth and weakest force is gravity. We are most familiar with this force from our everyday life; it has large scale effects and is always attractive. It is another inverse-square-law force and consequently has unlimited range.

It should be clear that each of these forces has special importance only within a limited range of the scale of dimensions. For example, although the gravitational force operates on the scale of atomic nuclei, its effects are quite unimportant compared with the strong nuclear force, which dominates the nuclear range of the scale. On the other hand, if the range is shifted to the cosmic end of the scale, such as in the study of galaxies or of the solar system, the gravitational force becomes completely dominant. The electromagnetic interaction is unique in that it has significant effects on both the microscopic and the macroscopic scales. Since the study of solids extends roughly between the macroscopic and the microscopic levels, the electromagnetic interaction is most important, and much of this book is a description of how this interaction is manifested in the metallic state.

Thus we see that limited ranges of the scale of dimensions have their appropriate interactions; they also have their own specialities in appropriate theories. In fact, a phenomenon may obey very different laws on different scales. The two outstanding examples of this, the discovery of which started the new era of modern* physics, are relativity and quantum theory. Motion on a cosmic scale is governed by the theory of relativity (on a macroscopic scale there is little difference from ordinary mechanics). However, on the microscopic scale with its processes involving distances typically of the order of 10^{-8} m or masses of the order of 10^{-30} kg, these processes are governed by the theory of quantum mechanics. It has been found that variables that on the macroscopic scale could theoretically take any of a continuous range of values are in fact restricted on the microscopic scale to a series of separated and discrete values. This is quantization: Nature has a graininess on the microscopic scale which is not obvious on the macroscopic one (rather as a newspaper photograph is resolved into discrete dots upon close inspection). But on the macroscopic scale the various effects that are a consequence of this quantization are completely obscured, and quantum theory dissolves into the classical theories of the macroscopic domain.

So we find that limited ranges have their own theories. A change of scale, therefore, usually requires a change of theory. This may seem straightforward enough—and it is if it is viewed as solely a problem of mathematical construction (no more problematical than, for example, extending the conceivable three-dimensional problem to the inconceivable n-dimensional one) but conceptually there are stumbling blocks. Take the quantum theory. It treats phenomena that occur on an atomic or

---

* The history of physics is divided into two great periods: The first, which began with Galileo and ended about 1900, is conventionally known as the classical period; the second as the modern. Roughly speaking, classical has come to mean nonquantal and nonrelativistic. Hence a nonclassical effect is one that cannot be understood without reference to either or both the quantum and relativity theories.

subatomic scale and are, therefore, not directly observable. To describe these phenomena we have only the concepts and language developed by our experience with the macroscopic world. When applied to the microscopic scale, macroscopic language and its concepts frequently seem inadequate or, worse still, downright misleading.

Perhaps the best known example of this is the wave-particle duality of quantum theory. On a macroscopic scale we have only two concepts— wave and particle—to describe the movement of matter in space and time. But when we use these concepts to describe the movement of matter on a scale where quantum theory is applicable, we find that they are inadequate in that they prove to be complementary: A particle can behave like a wave and a wave can behave like a particle, depending upon the circumstances. We shall encounter other cases of the inapplicability of concepts of macroscopic experience to the quantum scale; for example, the exclusion principle of classical mechanics does not carry over directly to become the Pauli principle (Sect. 3.2.2) and the classical concept of spin leads to complications when attributed to a quantum particle (Sect. 3.2.1).

All our concepts and explanations of phenomena that occur on a quantum scale (that is, on the atomic or subatomic level) are thus inevitably limited to analogies between the quantum phenomena and the more familiar macroscopic phenomena. The suitability of these analogies varies from case to case (as in the examples above) and it may even be quite undesirable to attempt such explanations at all. (See the remark by F. J. Dyson quoted in Sect. 3.2.1.) For example, the enigma of wave-particle duality arises from an insistence upon picturing quantum effects in terms of classical concepts. The correct approach is to accept that neither the wave nor the particle view is correct on the quantum mechanical scale; a more abstract entity is involved, one which is embodied and described by the rules of quantum mechanics but which is not describable classically. Put another way, a theory which requires such wave/particle complementarity has to renounce any attributes of their reality. Thus the initiate's typical question— "Which parts of the electron should I regard as wavelike and which as particlelike?"—shows that he has fallen into the pitfall of expecting real meanings for the wave and particle concept on the quantum scale. For some purposes the entity should be considered a particle, and for others a wave, but only the experience of applying the quantum mechanical rules can teach which of these macroscopic analogies is appropriate in a given instance. In fact, quantum theory is so at variance with intuition based on our macroscopic experience that even after nearly fifty years of successful application the experts still do not know what to make of it. (Struggling students may take some comfort from this!)

Although we must always put more reliance on equations than on any intuition drawn from analogy with our macroscopic world, it is a fact that the latter approach is well established and forms a large part of the physics of metals (with concepts like electron spin, the quasiparticle, and the exchange force, to list only a few examples). This book is written from a viewpoint which accepts this, and the aim in subsequent chapters will be to provide a descriptive coverage of some of these concepts and hopefully to make them available to a wider audience.

RELATED READING

Quantum mechanics has generated a vast and abstruse literature. The following two books give an entrance to this literature and are concerned with the meaning of quantum mechanics in the way it is touched upon above. Neither book is aimed at the novice, but both give generally nonmathematical descriptions.

J. M. Jauch, 1973, *Are Quanta Real?* (Indiana Press: Bloomington).

M. Audi, 1973. *The Interpretation of Quantum Mechanics* (Chicago University Press: Chicago).

A very readable description of quantum mechanical concepts given at a level appropriate to the present aim is

P. W. Atkins, 1974, *Quanta: A Handbook of Concepts* (Oxford University Press: Oxford).

2

INTRODUCTION

2.1 WHAT IS A METAL?

It seems pointless to try to start from a formal definition since metals mean different things to different investigators. As a physicist, I would define a metal as a material that possesses a Fermi surface. But metals can be liquid or solid, amorphous or crystalline, and what fraction of the people involved with the study or use of metals would in any case find such terminology meaningful? It is easy enough to write down a list of typical metallic characteristics that specify, in a general way, metallic behavior in a given context, but there is no reasonably succinct definition of metals that is acceptable to all the people who study or use them. A metallurgical engineer, for example, would stress that metals are materials that have particular malleability, ductility, and conductivity and that can be mixed and formed to vary these characteristics. A metallurgist, on the other hand, looks upon them as materials to be extracted from the earth's crust, purified, mixed together if necessary, and given some mechanical and thermal treatments to modify their bulk properties. To chemists, metals form a class of elements showing distinguishing tendencies in their interactions with other elements (and so occupying particular columns of the periodic table); in the solid state they provide the ultimate example of what is known as delocalized bonding between atoms. The physicist particularly stresses the atomic origins of a metal and tries to understand its behavior in terms of fundamental atomic parameters.

The outstanding characteristic of metals is good electrical and thermal conductivity. For example, the electrical resistivity (the specific electrical resistance) of the metallic elements at room temperature ranges from 1.5 (in the conventional units of 10^{-8} Ωm) for silver, the best conductor, to about 144 for plutonium, which is perhaps the worst. But even this latter has about 10^5 times better conductivity than the elemental semiconductors (such as silicon or germanium) which are the metals' nearest rivals. Moreover, the mechanisms by which an electrical current flows in a metal and in a semiconductor are fundamentally different; this is manifested in the temperature dependence of their resistances. In a semiconductor the electrical resistance decreases with increasing temperature, whereas, in a metal it increases—and at a roughly linear rate at moderate temperatures.

Consequently, even though we do not have a formal definition of a metal, these empirical features can serve to divide the elements into metallic and nonmetallic classes. In Fig. 2.1 we show that 68 of the 103 elements are conventionally classified as metals in their solid and liquid states at normal pressures. Four others are semiconducting in the solid state but become metallic when molten. The status of eleven others is as yet undetermined because of technical difficulties (although they are probably all metallic); at least four others are expected to become metallic if it ever becomes possible to apply sufficient pressure to a sample in the solid state.

Thus it is clear that among the elements metallic behavior is the rule rather than the exception.

2.2 AN ELEMENTARY VIEW OF A METAL: CORE AND VALENCE ELECTRONS

The aim of this section is to give, without attention to distracting details, an elementary physical view of a metal. Such an overall view at this point will hopefully provide the student with a useful framework within which to place the more specific aspects discussed in subsequent chapters.

First, let us recall the salient features of an isolated atom. It consists of a nucleus surrounded by a certain complement of electrons that are associated with it and localized about it in a specifically defined manner. It is a consequence of the Pauli exclusion principle (Sect. 3.2) that this cloud of electrons is not uniformly distributed around the nucleus but is concentrated in specific regions of space forming shells of charge each known as an atomic orbital. Each shell has a limited occupancy for electrons and corresponds to one of the strictly prescribed energy levels of an electron in the atom. We know that the total negative electronic charge surrounding the nucleus exactly balances the positive charge localized there; thus the isolated atom is electrically neutral.

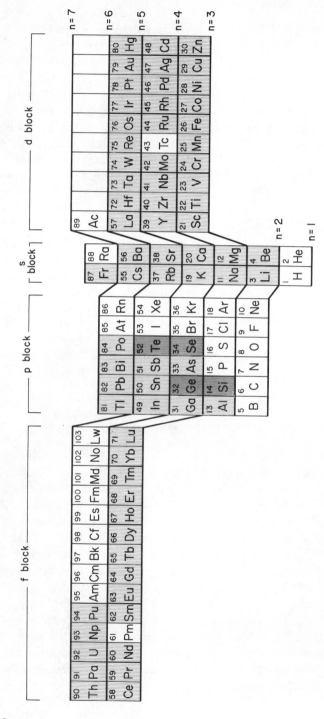

Fig. 2.1 Sixty–eight elements are classified as metals in their solid and liquid states at normal pressure (light shading). Four are semiconducting in the solid state but become metallic when molten (dark shading). The special cases of Tc, Pm, and the trans–Pu series (Am→Lw) probably have metallic phases too, but these radioactive elements do not exist in nature, and it is difficult to prepare them in adequate quantities. It is possible that other elements, such as H, P, C, and B, would form a metallic state if high enough pressure could ever be applied to them. Indeed, metallic H, which has significant military potential, has allegedly been produced recently in a Russian high–pressure laboratory (L. F. Vereshchagin et al. *JETP Lett.* **16**, 169, 1972. The significance of this particular arrangement of the periodic table is explained in Sect. 2.3.1.

8

If we imagine isolated atoms being brought closely enough together to form a lattice, clearly their outermost (or valence) electrons will interact most strongly with their neighboring counterparts as a result of the condensation. The inner (or core) electrons will be less affected, if they are affected at all. In a metal, these outermost electrons are, in fact, so affected by the change in their environment during the condensation that in the condensed state (a term which describes both the solid and liquid states) they are no longer effectively associated with any particular constituent atom. They form a group of electrons that belong to the metal as a whole; on the other hand, the core electrons are generally unaffected during the condensation and, therefore, remain bound to individual nuclei. Consequently, each constituent atom in the metallic state lacks the full complement of localized electrons that it had when isolated; in this state it is known as an ion and if we imagine it to be isolated, it will evidently have a net positive charge. We can, therefore, look upon a solid crystalline metal as having two structural units: The ions, which ideally are positioned to form a regular periodic structure of positive charges, and the group of delocalized electrons, which are distributed in some way throughout the interionic space.

Not only can these declocalized electrons move easily throughout the solid, but they can readily increase their energy in response to outside influences, such as an electrical field or temperature gradient.* In quantum language, we say that these electrons can be excited into energy states that lie above the normal unexcited levels of the system and that normally are vacant in the unexcited state. That such a group of electrons must exist in a metal is demonstrated by the universality of Ohm's law; if the electrons were even weakly bound to particular nuclei, Ohm's law would apply only above a certain threshold voltage, but this is not the case. It is this availability to the delocalized electrons of easily accessible, vacant, excited levels that is at the heart of metallic behavior.

The origin of a range of discrete energy levels into which the delocalized electrons can be excited lies in the quantization phenomenon referred to in Chapter 1. A particle that is restrained to move in a microscopic region of space has its kinetic energy quantized. For example, the kinetic energy

---

* If at this point the reader feels the need to have some picture of the underlying physical process, it is probably adequate to think of the electrons as hopping from interionic bond to bond in the solid under the impetus of the external influence. The energy transport is, in fact, a more complicated resonance process that involves the collective motion of all the delocalized electrons. Some consequences of attempts to retain the concept of an individual particle in dealing with such an electron liquid will be described in Sect. 4.2.

of an electron that is imagined to be confined in a box having impenetrable walls can take only those discrete values that are specified by the manner in which the corresponding wavelength of the electron fits into the space between the walls. It is instructive to consider here the specific example of an electron confined in a one-dimensional rectangular box (or square potential well) of length l. This is the situation of a delocalized electron in a linear hydrocarbon molecule, such as butadiene. Briefly, this molecule is a coplanar arrangement of four carbon atoms combined with six hydrogen atoms as shown in Fig. 2.2a. During the formation of the molecule most of the valence electrons become localized in the interatomic bonds lying in the plane of the molecule (and represented by the sticks in Fig.

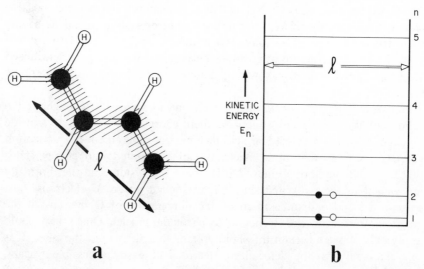

a   **b**

Fig. 2.2 (a) A ball and stick view of the butadiene molecule, which is a prototype metallic system. When the molecule is formed, each carbon atom (black circle) contributes four valence electrons, and each hydrogen atom (H) one. A total of twelve electrons from this pool are localized in the covalent C–C and C–H bonds (known as the σ–bonds, and here represented by the sticks). In an elementary picture these electrons can be regarded as distinct from the others (one per carbon atom) that are delocalized in the so-called π–bonds that extend both above and below the plane of the molecule (here represented by the shading). The π–electrons are confined to the molecule in the shaded areas by the attraction of the positively charged nuclei and, in fact, contribute significantly to the molecule's stability, but they are simultaneously free to range over the dimension l. The energy of these delocalized electrons is, of course, quantized. (b) A representation of the corresponding allowed energies. The availability to the delocalized electrons of vacant higher energy states is typical of the metallic state.

2.2). These are known as the σ-bonds; they are covalent bonds that are fully occupied when containing a charge equivalent to two electrons. But each carbon atom brings to the molecule one more valence electron than required by these σ-bonds. This electron becomes delocalized and occupies what are known as the π-bonds in the molecule. These are regions of space located over the carbon atoms, both above and below the plane of the molecule (shaded areas in Fig. 2.2a). The π-electrons are thus confined to the molecule by the attraction of the positively charged nuclei, but they are free to range over the linear dimension l spanned by the π-bonds. (Whether this dimension is straight, zigzag, or even a closed loop does not much matter.) A π-electron in this case is, therefore, just like one confined to a rectangular box, and we shall see that, as a result, the butadiene molecule forms a tiny prototypical metal.

The allowed energy levels E_n of an electron constrained by such a box can be derived straightforwardly as shown in many textbooks.* The result is

$$E_n = n^2 \left(\frac{h^2}{8ml^2} \right) \tag{2.1}$$

where n is a quantum number (Sect. 2.3.1) that can have only the integral values 1,2,3, . . . , m is the electron's mass, and h is Planck's constant. Figure 2.2b diagrams the arrangement of the first few energy levels in this system. Although it is premature at this point to discuss the occupation of these levels by the π-electrons, note that the Pauli principle (Sect. 3.2.2) allows no more than two electrons per level. Thus, if each carbon atom brings one electron to the delocalized pool in the molecule, their arrangement in the ground state will be that shown in Fig. 2.2b. As was emphasized above, the existence of the vacant, higher energy states available to the delocalized electrons is typical of the metallic state; thus we can look upon this molecule as a tiny one-dimensional metal. This does not imply that its metallic behavior can be demonstrated directly—an individual molecule is far too small for the attachment of electrical leads—but that the excitation of the delocalized electrons into the higher energy states can be inferred indirectly from the molecule's chemical reactivity and especially from its characteristic light absorption.†

---

* For example, L. Pauling and E. Wilson, *Introduction to Quantum Mechanics,* McGraw-Hill, New York, 1935, p. 95.

† The excitation by light of delocalized electrons in molecules of this class is responsible for pigmentation in biological systems. To give a few everyday examples, it makes grass green, butter yellow, and blood red.

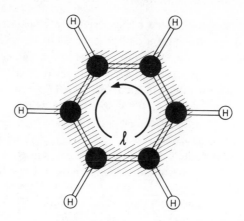

Fig. 2.3 Another ball and stick view of a molecule that is a prototype metal, in this case the familiar ring molecule of benzene. As in the case of butadiene (Fig. 2.2a) the equivalent of one electron per carbon atom is contained in the delocalized bonds, which are here shown as shaded areas. Again, these delocalized electrons are free to range over the dimension l and can be excited to vacant higher energy states represented in Fig. 2.2b.

There are other examples. In (Kekulé) benzene (Fig. 2.3) the molecular structure dictates that the π-bonds containing the delocalized electrons form a closed loop (these bonds are again shaded in the figure). The metallic circuit within the molecule is, therefore, also a closed loop, but this makes no difference to the overall argument: The allowed energy levels of the π-electrons are still given by Eq. (2.1) except that in this case l is the molecular perimeter. These electrons are effectively confined in a cylindrical box as contrasted with the rectangular one of a linear molecule.

We can imagine increasingly complicated linear molecules being built up as more and more atoms are added; the length l of the rectangular box would increase, and the corresponding energy levels in a diagram like Fig. 2.2b would also become increasingly compressed [compare Eq. (2.1)]. Ultimately the molecule would become large enough so that we could attach electrical leads in order to demonstrate directly its typically metallic behavior. This is one way of viewing any metallic crystal—as a gigantic one-dimensional molecule in which l has become so long that the energy levels are close enough together to form a practically continuous band of allowed energies throughout which the delocalized electrons are distributed. This is the elementary view of a metal known as the Sommerfeld model, which will be illustrated in greater detail in Fig. 2.7.

The group of delocalized electrons in a metallic crystal is known as either the itinerant or the valence group (the latter term emphasizes their atomic origin) or more simply as the conduction electrons. There are about 10^{29} valence electrons per cubic meter in a typical metal. They have an average separation of about 2×10^{-10} m and occupy up to 70 to 90% of the total volume of the metal. They move in all directions at high velocity, typically up to about 1.4×10^6 m/sec. However, because they interact with each other, with the ions, and especially with foreign atoms (inevitably present as impurities) and because there are other deviations from the ideal lattice structure, their lifetime between collisions is very short. (We remind the reader of the well-known result in intermediate theory that the deviations from the perfect periodicity of the ideal lattice are solely responsible for the scattering of the electrons. Thus if all the ions in an entirely pure metal could be perfectly static and located precisely at lattice points, there would be no scattering of the itinerant electrons by the structure and, therefore, no electrical resistance.) In silver of high purity, for example, the lifetime is about 4×10^{-14} sec at room temperature; this corresponds to a mean free path for the valence electron of 5.6×10^{-8} m, typically about 70 internuclear spacings. These figures give some idea of the typical magnitude of the different quantities, but we should note that they can vary enormously from metal to metal in different circumstances. For example, the mean free path at room temperature in a metal of normal commercial purity is typically 5×10^{-10} m, whereas it may be up to a million times greater in exceptionally pure metals cooled to the temperature of liquid helium.

In spite of all this activity among the valence electrons, there is no net transport of energy by them in a metal free from external influences. At any instant there are as many electrons moving in one direction as there are moving in exactly the opposite direction with the same momentum. It is only when an external influence upsets this balance that a net transport of energy occurs. An external electric field, for example, exerts a force upon all the electrons. Only those in the valence group are free to respond, however; as a result, if the applied field is steady, a certain mean velocity is superimposed upon the random motion of each valence electron. In copper, for example, a potential gradient of 10^2 Volt/m adds about 0.5 m/sec to the electron's velocity in the direction of the applied field.

The valence electrons in a metal collectively form what is known as a plasma. Because of its density, its particular resistance to compression, and the long-range, mutual interaction between electrons, this plasma behaves more like an electron liquid than like an electron gas. This behavior is the origin of the elementary view of a crystalline metal as a lattice of fixed ions immersed in a sea of itinerant electrons. Although this is a useful metaphor when considering electronic motion in metals, it does not

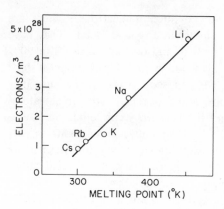

Fig. 2.4 The linear relationship between the melting points of the alkali series of metals and the density of their itinerant electrons illustrates the role played in the metal's cohesion by this group of electrons.

emphasize the important role played by the valence electrons in the solid's cohesion*; these electrons help to hold the ions in their lattice sites and, therefore, play a major role in holding the solid together. We become aware of the importance of this, at least for the alkali group of metals (Li, Na, K, Rb, Cs) which from an electronic point of view are the simplest of all metals, when we examine the relationship between the density of valence electrons in the metal and the temperature of the melting point. This examination reveals an essentially linear behavior (Fig. 2.4): The greater the density of itinerant electrons in the metal, the more rigidly the ions are bound in the solid, and the higher the melting point. (We should emphasize that, because other factors affect the melting point, this simple behavior is not generally observed for series of more complicated metals.)

This ability of the valence electrons in a condensed phase to be itinerant and simultaneously to be the bonding agents between contiguous atoms is expressed in the concept of the delocalized bond of organic chemistry. For example, in the butadiene and benzene molecules considered above, although the π-electrons are delocalized and can easily be excited to higher energy states, they also contribute significantly to the bonding energy of the molecule. In fact, they are largely responsible for the stability of the planar configuration, since the π-bonds resist any twisting of the atomic constituents out of the molecular plane.

The elementary view of a metal is, therefore, of a solid containing two distinctly different types of electrons: The itinerant (or valence) class, which have become effectively liberated from their atomic hosts during

---

* Note that the cohesive energy is defined as the energy required to separate the metal into isolated neutral atoms, but the binding energy is defined as that needed to separate it into isolated ions and isolated electrons.

the condensation process, and the ionic (or core) electrons, which are still localized about the atomic nuclei (more or less as in the isolated atom) and which are relatively immobile in the solid. Under normal circumstances the core electrons are unable to make long-range displacements, and their motion is restricted to the oscillatory thermal motions made by the ion as a whole about its equilibrium lattice site.

We should emphasize that this division of the electrons into valence and core groups is an oversimplification made for pedagogical convenience. We should not picture an ion in a metal as always surrounded by a definite boundary separating firmly bound electrons from entirely mobile ones; as we shall see in forthcoming chapters, the situation for many metals is not that simple. (Nevertheless, this approach is fairly widespread and is certainly applicable to certain metals such as the alkali group Li→Cs, for example. It leads to concepts such as the ionic radius, which is just a measure of the boundary referred to above.) Nor should we expect the valence electron fluid to be necessarily of uniform density throughout the interionic space. The valence electrons naturally seek out those regions of this space where their potential energy is a minimum, and we shall see (Sect. 5.3) that a nonuniform distribution of the more energetic electrons is a direct consequence of this behavior. Sometimes a slightly higher density of electrons is favored in regions centered on straight lines drawn between contiguous ions (just as with the build-up of charge in conventional chemical bonds); at other times the circumstances demand quite the opposite. The particular atomic properties of the constituent ions may favor denuding the regions mentioned above and building up charge in regions localized about the ions. In the chemical sense this is referred to as an antibonding configuration (the implication being that it is costly in energy and, therefore, detracts from the overall stability). However, in metals such a configuration can be just as bonding as in the more conventional configurations referred to above and can provide a link between ions that is just as strong.

2.3 FROM THE ATOMIC TO THE METALLIC STATE

The aim of this section is to give, again without unnecessary detail, a review of the important differences and similarities exhibited by metals that allow them to be further subdivided. The purpose is to provide a classification of metals that will be helpful, if rudimentary, in fitting into the overall scheme particular examples as they are cited in subsequent chapters. Just as in the elementary physical picture given in the preceding section, we shall continue to view a metal as a conglomeration of atoms brought together in such a manner that electrons belonging to the constituents are able to preserve their atomic characteristics to a greater or lesser extent. Although it is true that the metallic state is specifically characterized by the existence of a group of electrons sufficiently perturbed to have lost entirely their atomic characteristics, the relatively unperturbed ones (particularly the outer shells of the ion's core) play an equally important role in shaping the metallic properties. They provide most of the atomic legacy, and often singular features from the atomic state survive the condensation process and produce important and distinctive behavior in the metallic one. It should be clear from these remarks that our starting point will have to be the atomic state. Accordingly, we shall first review the salient features exhibited by metallic elements in the usual periodic classification of their atomic states, and then we shall turn (in Sect. 2.3.2) to a classification of the metallic state that will be based upon how these salient features fare in the condensation to the solid.

2.3.1 Trends in the Periodic Table.

Let us first review a few elementary features of the atomic structure of the elements. (Inevitably some of these remarks will anticipate results and concepts discussed in greater detail in subsequent chapters. Constant forward referencing seems unnecessary here, however; it is sufficient for present purposes to accept the various concepts at their face value.) We have already remarked in Sect. 2.2 that an atom consists of a nucleus around which are localized one or more electrons. In an extreme case of charge correlated motion, the electrons are controlled absolutely by the nucleus and are retained in a region of space that is known as an atomic orbital. Such orbitals can occupy only strictly prescribed regions of space, and these are dictated by the rules of quantum mechanics. They form shells of charge about the nucleus, and the orbitals of increasing effective radius generally correspond to electron states of increasing energy. (There are exceptions to this statement, the most notable being the $4f$ shell of the rare earth lanthanide series, which is discussed in Sect. 2.3.2.) The energies of the orbitals vary in steps as the radius is changed; within a shell there is a fairly narrow range of energies between the two extreme orbitals, but larger differences (typically of the order of 1.6×10^{-19} J (1 eV); see Table 1) exist between the individual shells. Any orbital can accomodate up to two electrons, providing they are of opposite spin.*

---

* Rather than anticipate the discussion of Sect. 3.2.1, let us say simply that this spin expresses one of the two possible internal alternatives that the electron has chosen.

Table 1 Scale of Energies in Solid State Physics

Description	Order of Magnitude	
	eV	10^{-19} joules
Average energy of a plasmon	10 to 20	16 to 32
Intra-atomic Coulomb repulsion U of eqn. (3.22)	up to 10	up to 16
Intra-atomic exchange energy. (Type 1 of Fig. 3.9)	$\sim$0.3 to 3.0	$\sim$0.48 to 4.8
Widths of conduction bands in metals	$\sim$1 to 5	$\sim$1.6 to 8.0
Energy per itinerant electron in a metal due to Coulomb interactions (Coulomb correlation energy)	$\sim$1	$\sim$1.6
Separation between energy levels in an atom	$\sim$1	$\sim$1.6
Energy gap between first and second energy bands in a metal	$\sim$1	$\sim$1.6
Direct inter-atomic exchange energy (Type 2 of Fig. 3.9)	$\sim$0.1 eV and 1.0 eV for $3d$ and $4f$ electrons, respectively.	
Spin-orbit coupling (Sect. 3.2.3)	As low as 0.002 eV for elements lighter than the rare earths, but increases rapidly with atomic number thereafter reaching $\sim$1.0 eV for heavy elements like Th or U.	
Typical phonon energy ($k_B\theta_D$)	$\sim$0.025	$\sim$0.04
k_BT at room temperature	$\sim$0.025	$\sim$0.04
Average energy of a phonon	10^{-2}	0.016
Superexchange energy (Type 4 of Fig. 3.9)	$\sim$7 $\times$ 10^{-3}	$\sim$0.011
Energy change per electron upon superconducting condensation	10^{-3}	1.6 $\times$ 10^{-3}
Zeeman energy of an ion in 3T(30 kG)	10^{-3}	1.6 $\times$ 10^{-3}
Energy of an itinerant electron in a metal due to electron-phonon interactions	10^{-4}	1.6 $\times$ 10^{-4}
Binding energy of a Cooper pair	10^{-8}	1.6 $\times$ 10^{-8}
Separation of the delocalised electrons' energy states in a metal	10^{-10}	1.6 $\times$ 10^{-10}

An electron's assignment among these orbitals is designated by the four labels n, l, m_l and m_s. The first, known as the principal (or main) quantum number, relates to the total energy of the corresponding orbital state; roughly speaking, it can be considered a measure of the orbital's effective radius. In the ideal one-electron system (Sect. 3.2.3) without external electric or magnetic fields, this quantum number alone is sufficient to specify the state's energy. In these circumstances a given value of n, which can be one of 1,2,3. . . , designates a particular shell of orbitals of equal energy; the other three quantum numbers specify the various possibilities existing within this shell. (Table 2 summarizes the various orbitals available in successive shells.)

Table 2 Orbitals Available in Atomic Shells

| | l Subshell | | | | | |
| | $l = 0$ | 1 | 2 | 3 | Total Number | Total Number of |
n Shell	s	p	d	f	of Orbitals	Electrons Permitted
1	1	–	–	–	1	2
2	1	3	–	–	4	8
3	1	3	5	–	9	18
4	1	3	5	7	16	32

These different possibilities that are available to an electron characterized by a given n arise because it has angular momentum: It has orbital angular momentum because of its orbital motion about the nucleus (in a corpuscular picture), and it has an intrinsic angular momentum because of its internal degree of freedom, which we have called spin. Furthermore, since it is a charged particle and since a moving charge generates a magnetic field, each of these angular momenta will have an associated magnetic dipole moment. The magnitude of this moment will be proportional to that of the angular momentum that produces it, and in a vector representation the two quantities will be colinear. (Note, however, that the electron carries a negative charge so that in terms of the usual convention the dipole moment and angular momentum vectors are opposed in both the orbital and intrinsic cases.) On this microscopic scale, angular momentum—whether orbital or intrinsic—is quantized: Its magnitude can have only certain discrete values, and its vector is permitted to take only certain discrete configurations. Obviously, the concomitant magnetic dipole moments are correspondingly restricted. These different quantum configurations make up the different states possible for a given n. They form what has been called a shell of orbitals; each configuration corresponds to the same energy in the field-free case, but each can be resolved in energy by applying electric or magnetic fields.

The second label l in the above list is known as the angular momentum (or azimuthal) quantum number since it is a measure of just that. The smaller the value of l, for example, the less the angular momentum of the electron, and the greater the probability that the electron will be located near the nucleus. Therefore, l determines the form of the electron's probability distribution about the nucleus; in other words, it determines the shape of the atomic orbital.* l can have one of only $0, 1, 2, \ldots, n - 1$, and those particular orbitals having $l = 0, 1, 2$, and 3 are known respectively as the s, p, d, and f orbitals† (this usage having been carried over from an older spectroscopic notation; for $l > 3$ alphabetical order is followed except that j is omitted). The set of all possible l values for a given n forms what in earlier descriptions was often called an l subshell; however, the use of this pedanticism is declining so that today we often see reference to s shell, p shell, and so on—even though they are strictly subshells of a given n shell. Henceforth we shall follow this modern trend.

In the case of a multielectron system (as opposed to the one-electron system cited above) this variation in the probability distribution leads to a hierarchy of energies among the various l values within a given n shell. This is caused by the mutual interaction of electrons in the atom; the energy of an electron is lowered as it penetrates the charge cloud of the other electrons surrounding the nucleus. Consequently, for a given shell we can expect orbitals of lower angular momentum to be more stable (that is, to have a lower energy) than those of higher momentum. Thus an s level has a lower energy than a p; a p level is lower than a d, and so forth. Generally speaking, energy differences specified by the third and fourth quantum numbers described below are smaller than that produced by a variation of l. In fact, the destabilizing effect of an increasing l is evident in the periodic structure of the elements, as will be shown below.

The third label m_l is known as the magnetic quantum number. It specifies that component of the electron's orbital angular momentum that appears along the (arbitrary) direction of an externally applied magnetic field. This is generally pictured at an introductory level as resulting from

---

* The shapes of atomic electron distributions are discussed and illustrated in Appendix A.

† This s should not be confused with the spin quantum number to be introduced below.

the classical precession of the orbital's magnetic dipole (and, hence, its orbital angular momentum vector) about the applied field's direction. The time-averaged result for both the angular momentum and the magentic moment is zero for all directions except that defined by the field. Furthermore, quantization restricts the inclination of the precessing moment. It is forced to make discrete and fixed angles with the field's direction, and this produces the discrete values of the magnetic moment (and hence the orbital angular momentum) which are observed along that direction. (Compare Fig. 2.5.) This quantization is such that m_l can have values $\pm l, \pm (l - 1), \pm (l - 2), \ldots, 0$. The magnitude of the orbital angular momentum observed along the field's direction is actually $m_l \hbar$ (where $\hbar = h/2\pi$), and that of the corresponding magnetic moment is just $m_l \mu_B$ (where μ_B is the Bohr magneton).

Finally, s, the electronic spin quantum number, is discussed at some length in Sec. 3.2.3. It is sufficient to note here that it is always equal to 1/2 and that it characterizes the intrinsic angular momentum possessed by the electron. The origin of this momentum is not yet fully understood, but it is a manifestation of certain internal alternatives available to the electron. Again, an intrinsic magnetic moment is associated with this spin angular momentum, and this moment, together with the associated angular momentum vector, can be imagined to precess about the direction of any externally applied magnetic field. The inclination of these vectors to the applied field is again restricted by quantization, and the quantum number m_s, known as the spin projection quantum number, is introduced to characterize the component of the spin angular momentum appearing along the applied field's direction. Thus the magnitude of the spin angular momentum observed along this direction is $m_s \hbar$ and the corresponding magnetic moment is $m_s g \mu_B$; g is the so-called g-factor, which, if the classical relationship between the spin angular momentum and its corresponding magnetic dipole moment held, would be unity and, therefore, a redundant parameter. However, unlike the orbital case, g is not redundant here and is approximately 2. The particular feature of the electron's quantization is that m_s can only have the values $\pm s$, that is $\pm 1/2$; the spin can only be either up or down with respect to the applied field's direction, and these two cases correspond to the two internal degrees of freedom possessed by the electron. (Figure 2.5 is a typical illustration of the macroscopic analogy of these quantum effects.)

When considering the interaction of atoms to form a solid, we have to take into account the nature of their interacting orbitals. The most important orbitals directly involved in chemical bonding (and, therefore, those of importance in the formation of the solid state) are the $s, p,$ and

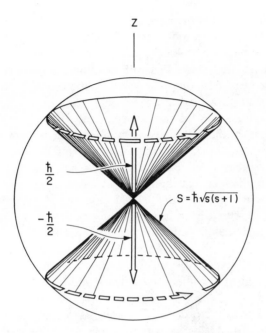

Fig. 2.5 An example of the macroscopic analogy made to describe the quantization of angular momentum—in this case the intrinsic spin angular momentum of an electron. The magnitude of this quantity, here designated S, is found from quantum mechanics to be $\hbar[s(s+1)]^{1/2}$ where s is the electronic spin quantum number. A spinning charge has inevitably a net magnetic dipole moment; as the text explains, for an electron this moment points exactly opposite to **S** and has a magnitude of $g\mu_B[s(s+1)]^{1/2}$. When an external magnetic field is applied to the system to define a particular direction (conventionally the z axis), this magnetic moment—and its precursor angular momentum vector—precesses about the applied field's direction. The time-averaged value of these quantities is a zero along all directions except the z axis, so that to an external observer both the angular momentum and the magnetic moment only exist along the applied field's direction. Furthermore, quantization restricts the angle of inclination of these vectors to the z direction such that it can take only constant and discrete values. The components of these vectors along the z direction are consequently limited to discrete magnitudes. For the angular momentum and magnetic moment these are $m_s\hbar$ and $m_s\mu_B g$, respectively, where m_s can be only either $\pm 1/2$. The diagram shows the two possible configurations of the spin angular momentum when a magnetic field is applied along the z axis to resolve them.

(to a lesser extent) d orbitals, since for many elements one or more of these types forms the outermost shells of the atom. Figure 2.1 shows an appropriate form of the periodic table* in which the elements are divided into four blocks (s block, p block, and so on); these are determined by the l subshell currently being filled with electrons as the atomic number is increased to the element in question.† We shall see below that the subshell being so filled is not in all cases the outermost shell of the atom.

The vertical separations between the horizontal n lines forming the basis of Fig. 2.1 represent the potential energy of the electrons in the respective n,l subshells. For example, the effect of the destabilization caused by the occupation of orbitals of increasing l values, which was referred to above, is manifested by the fact that the energy of the $n = 3$, $l = 2$ level (which is being filled continuously in the Sc→Cu sequence) is slightly higher than the $n = 4$, $l = 0$ level of the K→Ca sequence. In other words, the influence of the angular momentum is sufficient to put the energy of an $n = 4$ level below that of an $n = 3$. In fact, this is but one example of a general rule among the elements (see Fig. 2.1): For a given value of n the n,d energy level is higher than the $n + 1,s$ level, and the n,f level lies between the $n + 2,s$ and $n + 2,l$ levels.

Let us turn now to the order of filling the atomic orbitals as the atomic number is increased. It is shown in Appendix A that an s orbital is spherically symmetrical. It, therefore, has only one spatial form, and it can accommodate a maximum of two electrons, and then only if they have antiparallel spins. There are, however, three spatially distinct p orbitals (Fig. A.1 of Appendix A) each of which can accomodate two electrons of opposite spin. As these are filled, the Coulomb repulsion between the electrons spaces them to the greatest extent possible. This is achieved by avoiding, as far as possible, full (that is, double) occupancy of any orbital. Furthermore, the intraatomic exchange force (which is described in Sect. 3.3.3) dictates that the electrons shall have the maximum spin multiplicity that is consistent with the Pauli principle, which allows the double occupancy. Thus if we imagine the six p orbitals to be divided into two groups of three, each of a given spin direction, the first three electrons will occupy the spatially different orbitals of a given spin orientation. Only the addition of more electrons will populate the orbitals of the other spin direction.

---

* After D. G. Cooper, *The Periodic Table,* Butterworth, London, 1968 and W. F. Sheehan, *Physical Chemistry,* Allyn and Bacon, Boston, 1961.

† The allocation to a given block is somewhat arbitrary in certain cases. For example, it might seem illogical to put the Hg→Zn column in the d block, but it corresponds to the inclusion of the Rn→Ne column at the end of the p block.

A consequence of all this is that a half-filled shell and a completely filled shell have special stability.

This can be seen in Fig. 2.6, which shows, in terms of the arrangement of Fig. 2.1, the distribution of the higher-energy electrons in the neutral atoms. In the s and p blocks the valence shells are outermost and the only incomplete shells in the atom. Without exception they are systematically filled as the atomic number increases. But in the five groups of elements forming the d and f blocks the situation is complicated by the presence of more than one shell of roughly the same energy; in the d block it is the d and s shells, and in the f block it is the $d,f,$ and s shells. In these five groups the s shell has generally the lower energy and is, therefore, filled before any electrons occupy the d (or f) levels. We, therefore, find vacancies in a shell inside the valence electrons' s orbitals in these cases. After the s shell is filled, however, these vacancies are progressively filled as the atomic number increases—although with the interruptions described below. Specifically, these five groups are the ion group (Sc→Cu; incomplete $3d$ shell) the palladium group (Y→Ag; incomplete $4d$ shell) the rare earth lanthanide group (Ce→Yb; incomplete $4f$ shell) the platinum group (La→Au; incomplete $5d$ shell) and the actinide group (Th→Md; incomplete $5f$ shell). The interruptions in the systematic filling of the d and f shells in these series occur notably in the middle of the sequences, and they reflect the particular stability of the half-filled shell, which we commented upon above. For example, in the iron group the lowest energy configuration for Sc (atomic number 24) has five $3d$ electrons (which just half fill the shell) but only one $4s$ electron. Further along the same series, Cu (29) takes ten $3d$ electrons (which just fill the shell) and only one $4s$ electron; this is another example of the same effect. In the f block, Gd (64) prefers to promote an electron to the $5d$ level (and so retain the half-filled $4f$ level) rather than to accommodate all eight electrons in the $4f$ shell; Cm (96) is believed to behave similarly.

The important lesson that we should learn from all this is that we can expect complications when we condense a group of the d and f block elements to form a solid. It is evident that these atoms have more than one shell containing electrons of roughly equivalent energies, even though these shells generally have quite different shapes and spatial distributions. When we bring the atoms together in such a way that their outer orbitals interact and, in the case of metals, result in the delocalized bonds, it is unlikely in the cases of the d and f blocks that we can consider the interaction of the valence s (or p) orbitals in isolation. For example, if the $4s$ orbitals in Cu (29) form the delocalized states in the metal, we should not be surprised if the $3d$ orbitals are also influential, for we know from the atomic

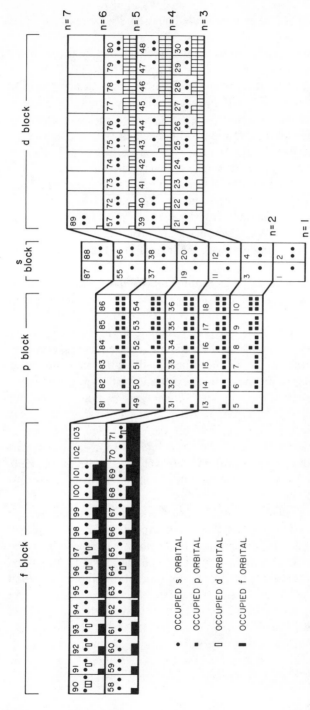

Fig. 2.6 The atom's valence electrons, which are those in the outermost incomplete subshells, are responsible for most of its distinctive chemical and physical properties. Since they suffer the greatest interaction with neighboring counterparts in the solid, they also control to a large extent the type of bonding adopted. Here the occupancy of the higher energy states of the natural atoms is illustrated in the same layout as Fig. 2.1 (refer to it to identify the elements). In the *s* and *p* blocks the valence shells are systematically filled as the atomic number increases, but in the *d* and *f* blocks the situation is complicated by the fact that the *s* and *d* (or *s*, *d*, and *f*) orbitals have roughly equivalent energies in some cases, particularly in the middle of the block. The adoption of the alternative of lowest energy by the electrons leads to interruptions in the systematic filling of the *d* and *f* shells. Permanently filled shells in the various blocks are, of course, not shown. There is considerable uncertainty about the outer electron distributions in the actinide series (89 → 94; Ac → Pu). There is some evidence suggesting that no 5*f* electrons at all are present in the earlier elements (89 → 92), and only one in Np (93).

state that these orbitals have somewhat similar energies—and this is the first requirement for resonance interaction between any two systems. (Compare Fig. 2.9.) We shall not now provide any further examples, since the important ones will be discussed as they arise in subsequent chapters. Let us turn instead to the main objective of this chapter.

2.3.2   An Elementary Classification of Metals.   It is convenient to think of a metal as being divided into atomic cells. Each cell is imagined to be centered upon a nucleus and is defined to contain all the points in the metal that are closer to that nucleus than to any other. Such a cell is known as a Wigner–Seitz cell or an atomic polyhedron. (It obviously has to be a polyhedron since a sphere would not lead to a space-filling assembly, although theoreticians often for convenience assume a sphere.) The cell is constructed by drawing the planes that perpendicularly bisect each line joining the chosen nucleus with its nearest equivalent neighbors. Clearly, the actual polyhedron obtained depends upon the symmetry of the ionic lattice, and standard textbooks show the commonly encountered examples.*

Sect. 2.2 has given us an idea of the occupancy of an atomic cell in a typical metal. First, it will contain the core electrons. These are localized relatively close to the nucleus, controlled absolutely by it even in the metal, and are thus effectively unchanged from the atomic state. They occupy typically about 5–10% of the cell's volume, but they are not of great importance in the present context since they make no direct contribution to the delocalization of the electrons caused by condensation. We shall henceforth ignore them. This core is surrounded by what in the atomic state are the outermost and valence electrons. These are the electrons that range far enough from the nucleus to interact appreciably in the metal with their counterparts around neighboring nuclei. They are our exclusive concern in the present context.

Except near the boundaries of the atomic cell, these electrons move in an environment that is, in fact, not greatly different from the atomic state. But near the boundaries they interact to varying degrees with those of contiguous cells, and in that region their environment changes markedly from the atomic situation. Put into more technical language, we can say that near these boundaries the electrons experience a different potential (and their wave functions have to satisfy different boundary conditions) from the atomic case. Well away from the boundaries, however, they experience a potential that is similar to that in an isolated atom.

---

* For example, S. Raimes, *The Wave Mechanics of Electrons in Metals,* North-Holland, Amsterdam, 1961, p. 247.

The degree of interaction between electron states in adjacent cells essentially depends upon the density of the corresponding atomic states near the cell boundaries, and this in turn clearly depends upon the shape and range of the relevant atomic orbitals. If these parameters for a particular outer orbital concentrate the electron near the nucleus (thus leaving only a small probability of finding it at the cell's boundary) there will be correspondingly little modification of the electron's state through overlap and interaction with those in surrounding cells. The electron's environment in this case will still be dominated by the influence of the nucleus and the core electrons. On the other hand, if a valence electron's orbital concentrates it generally outside the core, giving a high probability density in the region of the cell's boundaries, then in the metallic state that electron will be dominated by the new boundary conditions referred to above. The influence of the nucleus and the core electrons in this case becomes incidental, or at least only a weak perturbation.

This variation in the degree of interaction between the outer electrons in adjacent atomic cells can be used to devise a useful classification that also ties in directly with the separation of the periodic table into the blocks shown in Fig. 2.1. However, for this purpose we can combine the s and p blocks (for reasons outlined below) so that we have just three categories: d, f, and s/p (s or p). The unfilled shells in each of these classes are composed of atomic states that interact to markedly different degrees in the metallic state. Accordingly, we shall distinguish the following three types of atomic state(s):

Type I   States that concentrate the electron density deep within the atom's core, leaving correspondingly little density in the atom's outer regions. In the metal, the electron density produced by these states is very low at the boundaries of the atomic cell.

Type II   States whose characteristics are the opposite of Type I; namely, those leaving a large electron density at the furthest extremity of the atom and correspondingly little within its interior. In the metal, the electron density produced by these states is very high at boundaries of the atomic cell.

Type III States whose characteristics are intermediate between Types I and II. These produce a relatively moderate electron density at the boundaries of the atomic cell.

The manner in which these different types of orbital behave when identical atoms are condensed to form a metal is illustrated in Fig. 2.7. The Type I states experience no appreciable interaction between their

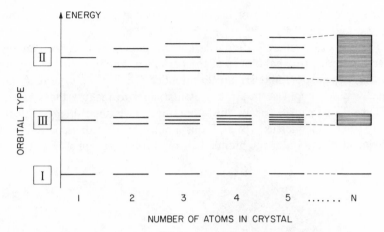

NUMBER OF ATOMS IN CRYSTAL

Fig. 2.7 The formation of energy bands through the repeated addition of identical atoms to form a metallic crystal. Three types of atomic orbital are distinguished in the text depending upon their electron density at the boundaries of the atomic cell. In a Type I orbital, which is typically one deep within an atom's core, there is no appreciable interatomic interaction and, therefore, no shifting of the corresponding energy level as atoms are added. In the crystal these N levels are degenerate in energy. The other two types, however, suffer interaction to differing degrees with the result that the energy levels are shifted slightly as each atom is added, the number of separate states in the band being equal to the number of atoms present. In the metallic crystal these discrete but closely separated levels range between upper and lower limits, and this range of energies is known as the energy band. (After F. Seel: "Atomic Structure and Chemical Bonding", 1963.)

orbitals in adjacent cells, and consequently the energy of these orbitals remains essentially unaffected as atoms are added to the crystal. In the solid, these levels all have the same energy.* The Types II and III, on the other hand, do experience appreciable interaction with adjacent cells. In these types the energy of every interacting atomic orbital is shifted slightly from its unperturbed value so that the number of discrete states in the system is equal to the number of atoms contributing an interactive orbital. These discrete states span a range of energy that has upper and lower limits. As more atoms are added, more orbitals are added to the interacting set, and the spacing of these discrete levels becomes less and less. Eventually, since a typical metal has 10^{29} atoms per m^3, the spacing becomes so small (about 10^{-29}J or 10^{-10}eV; see Table 1) that the energy levels form what is effectively a band of continuously varying energy between the prescribed limits. This is known as an energy band; its width

* In the jargon of the subject they are said to be degenerate in energy.

is a measure of the degree of interaction between the constituent orbitals (Fig. 2.8 and Fig. 2.9).

To return to the classification of states introduced above, we can relate this to the three principal blocks of the periodic table by noting that atomic states of Types I, II, and III are exemplified by the f, s/p, and d states, respectively. It might seem strange at first sight to include the s states in the Type II category, for in the atom they have their highest electron density near the nucleus. But at this point we are concerned with the behavior of electrons in the atomic cell of a condensed phase, as opposed

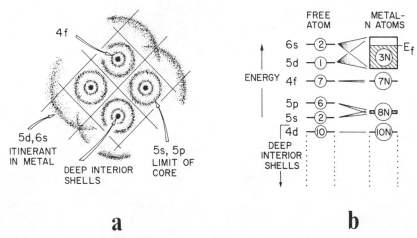

a   b

Fig. 2.8 Three classes of atomic orbital are distinguished in the text depending upon whether their electron density at the boundaries of the atomic cell is high, low, or intermediate. This simplified sketch shows the salient features of these three classes and their consequences for the metallic state. (*a*) Pictorial view of the distribution of electron density in part of an imaginary two–dimensional metal that could be from the lanthanide series. (A square atomic cell is chosen for simplicity.) The 5s,p orbitals form the outer shells of the atom's cores; their mutual repulsion between contiguous ions in the metal sets the interionic separation. The 4f orbitals are assumed to be pulled well within this core by the same mechanism that operates in real lanthanides and is discussed in the text; the valence electrons in the 5d and 6s orbitals in the atomic state are assumed to have the range indicated. Electrons from the latter orbitals are, of course, itinerant in the metal and not bound in defined orbitals as suggested by the figure. (*b*) Showing schematically how these various orbitals fare during the condensation of atoms to form the metal. The 5d and 6s states are Type II orbitals (Fig. 2.7) and broaden and mix to form a band of closely spaced states containing delocalized electrons up to the level E_f dictated by Pauli's principle. The 4f states (of Type I orbitals) have virtually no direct interaction with their counterparts on contiguous ions; the 5s,p states (of Type III orbitals) produce narrow bands that are completely filled in this case and represent electrons localized in the ion's vicinity.

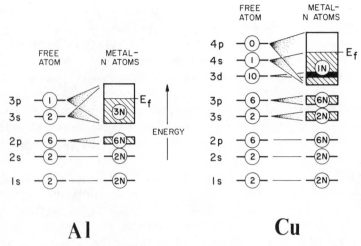

Fig. 2.9 As free atoms are condensed to form a metal, their outer electrons are influenced to varying degrees by the electrostatic potential of surrounding nuclei. This shows, in the fashion of the hypothetical example of Fig. 2.8*b*, the consequences for two real metals. In Al the 3*s* and 3*p* levels form an *s/p* band that in a sample of *N* atoms accommodates 3*N* electrons in delocalized states. Since this band is relatively isolated in energy from other energy levels, the itinerant electrons behave largely as free particles. Consquently, Al is a simple metal for which its electrons' dynamics are quite well described by free electron theory. In Cu the same effect is observed for the atoms' valence electrons—the 4*s* and 4*p* levels form an *s/p* band— but the presence of a narrow, filled band formed from the 3*d* orbitals complicates this situation. The fact that electrons in these *s/p* and *d* states interact with each other to varying degrees means that the dynamic properties of the former are modified from the simple unfettered *s/p* behavior. Cu is thus not a simple metal. This point is considered in more detail in Sect. 6.2.

to their behavior in the atomic state. A valence *s* electron loosely bound outside an ionic core of filled shells can range appreciably to the extremities of the atomic cell when the ion is part of the condensed phase.

The pure metals, therefore, fall into three classes, depending upon their position in the periodic table, and that position in turn specifies generally the type of atomic valence state that leads to dominant effects in the metal. Frequently these classes are called the simple or normal metals (when the valence states are of the *s/p* type) the transition metals (when the valence states are of the *d* type) and the rare earth metals (when the valence states are of the *f* type). These titles reflect conventional terminology rather than any rigid classification, however; for example, the actinides have 5*f* valence electrons, but these electrons are so much less localized in the atom than the corresponding 4*f* examples that this series is more appropriately included in the transition metal's class. However, it is not appropriate at this point to discuss in detail the various exceptions to the above classification. Rather, we can accept it as a useful framework in the present context. The following sections build upon it with a description of the salient features exhibited by the members of each class.

2.3.3. The Origin of the Energy Bands.

Although the preceding paragraphs have glibly referred to the formation of energy bands by the interaction of orbitals on adjacent atoms in a metal, there are, in fact, three main conditions that have to be fulfilled by atomic orbitals before they can produce such bands in the solid state. First, the orbitals of adjacent ions in the solid must overlap in space; in other words, the participating orbitals must have an appreciable electron density in the neighborhood of the atomic cell's boundary. Second, the participating orbitals must also overlap in energy; as is well known, if an efficient resonance interaction is to occur between any two physical systems, they must have comparable energies. Third, the participating orbitals must have the proper symmetry.

The first two requirements have already been referred to in preceding paragraphs. They are clearly and intuitively reasonable in terms of our macroscopic experience and language (Chapter 1); taken together they state the obvious fact that unless the ranges of the participating electrons overlap in space and energy, each will be unaware of the others' existence and there will be no direct interaction between them. The third condition, on the other hand, has no immediately obvious interpretation since it stems from a quantum mechanical requirement that is the physical reason for the existence of energy bands in solids.

The wave function Ψ corresponding to a given atomic orbital is a solution of Schrödinger's equation. As such, it can have both positive and negative parts. (See Appendix A for illustrations of the s, p, d, and f cases.) In conventional solid state textbooks this attribute is frequently underemphasized—perhaps because $|\Psi|^2$ is the physically meaningful quantity—but it is all-important when considering the interaction between orbitals. For example, the concept of hybridization between different orbitals in atoms, which is widely used in chemistry, is simply a convenient expression of the influence of this attribute (Fig. 2.10a). In the present context the importance of the signs of overlapping wave functions can be seen from the simplified case considered in Fig. 2.10b. This shows two situations in which otherwise identical orbitals overlap. (These are typical Type II orbitals, having appreciable electron density at the atomic cell's boundary.) When orbitals of like sign overlap, as in the lower part of the figure, the resulting hybrid wave function, which is also frequently known

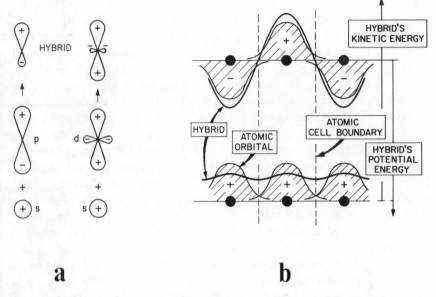

a   b

Fig. 2.10 (*a*) Two simple examples that show the effect of hybridization upon the angular distribution of the resultant atomic orbital. At left atomic *s* and *p* orbitals are hybridized; at right *s* and *d* orbitals. In each example the orbitals should be imagined to be centered upon the same nucleus. The angular distributions of the atomic orbitals are illustrated in Fig. A.1 and Fig. A.2 of Appendix A. (*b*) Showing the hybridization of equivalent *s* orbitals in an imaginary one-dimensional metal. The resultant hybrid wave function, which is also known as the crystal's wave function, has no nodes when the interacting orbitals have the same sign; it has one node per atomic cell when they have opposite signs. The text explains that, since the kinetic energy of the electron in the corresponding state is proportional to the number of nodes in the wave function, the lower configuration corresponds to the lower kinetic energy. On the other hand, it corresponds to the higher potential energy for the electron because the spaces between the ions are regions where the electrostatic potential is high, and, since the configuration with nodes excludes the electron density from these regions, it, therefore, corresponds to the lower potential energy.

as the crystal's wave function, is entirely of one sign; it is a relatively smooth function and possess no nodes (points at which it changes sign). When orbitals of unlike sign overlap, on the other hand, the hybrid function has a node wherever it crosses a cell boundary—that is, midway between each atom. The significance of this detail is that the number of nodes in the function is roughly proportional to the kinetic energy of the delocalized electron in the corresponding state. (We are temporarily ignoring any influence upon the electron's motion which might be produced by the periodic potential of the ionic lattice.) In simple language, as more constraints are imposed upon the charge distribution of a delocalized electron in a system, its kinetic energy is increased. A node in a crystal wave function illustrates just such a constraint, since it represents a point in space where the electron in that state has a zero probability of being found. Consequently, the more nodes that appear in a delocalized electron's wave function, the higher, in general, will be the electron's kinetic energy.* Therefore, the kinetic energy of an electron in the hybrid state in the upper part of Fig. 2.10b is greater than that in the lower part.

The form of the crystal's wave function also affects the delocalized electron's potential energy, and this has to be taken into account when we consider the effect of hybridization upon the electron's total energy. Such an effect arises because the hybridization redistributes the electron density; it alters the electron's probability of being found in regions subjected to electrostatic fields of different strengths, and, as a result, the potential energy of the electron is changed. For example, in the upper part of Fig. 2.10b the charge density corresponding to the hybrid wave function is centered upon the nuclear sites where the attraction for the electron is greatest and its potential energy is, therefore, lowest. In the lower part of the figure, the redistribution of the hybrid state prevents some of this charge density from occupying the region of low potential and, in fact, forces it into the region between the nuclei at a corresponding cost in energy. Consequently, the potential energy of the delocalized electrons in the lower part of the figure is greater than that in the upper part. Note that in this case the situation of lower potential energy has the electrons distributed in what in chemical parlance is called an antibonding configuration. This is an example of the situation referred to at the end of Sect. 2.2, where we pointed out that in metals the antibonding electron distribution can be just as bonding as the more conventional bonding (or covalent) distribution that has an appreciable charge density between the ions.

The interplay of these competing contributions to the delocalized electrons' energy in a slightly more complicated situation is illustrated in Fig. 2.11. This shows an imaginary metal being formed from a row of identical atoms. It is assumed that two outer orbitals in the atom—one each of s and p symmetry (see Appendix A)—overlap sufficiently in space with their counterparts on adjacent atoms to produce an appreciable interaction in the metal. Since the atoms are identical, the overlap in energy of equivalent orbitals on adjacent atoms is, of course, automatically assured. The

---

* This effect is also evident, for example, in the solutions to the standard problem of an electron constrained to a rectangular box, which is cited in Sect. 2.2.

figure shows the influence of the third condition cited above: The symmetry of the atomic orbitals.

For simplicity, let us consider that the metal is just four atoms long. As each atom is added, a further alternative combination of interacting atomic orbitals becomes possible, and this leads in turn to an additional energy level for the delocalized electrons in the metal. Thus the four atoms in this example produce four alternative symmetry combinations for each of the s and p orbitals; N atoms would likewise give rise to N separate possibilities for each orbital (compare Fig. 2.7). These alternatives produce separate energy states for the reasons outlined above: The kinetic energy of an electron in the state is proportional to the number of nodes in the hybrid wave function, but its potential energy depends upon the details of the corresponding charge distribution. Figure 2.11 shows the form of the hybrid (or crystal) wave function in each case; it is clear that the influence of the kinetic component of the total energy is dominant in setting the hierachy of the energy levels. Thus, the sequence of states $s1 \rightarrow s4$ corresponds to levels of increasing total energy for the resident electrons and is reflected in the increasing number of nodes per unit atomic cell across the series (from $0 \rightarrow 1$ for the s hybrids; from $1 \rightarrow 2$ for the p ones). The influence of the potential component is also evident, however, notably in the separation between the levels $s4$ and $p1$. These states have the same number of nodes per unit cell; however, electrons in $p1$ have the higher potential energy because the nodes in that state occur at the atomic nuclei, but the nodes in the $s1$ state appear midway between them. Thus, since electrons in state $p1$ are prevented from being near the nuclei, the total energy of the state is lifted with respect to $s4$. This is the physical cause of the separation of the energy bands in solids.

By analogy with the symmetry designations of an atomic orbital (viz., s, p, and so forth), the hybrid states in the metal are said to have s-like or p-like character depending, respectively, on whether the electron density is concentrated in, or repelled from, the region around the nuclei. In the particular case of Fig. 2.11, where the s and p states in the atom are assumed to be so widely separated in energy that no hybridization occurs between them in the solid, the hybrid states retain their separate s and p character throughout the s and p bands. Thus the state $s1$ at the bottom of the band is an s-like bonding one, since it permits the pile-up of charge in the regions between the atoms. Similarly, the state $s4$ at the top of the band is an s-like antibonding one, since the electrons are moved away from the internuclear regions to add to the charge density about the nuclei ($p1$ and $p4$ are likewise of p-bonding and p-antibonding character, respectively).

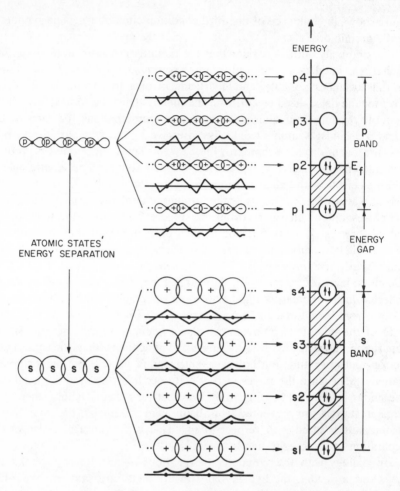

Fig. 2.11 Schematic view of the role played by the atomic orbital's symmetry in the formation of energy bands in an imaginary one-dimensional metal. For simplicity the situation shown is that when the metal is just four atoms long. It is assumed that two outer orbitals in the atom—one each of s and p symmetry—interact appreciably only with their counterparts on adjacent atoms. Furthermore, these orbitals are taken to have the occupancy s^2p^1 in the atomic state and to have the energy separation indicated. As each atom is added, a further alternate combination of interacting atomic orbitals becomes possible, leading in turn to an additional energy level in the metal. Each of these is illustrated together with its corresponding hybrid (or crystal) wave function. The kinetic energy of an electron in the state described by this function is proportional to the number of nodes the function possess per atomic cell; hence, the energy hierarchy of the states within a band. The potential energy of the electron is determined, on the other hand, by the distribution of the wave function in the interionic regions of high and low electrostatic potential. For example, $s4$ and $p1$ have the same number of nodes per atomic

In a real metal the energy bands are generally formed from the hybridization of atomic states of different symmetries: Thus, hybridizations of s/p, $s/p/d$, and $s/p/d/f$ types are known, but the first—which is understandably the simplest—is the one we will consider here (Type II of the next section). The dominant character, then, of the hybrid can change across the band because of the different degree of mixing between the precursor states. The details depend upon the relative energies of the s and p states in the atom. In the situation depicted in Fig. 2.11 the hybridization between the s and p states in the metal leads to an energy band that has a purely s-like bonding state at its lower extremity (like $s1$ in Fig. 2.11) and that changes to a purely p-like bonding state at the top of the band just below the energy gap (corresponding to $s4$ in Fig. 2.11). We can visualize this as a physical process in which electrons that are added to the available states are forced by the exclusion principle (Sect. 3.2.2) to occupy states of increasing kinetic energy; however, they are able to minimize this inevitable increase by modifying their distribution. In going from the bottom to the top of the s band of Fig. 2.11, for example, the electrons' energies can be reduced by progressively mixing a p-character into their distribution —to the extent that at the top of the band it is entirely p-like, as we have already said. This admixture, and the corresponding reduction in the electrons' total energy that it produces, is thought to be an important determinant of the crystal structure of metals possessing s/p energy bands. This point is considered further in Sect. 5.3.

Turning to the ocupation of the hybrid states, we know in anticipation of Sect. 3.2.2 that each of them can accommodate two electrons, providing they have opposite spins. If, for the sake of illustration, the electron configuration of the outer orbitals in the atom is taken to be s^2p^1, then each atom contributes two electrons to the s band of energies and one to the p band. This term is used here with some license, since four levels can hardly be considered to constitute a band. But as we imagine more atoms being added and the number of levels increasing correspondingly, the term becomes strictly applicable (see, for example, Fig. 2.7). In this case the s

---

cell, but the former corresponds to the lower potential energy because the electron density is there concentrated upon the nuclei, where the attraction of the electron is greatest and so its potential energy is lowest. In $p1$ the electron density is conversely prevented from being near the nuclei and is concentrated in the spaces between the ions where the electron's potential energy is greatest. This difference in the electron density distributions corresponding to the crystal wave function is the physical cause of the separation of the energy bands in solids. The corresponding case of overlapping d levels is rather more complicated and is reserved for Chapter 6 (Fig. 6.14).

band will be completely filled in the manner illustrated, but the p band will be fully occupied up to the level $p2$ at absolute zero. This level is known as the Fermi level, E_f. The remaining unoccupied levels in this band are thus available for occupancy should the kinetic energy of a delocalized electron* be increased by some outside influence, such as an applied electric field. This availability of vacant, delocalized states is what characterizes a metal (as is emphasized in Sect. 2.2); if the atomic configuration is s^2p^2 (that is, if both s and p bands are completely filled) then the situation corresponds to that in a classical insulator. A partially filled band, like that described above, is known as a conduction band.

2.3.4   The Expected Forms of the Energy Bands.   In Sect. 2.3.2 three degrees of interaction were distinguished between orbitals localized about atoms that can be condensed to form a metal. These were exemplified by the behaviors of the f, d, and s/p atomic orbitals during the formation of a pure metal. Staying within that classification, we will now examine in slightly more detail the salient features we can expect in the three resulting types of energy bands.

BANDS OF TYPE I: F BANDS.   Partly because their shape (see Appendix A) does not very effectively screen the f electrons from the nuclear positive charge, and partly because they occur in elements of high atomic number, the f states in atoms are drawn disproportionately close to the interior of the atom. (The high atomic number emphasizes this inadequate screening because the atom contains a correspondingly large nuclear charge.) Thus at the beginning of the lanthanide series (conventionally La→Lu) the empty $4f$ shell of La lies outside the Xe core shells (which are $5s^25p^6$). But throughout the balance of the lanthanide series (Ce→Lu in the f block of Fig. 2.1) the $4f$ shell is pulled inside this core by the factors mentioned above; in fact, it is pulled increasingly within the atom's interior as the sequence Ce→Lu is traversed.† The typical radius of the $4f$ shell in this series is about 0.3×10^{-10}m, and, since the interatomic spacing in the metal is about 3×10^{-10}m, there can evidently be very little overlap of the $4f$ states.

---

* Since there are only four atoms in the present example, we are influenced by the molecular connotation and, therefore, refer to the hybrid states as containing delocalized (as opposed to itinerant) electrons, but this is a pedantic distinction.

† This is the origin of the lanthanide contraction, which is the progressive decrease observed in the atomic radii of the sequence La→Lu (with the notable exceptions of Eu and Yb). See, for example, K. N. R. Taylor and M. I. Darby, *The Physics of Rare Earth Solids*, Chapman and Hall, London, 1972, Fig. 2.1.

Although the $4f$ shell lies inside the $5s$ and $5p$ shells and is, therefore, closer to the nucleus, the $4f$ electrons have the highest energies. This may easily cause confusion, particularly when comparing energy level diagrams with physical pictures of the electron distributions (as is done in Fig. 2.8). The higher energy of the $4f$ electrons is caused by the shape of their orbitals (see Appendix A): Their complicated angular nodes make it difficult for their electrons to avoid coming close to other electrons in the atom. Consequently, the $4f$ electrons experience unusually high Coulomb repulsions from other shells. These repulsions have to be overcome in order to fill a vacancy in a $4f$ shell, and they push that shell's energy above that of even the $5p$ shell.

Since in the metal there is very little interaction between these Type I states and their counterparts in adjacent atomic cells, the atomic legacy that they bring to the metal is very great. For example, such features as a vacancy in the atomic states can survive the condensation process to appear with little modification in the metallic state; the magnetic moments possessed by the ions of the rare earth lanthanide metals are a direct consequence of this.

BANDS OF TYPE II: S/P BANDS. Since they are loosely bound the outermost s or p orbitals of an atom have a high electron density at the boundaries of the atomic cell (Fig. 2.8) and, therefore, their interaction with counterparts in adjacent cells is very great indeed. This results in the formation (Fig. 2.11) of wide bands of closely spaced delocalized energy states that are occupied in accordance with Pauli's principle (Sect. 3.2.2) by the s or p electrons liberated from the atomic hosts. These electrons form the itinerant class described in Sect. 2.2. They have completely lost their atomic features and move in states that are dominated by the conditions at the boundaries of the atomic cell; the effect upon them of the atomic nucleus or the core electrons in the cell is relatively very slight. This means that, as an itinerant electron travels through an ideal sample of such a metal, its probability of being scattered is very small. Thus metals formed from elements of the s and p blocks generally behave very much as predicted by the simple theory that treats itinerant electrons as free particles. The alkali metals of the s block (Li→Cs) and the remaining metals of the s and p blocks are generally referred to as the free and nearly free electron (NFE) metals, respectively.

Typical examples of the behavior under discussion are provided by the band of delocalized s states formed in any alkali metal or the band formed from the $3p$ and $3s$ states in Al (Fig. 2.9). But we should realize that such behavior is not restricted to the s/p blocks. It is only that they provide distinctive examples in which the s or s/p bands exist in the metal in relative isolation. In fact, the outer s or p states of atoms in all metals form bands of delocalized states of various widths, but these bands are not necessarily so isolated. Cu is a good example of this: The $4s$ electrons populate a wide band of delocalized states formed from the $4s$ and $4p$ orbitals, but the $3d$ states (which we know have roughly similar energies in the atom—see Fig. 2.6) also interact and form a band, albeit narrow and completely filled, that is superimposed upon the $4s$ band and lies in energy roughly in its middle. The presence of this $3d$ band significantly affects the dynamic properties of the itinerant electrons, and in a detailed analysis it cannot be ignored.

BANDS OF TYPE III: D BANDS. A typical $3d$, $4d$, or $5d$ orbital in an atom of the d block is very localized. Its electron density does not extend to large distances from the atom (the d orbitals are generally well inside the s/p ones of comparable energy), at the other extreme the electron density falls off parabolically as the atom's nucleus is approached. Because of this shape, the d electrons screen the nuclear charge rather badly—somewhat like the f electrons—and this increases their compact and well-defined distribution in the atom. (In fact, there is evidence of a corresponding mini lanthanide contraction of the d shell as the atomic number is increased.) As a result, the d orbitals produce only a moderate electron density at the atomic cell's boundary and have correspondingly an intermediate degree of interaction with similar states in adjacent cells. We, therefore, expect that narrow bands of such states will be formed and that the electrons in them will have some degree of itinerancy. But such electrons are still very much dominated by the electrostatic potential of the nucleus and the core electrons, and they, therefore, retain their atomic characteristics to a marked degree.

Figure 2.12 shows the results of some calculations* leading to the relative positions of the $3d$ and $4s/4p$ bands along a fixed crystallographic direction in the metals of the first transition series. (This is concerned only with the paramagnetic state of the solid; we reserve the effect of d band splitting and other complications for Sect. 6.4.) We can see certain systematic tendencies in these results that reflect clearly the coresponding atomic characteristics referred to above. First, the d bands are relatively narrow compared with the s/p ones: In the Ti$\rightarrow$Cu series the d bands range approximately between 6.4 and 12.8 $\times$ 10^{-19}J (4–8 eV) in width. Second, as this series is traversed and the d levels fill up, there is a generally system-

---

* L. F. Mattheiss, *Phys. Rev.* **134**, A970, 1964.

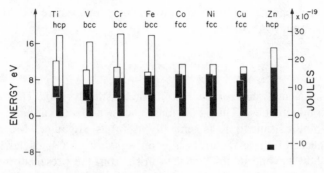

Fig. 2.12 Schematic form of the $3d$ and $4s/p$ energy bands in the paramagnetic form of metals of the first transition series of the d block. In each case the $3d$ band is that having the narrower energy spread. The solid regions illustrate the extent to which the bands are filled.

atic narrowing of the d band. This indicates a decreasing overlap of the d wave functions between neighboring ions in the solid and reflects the increasing compactness of the wave functions about the nucleus as a result of the lanthanide-type contraction referred to above. At Cu, the d band is completely filled (see also Fig. 2.9). Although the situation for the p block series following Zn (Ga→As) is not shown in Fig. 2.12, it is known that the d band becomes progressively narrower and falls increasingly and substantially below the energy range of the s/p band. In Zn, as Fig. 2.12 shows, it has already dropped to about 11×10^{-19} J below the conduction band, and its width is down to about 1×10^{-19} J.

As in the case of the f electrons of Type 1, a vacancy in an atomic state can survive the condensation process and produce a magnetic moment on the ion in the metal; hence the magnetism of the $3d$ transition metals. But there the similarity between the two cases ends, for here we have the vexing and still unresolved question of itinerancy in the d bands. Is it a significant feature which we should always allow for when considering d bands in metals, or is it enough in most cases to regard the d electrons as localized? There is no general answer to this question*; each metal has to be treated in whatever fashion seems appropriate in the light of experimental evidence. We shall comment upon individual cases as they arise

* C. Herring in *Magnetism*, Ed. G. T. Rado and H. Suhl, Academic Press, New York, V. IV, p. 118; this is a descriptive discussion of this question at an advanced level. J. Hubbard, *Proc. Roy. Soc.* **A276**, 238, 1963; this is a lucid description of how such electrons could show simultaneously both itinerant and localized characteristics.

in subsequent chapters, but to go into further details here would take us outside the required scope.

2.4 CHAPTER SUMMARY

A metal is a substance which has characteristically good electrical and thermal conductivities. This is obviously not a perfect definition, but as a starting point it seems to be as generally useful as any other. More quantitative aspects of these features—the universality of Ohm's law, for example—show beyond doubt that they arise from the presence in the substance of electrons that are itinerant, that must belong to the sample as a whole and not to particular atoms, and that must be very easily excited to higher energies by external influences such as electric fields or temperature gradients. Thus we could also, in the jargon of the subject, define a metal as a substance that has some itinerant electrons in delocalized states.

Just how such a group of electrons comes to be present in a solid is rooted in the quantum mechanical requirements of the situation and is, therefore, not directly amenable to exact classical analogy. However, the same can be said of the specific distribution of electrons in a free atom, for example, and yet in that case a working science, such as chemistry, has been developed with the aid of specious but not entirely satisfactory concepts like exchange, the atomic orbital, hybridization, covalent and resonating bonds, and so forth. A similar situation obtains in the present context: A feeling for the origin of the itinerant electrons can be obtained by considering the solid as a conglomeration of atoms and by falling back on concepts like those listed above.

In this picture, as free atoms are brought increasingly close together to form a pure metal, the electrons in the outer orbitals are influenced sooner or later by their counterparts in adjacent atoms. Hybridization of the corresponding energy states forms closely spaced energy levels that occupy well-defined ranges of energy and that are populated by the previously localized electrons from the atoms' interacting shells. The degree of this interaction, and hence the amount by which the circumstances are shifted from the free atom's situation, varies enormously for different electrons in the atom. But it is convenient to separate the range covered into three categories, corresponding simply to large, small, and intermediate degrees of interaction. This allows the metals to be similarly divided into three classes and provides a start in understanding how the various atomic charactertistics fare during the condensation of atoms to form a metal. Thus we are able to see the origin of specific traits by which one homologous group of metals is distinguished from another in the periodic table.

RELATED READING

Specific research papers are referenced where cited in the text.

SOME OF THE MORE ELEMENTARY STANDARD TEXTBOOKS.

N. F. Mott and H. Jones, 1936, *The Theory of the Properties of Metals and Alloys* (Oxford University Press: Oxford). Still useful, this text gives generally an intermediate (and mathematical) discussion of the quantum mechanical principles and the formation of energy bands.

S. L. Altmann, 1970, *Band Theory of Metals: The Elements* (Pergamon Press: Oxford). With a useful introductory review of quantum mechanims, this text deals exclusively with the energy band picture of metals.

F. Seel, 1963, *Atomic Structure and Chemical Bonding* (Methuen and Co.: London). A very readable description of the place of metallic bonding in the general scheme of bonding in solids.

L. Solymar and D. Walsh, 1970, *Lectures on the Electrical Properties of Materials* (Clarendon Press: Oxford). Chapter 7 gives a clear description of the different approaches that lead eventually to the realization of energy bands. The approach we have followed is called the Feynman model by Solymar and Walsh.

J. M. Ziman, 1963, *Electrons in Metals* (Taylor and Francis: London).

MORE SPECIALIZED REVIEW ARTICLES

N. F. Mott, "Electrons in Transition Metals." *Adv. Phys.* **13**, 325 (1964). Reviews in a generally descriptive way the evidence relating to the form of the d bands.

J. Friedel, 1969, *The Physics of Metals*, Ed. J. M. Ziman (Cambridge University Press: Cambridge). Article entitled: "Transition Metals: Electronic Structure in the d band." A generally more mathematical description of the same subject covered by Mott.

ELEMENTARY REVIEW ARTICLES

N. F. Mott, "The Solid State," *Sci. Amer.,* September (1967).

A. H. Cottrell, "The Nature of Metals," *Sci. Amer.,* September (1967).

3

THE ELECTRON LIQUID

I

GENERALLY SINGLE-BODY EFFECTS

3.1 INTRODUCTION

The preceding chapter describes how a crystalline metal can be regarded from an elementary point of view as a lattice of positive ions immersed in a fluid of itinerant electrons. In this and subsequent chapters this view will be developed by a more detailed discussion of selected points. To begin let us examine the electron fluid, which is discussed in this chapter and the next.

Two facts are the prime causes of the characteristic behavior of such a fluid: First, the electron is a charged particle and, second, it possesses spin. Being charged particles, electrons experience a mutual interaction through the Coulomb force. This makes the electron fluid a many-body system because anything done to one constituent will sooner or later be felt to some extent by the others. Consequently, we cannot study the behavior of an individual constituent in isolation.

However, the full many-body problem is presently intractible, as is emphasized in Sect. 4.2, and if we are to make any progress, we must make some simplifying approximations. The most sweeping one of all is simply to ignore the mutual interactions between all electrons in the fluid except the one (or pair) whose behavior we are examining. In effect, we are assuming that the electron–electron interaction between the constituents in general is weak enough to be negligible—even though we know that it is not. The resulting model is called the independent particle (or, in this case, the independent electron) model. Ideas and suggestions that arise from this approach are the subject of this chapter; the next chapter is concerned with the inevitable extension to the more realistic situation where the mutual interactions between the electrons are not assumed to be weak.

We have said that charge and spin are the two important attributes possessed by the constituents of an electron fluid; they also make convenient pivots for this chapter and the next. The Coulomb force that is discussed in the next chapter serves as a useful introduction to several concepts of field theory that are used in the contemporary description of a metal; electron spin and its ramifications are our starting point in this chapter.

Electron spin is a quantum mechanical concept that is difficult because it has no clear-cut physical interpretation in terms of our macroscopic language. The consequences of spin can be made evident enough, however, and they are all-important for electrons in metals. It will be shown that the possession of spin leads to the Pauli Principle (Sect. 3.2.2) the exchange interaction (Sect. 3.3) and, ultimately therefore, to several characteristic features of metallic behavior (Sect. 3.4).

3.2 ELECTRON SPIN AND THE PAULI PRINCIPLE

3.2.1 Electron Spin. For those whose understanding of the physics of solids leans towards a corpuscular or classical* interpretation, the idea of electron spin is complicated by conceptual difficulties, for it is a quantum mechanical concept that is partially obscure and will probably remain so until the full significance of quantum theory is understood. Indeed, some authorities take the extreme view that it is part of quantum mechanical theory that questions (such as What is electron spin?) should not be asked† and that to do so inevitably leads the student into further confusion since he is conceptually handicapped by his experiences with the Newtonian mechanics of the macroscopic world. F. J. Dyson says in this context:‡ "The reason why new concepts in any branch of science are hard to grasp is always the same; contemporary scientists try to picture the new concept in terms of ideas which existed before."

This may be so, but, generally speaking, the commonsense and everyday experience of the working scientist make such an approach almost inevitable; most of us need to have some picture of a given physical concept in order to have an effective way of working with it, and we have only

---

* As pointed out in Chapter 1, this has come to mean nonquantal and nonrelativistic.

† C. W. Kilmister, *Quantum Theory and Beyond,* Ed. T. Bastin, Cambridge University Press, London, 1971, p. 124.

‡ F. J. Dyson, *Sci. Amer.,* September, 1958, p. 74. (Quoted with permission.)

previous experience to draw upon. Consequently, in a case such as electron spin, it is not surprising that the search for a classical or semiclassical analogy is as old as the concept itself and remains an active field of research. But before we look at the product of some of these efforts, it is instructive to consider briefly the well-known early history of the concept.

Pauli* had discovered when analyzing certain alkali spectra that the anomalous results could be understood if the electron were given a new property, a certain "two-valuedness not describable classically"; this property lead to the introduction of an additional quantum number (Sect. 3.2.3). Shortly afterwards Uhlenbeck and Goudsmit† took the explanation a step further. They were examining results of a study of the anomalous Zeeman effect in the spectra of alkali atoms. Zeeman had observed that, when the radiating atoms were subjected to a magnetic field, the spectral lines were not simply shifted in wavelength in one direction or the other, as might have been expected, but that each normal line was split into a multiple set of sharp and distinctly separate lines. Uhlenbeck and Goudsmit postulated that, if in addition to its corpuscular attributes of mass and charge an electron were given some properties of a spinning charged body, then they could provide a satisfactory explanation of the anomalous Zeeman effect. They thus associated Pauli's two-valuedness with an intrinsic angular momentum possessed by the electron (later known as spin angular momentum or spin) and also with an associated magnetic dipole moment.‡ The spin moment gave the additional quantum number required, and the magnetic moment came automatically; since it constitutes a rotating cur-

---

* W. Pauli, *Z. Phys.* **31**, 373, 1925.

† G. E. Uhlenbeck and S. A. Goudsmit, *Nature* **117**, 264, 1926.

‡ Nearly fifty years later, in an address to an American magnetism conference, Goudsmit recalled that it was he who felt that *some* new property had to be associated with the electron, but it was Uhlenbeck who had the idea that this should be an intrinsic angular momentum. Initially they were not aware that A. K. Compton had suggested the idea of a quantized spinning electron some five years earlier (*J. Franklin Inst.*, **192**, 145, 1921). Compton's idea was not applied to the Zeeman effect, however, and its neglect seems to have been of no direct detriment to the topic's development.

It is also interesting to note that, although Pauli had established the "two-valuedness" and the early form of his exclusion principle (Sect. 3.3.2, Sect. 3.3.3) it did not occur to him to ascribe these effects to an intrinsic angular momentum of the electron; it would seemingly have been a simple matter for him to be the first to do so. R. Kronig, a contemporary of Pauli, had actually suggested the idea of a spinning electron to Pauli a year before Uhlenbeck and Goudsmit's paper appeared. But Pauli, who could be very autocratic and domineering, had so ridiculed the idea that Kronig had not dared show it to anyone else.

rent, any rotating charged body would be expected to have a magnetic moment.

Uhlenbeck and Goudsmit found that the value of the spin angular momentum required for the electron was $\hbar/2$, which was exactly half the orbital angular momentum of an electron in the lowest Bohr orbit in the hydrogen atom. The spin was pictured as having two possible orientations: Either parallel or in opposition to any local field that is present. What was puzzling, however, was that in order to agree with experimental evidence the associated dipole moment apparently had to be just twice that expected from the assigned spin momentum. Specifically, the intrinsic magnetic moment had to be $eh/2m$, this being the moment produced by the orbital motion of an electron in the lowest Bohr orbit in the hydrogen atom. The gyromagnetic ratio g for the electron (which is the ratio of the magnetic moment to the angular momentum) was thus e/m (or 2 in the natural units of $e/2m$, which were used conventionally at that time). However, in classical theory any rotating charge distribution of a single sign and constant charge/mass ratio throughout is known to have a fixed gyromagnetic ratio; in the case of an orbital electron this would be $e/2m$ (or 1 in natural units). It was necessary to accept, therefore, the apparent contradiction that although the classical electron has an apparent g factor of 1, the concept of Uhlenbeck and Goudsmit required the electron to have an apparent g value exactly twice as large.* In spite of this, it soon became obvious from the results of various experiments, such as the fine structure of optical spectra and the scattering of β rays by electrons, that the concept of intrinsic spin was somehow fundamental. Its manifestations were accountable in the manner suggested by Uhlenbeck and Goudsmit, even if classically some aspects were incomprehensible.

Dirac† apparently took a big step forward when he showed that the manifestations of spin possessed by the electron, including the anomalously large magnetic dipole moment required by Uhlenbeck and Goudsmit's interpretation, are a direct result of the incorporation of the principles of

---

* It was thought for at least 20 years afterwards that the electron's g value was exactly 2, but it has since been shown both experimentally and theoretically that it is in fact about 0.1% greater than this. The difference is produced by the electron's interaction with the physical vacuum of space—one of the key ideas of quantum electrodynamics. This will be referred to briefly in Sect. 4.3.2; it can safely be neglected in this chapter. An elementary description of the determination of the electron's g value has been given by H. R. Crane (*Sci. Amer.*, January 1968, p. 72).

†P. A. M. Dirac, *Proc. Roy. Soc.* A117, **610** (1928); a more popular account of this period is given by the same author in *The Development of Quantum Theory*, Gordon and Breach, New York, 1971.

relativity into quantum mechanics. Briefly, when relativity is introduced and all equations are made invariant under the Lorenz transformation, there is an additional internal degree of freedom for the electron that is not evident in its normal translational and rotational motion. Since the property is peculiar to quantum theory, this extra degree of freedom has no strictly classical interpretation, although historically it was called spin and the name has stuck. This position has become the standard pedagogical approach to electron spin, and it essentially provides the conventional answer to the question in the opening paragraph. But it is a misconception to deduce from this that spin is a purely relativistic effect (as in some textbooks). This is not so, for a theory that produces the correct spin and intrinsic magnetic moment can also be obtained without the Lorenz invariance required by relativity.* (This contrasts with spin-orbital coupling, for example, which is a purely relativistic effect. It is referred to in Sect. 3.2.3.)

All theories agree, however, that spin is a quantum mechanical effect (as it disappears when $h \rightarrow 0$) and it can, therefore, not be understood classically. But Dirac's arguments are not universally accepted without reservation,† and his conclusions have not eliminated the search for a semiclassical analogy of electron spin. "Semiclassical" here means a corpuscular view of the electron combined with the consequences of relativity, for it is easy to show that an entirely classical and corpuscular view is untenable. Indeed, it may be instructive to do so for the benefit of any reader who still retains the idea of an electron as a spinning, rigid body of charge.

A purely classical view requires a rigid charge distribution of a given size and form that, under the application of the normal rules of mechanics, gives quantitatively all the observable properties of the electron, including those of spin. The first question that this approach raises is whether it is realistic to represent an electron by anything but a point of charge. It is true that in theoretical studies involving interaction between moving electrons there is a factor that occurs repeatedly and that has the dimensions of length. In the usual terminology this factor is e^2/mc^2, and it has a value of 2.8×10^{-15} m. It has become known as the classical electron radius (it can also be deduced by equating the relativistic self-energy of the electron m_0c^2 with the Coulomb self-energy of the electron's electrostatic field) but experiments give no justification for interpreting this quantity as the probable radius of the electron. Indeed, at the time of writing, experiments involving high-energy electron collisions all indicate that the

---

* A. Galindo and C. Sanchez del Rio, *Am. J. Phys.* **29**, 582, 1961.

† *Quantum Theory and Beyond*, Ed. T. Bastin, Cambridge U. P., Cambridge, 1971, p. 123.

electron's radius is less than 10^{-17} m. In other words, there are no experimental grounds for taking an electron to be anything but a point charge.*

Even if we ignore this difficulty and attribute to the electron a certain radius r and a spin momentum $\hbar/2$, in order to reach Uhlenbeck and Goudsmit's interpretation we must allow somehow that the range of the charge itself can extend further than r; far enough, in fact, to give an intrinsic dipole moment of $\hbar e/2m$, exactly twice as large as that expected from the range of the mass. This is clearly not an acceptable concept, and we are straining the credibility of the classical picture beyond its usefulness.

Recent attempts at a semiclassical approach have not been much more productive. No suitable mass distribution has yet been found† that when attributed to a relativistic electron, can give the intrinsic spin of $\sqrt{3}\ \hbar/2$ required by wave mechanics. Moreover, there is serious doubt as to the reasonableness of attributing to the electron the radius $\hbar/m_0 c$ ($=3.86 \times 10^{-11}$m) required by this semiclassical approach.

The picture of an electron as a spinning body of charge having an intrinsic spin pointing along permissible directions in space must be accepted as just a symbolic, if specious, analogy. In fact, it has been argued‡ that such directions attributed to the electron's two-valuedness unnecessarily build into the theory classical concepts of space and direction that can only lead to further confusion. For example, take an electron (or any particle) with an observable spin of $\hbar/2$. If the ideas of space and direction are introduced someone may ask: "Well, does the electron's spin point up or down in space?" The answer is that one cannot say; the electron has no way of knowing which way is up and which is down. The best answer to this question, as we mentioned earlier in this section, is that a direction is defined by some local magnetic field to which the observable spin may be either parallel or opposed. But this complication has been forced by the unnecessary introduction of the idea that the spin can take up directions in space in the ordinary sense.

---

* The discovery of the muon, which is possibly an electron in a very excited state, has lead to the revival of the view that perhaps the electron is an object of finite or even variable size. Dirac himself has worked on this problem and gives an account at a popular level in the *Sci. Amer.*, May, 1963, p. 45. The problem with a point electron is how to attribute excited states to it; the problem with an extendable body is how to incorporate the correct manifestations of spin.

† A. R. Lee and J. Liesegang, *Phys. Letters* **37A**, 37, 1971.

‡ R. Penrose, in *Quantum Theory and Beyond*, ed. T. Bastin, Cambridge University Press, Cambridge, 1971, p. 152.

Quantum particles should be regarded instead as having discrete internal alternatives open to them within their own frame of rest. These do not refer to any preexisting directions in macroscopic space but can be thought of as the only available choices known to the particle as regards its state of what we have called spin. For example, a particle* with zero spin has only one alternative, whereas one with spin $\hbar/2$ has two. A spin $\hbar$ particle will have three, and as a rule for a spin quantum number s there will be $2s+1$ alternatives. The manifestation of these internal alternatives on a macroscopic scale may vary depending upon the spin value of the particle. For example, the single alternative of the zero-spin particle is manifested by the fact that such a particle has to be spherically symmetrical; there is no question here of associating the concept of direction with the single alternative. On the other hand, in the case of a particle having two internal alternatives (such as an electron) it is conceptually quite useful to identify these alternatives with directions of spin in real space (the above complications of defining a preferred direction being accepted). However, the main point is that we should not necessarily think of the particle's internal alternatives as being directions of spin in all cases, although in the case of our particular concern—the electron—the concept of an angular momentum existing with respect to some specified direction is very firmly entrenched, and it underlies all subsequent considerations.

3.2.2   The Pauli Principle.   From everyday experience there is no difficulty in accepting a principle that states that no two impenetrable bodies can be in exactly the same position in ordinary space at the same time. This self-evident exclusion principle of classical mechanics can be regarded as a forerunner of the quantum mechanical equivalent: The Pauli exclusion principle.† We will show that the temporal and positional coordinates of a classical body specify the equivalent of the state of a particle on the quantum mechanical scale. When it is said that no two quantum particles in a system in which the Pauli principle operates can occupy the same state, the

---

* All elementary particles have an intrinsic angular momentum of total observable magnitude $n\hbar/2$, where n is either zero or another small integer: $n = 1$ for the electron and for most leptons and baryons; $n = 2$ for the photon; and $n = 4$ for the graviton.

† In a lighter vein, we can distinguish the *Pauli principle* from the *Pauli effect*. The latter can be defined as the uncanny ability of some individuals to cause havoc among experimental apparatus by means of either direct or vicarious contact. We are told that in Pauli this talent was developed to a very high degree (G. Gamow, *Sci. Amer.*, July, 1959, p. 74) and that he was very proud of it. Gamow implies that such ability is particularly a prerogative of theoretical physicists, but I doubt that this is so. Some experimentalists certainly possess it, although for obvious reasons they prefer to conceal the fact.

usefulness of the analogy is obvious. But on the quantum scale there is more to it than this, for not all types of quantum particles obey the Pauli principle and multiple occupancy of quantum states is permitted in some cases. When this fact is accepted, a rigid application of the analogy with the classical picture leads to confusion, for how does one picture multiple occupancy in the classical context? Do some particles on the quantum scale have penetrability, and can they effectively occupy the same position at the same time? The answer is that here is another example of an analogy drawn between phenomena occurring in vastly different scales that can easily be pushed too far—just as we warned in Chapter 1.

The aim in this section is to describe the Pauli principle and its important manifestations. We will eventually describe how the principle is rooted in the indistinguishability of quantum particles in a many–particle system, but it is convenient first to consider specifically a one-electron system; for, although the question of distinguishability of the electron is not prominent in that situation, it will allow the introduction of useful nomenclature as well as a statement of the principle in a form appropriate to electrons in atoms. This we do in Sect. 3.2.3. In Sect. 3.2.4 we turn to an alternative statement of the principle in the context of an N-particle system. Because of inevitable approximations, an N-particle system becomes really no more than a collection of N single-particle systems, but it does bring the description closer to the operation of the principle among itinerant valence electrons, which is the principal interest, and we see how the operation of the principle leads to what is known as the exchange interaction. This important effect has many consequences, and we give a descriptive review of those relevant to the metallic state in Sect. 3.3.

3.2.3 The Pauli Principle in the One-Electron System.

On the microscopic scale, where quantization effects are important, only certain well-defined configurations are permissible in a given system. Each of these is known as a state and is characterized by a wave function Ψ known as the state function. In the most general cases Ψ will be a function of the positions of all the particles in the system, of time, and perhaps of other coordinates that it may be necessary to introduce in particular circumstances. This mathematical construction Ψ contains all the information about the motion of the system that it is possible to know from experiment. To extract this information it is in principle only necessary to know the detailed form of Ψ in any given context and to solve the equations generated by the application of the theory in which, we remind the reader, observables are represented mathematically by operators chosen in a particular way.

This wave function, or the state that it represents, can also be specified conveniently by a series of labels, known as quantum numbers, that are chosen to suit the particular system or context. (We have already encountered examples in Sect. 2.3.) For example, the elementary particles in nature have as appropriate labels numbers specifying such things as the value of their electrical charge and spin angular momentum. For itinerant electrons in a metal the appropriate quantum numbers refer to the particle's linear momentum and spin angular momentum. Those quantum numbers appropriate to the motion of an electron in a central field of force will be presented in the following paragraphs.

As a matter of notation, the quantum numbers are frequently written as a subscript to the wave function: Thus $\Psi_{a,b,c}$ . . . is the wave function of the state characterized by the quantum numbers a,b,c. . . . In Dirac's bracket notation this is put even more economically by writing only the quantum numbers: Thus $\Psi_{a,b,c}$ . . . is denoted by $|a,b,c \ldots >$ or just $|a>$ for short. This is the ket; the bra is written $<a|$ and stands for the corresponding complex conjugate $\Psi_a{}^*$, for in general the wave function is complex and has real and imaginary parts.

The conventional one-electron system is one where an electron is assumed to be moving under the influence of a central field of force. This could be, for example, just the field of a nucleus together with some average field induced by any other surrounding electrons that might be present as in the case of an electron bound in an outer shell of an ion in a solid, which was considered in Sect. 2.3.1. It is a standard result of quantum theory that in this system (of which the hydrogen atom is a prototype) the electron is represented by a quantity which, in the absence of spin angular momentum, is a function of only the spatial coordinate $\mathbf{r}$. This quantity is called the orbital wave function (atomic orbital or positional coordinate wave function) and is represented by ψ ($\mathbf{r}$). It is a solution of the Schrödinger equation for the one-electron system, and ψ ($\mathbf{r}$) ψ^* ($\mathbf{r}$) turns out to be proportional to the probability that the electron will be at the point $\mathbf{r}$ in ordinary space. In terms of appropriate quantum numbers we can write

$$\psi(\mathbf{r}) \equiv \psi_{n,l,m} \tag{3.1}$$

The orbital state of the electron is thus characterized by three quantum numbers, which conventionally have the subscripted symbols. Strictly speaking, these are simply labels defining a particular state, but if we wish to retain a part of the classical picture of electrons in fixed orbits about the nucleus, n can be thought of as a measure of the distance of the electron from the nucleus, l as a measure of its angular momentum, and m as a number that fixes the orientation of the electron's orbit in space.

There are other one-electron systems that can be introduced at this point, since they will be useful in later discussions. Suppose that instead of being restrained in a central field of force the electron is confined in a box of dimensions l. (Neither related to, nor to be confused with, the quantum number above.) If the electron has fixed velocity u along the direction of this dimension, its wave function is periodic and fits exactly into l as illustrated in Fig. 3.1a. The wavelength λ associated with this period is related to the momentum by $\lambda = h/mu$ (providing the velocity is nonrelativistic). Alternatively, if the electron is known to be localized at a point $\mathbf{r}$ at some instant, it cannot usually be described by a plane wave function like that of Fig. 3.1a since this implies an equal probability of finding the electron anywhere within the range l. A more appropriate wave function in this case is that shown in Fig. 3.1b, which is known as a wave packet. Technically, it is constructed by superimposing a group of monochromatic waves like that of Fig. 3.1a whose momenta lie in a small interval about $\mathbf{r}$. This procedure is permissible because any linear combination of the solutions of Schrödinger's equation is itself an equally good solution.

Returning to the electron in the central field of force, the use of the word orbital in this context is carried over from the days of the older theory when, as indicated above, electrons were pictured as rotating about the nucleus in well-defined orbits—somewhat like a minute planetary system. But modern quantum mechanics teaches that it is hopeless to try to follow an individual electron in its path; therefore, it is wrong to retain the above planetary picture. Instead, an atomic electron is considered to have a certain probability of being found in a given region of space about the nucleus, and all the information about this probability is contained in its orbital wave function. This varying probability in space (known as the particle's probability density) refers to the presence of a material particle,

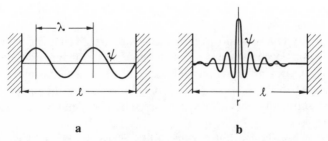

a   b

Fig. 3.1 Illustrating the wave function ψ of an electron confined within the linear dimension l. (a) An electron moving along l with a nonrelativistic velocity. (b) An electron localized at the point $\mathbf{r}$.

but, since the electron is charged, there is inevitably a corresponding charge distribution; a higher charge density is found in those regions of higher probability density. This leads to what is known as the charge cloud picture of an orbital in which the electron's charge is imagined to be smeared out in space to give the appropriate charge distribution. Although this is technically an incorrect viewpoint (assuming an electron cannot divest a portion of its charge), it is convenient and particularly suitable to describe the interaction of atoms. We shall fall back on it in the discussion wherever it suits our purpose.

Returning to the exclusion principle, what Pauli observed in 1925,[*] before the concept of intrinsic spin was established, was that there is never experimental evidence that more than two electrons have the same set of quantum numbers n, l, and m_l. He concluded that a systematic description could be given of the spectroscopic data he was examining if it was admitted that an atomic electron's state was characterized by, rather than three, quantum numbers. Accordingly, a fourth quantum number m_s was introduced which could take only two values $\pm 1/2$. This was the two-valuedness of the electron referred to in Sect. 3.2.1 and later identified with intrinsic spin.

When this is incorporated into the one–electron system, it can be shown that, under certain conditions specified below, the electron's wave function can be written as the product of two independent components.

$$\phi_{n,l,m_l,m_s} = \psi(\mathbf{r})\chi(\sigma) \tag{3.2}$$

In this equation $\psi(\mathbf{r})$ is the spatial part [that is, the orbital wave function of Eq. (3.1)] and is a continuous function of $\mathbf{r}$, but $\chi(\sigma)$, the so-called spin wave function, can take only two values. These are specified by the two-valued spin coordinate σ, which is introduced conventionally as $\sigma = \pm 1$ corresponding to $m_s = \pm 1/2$. ($\sigma = 2m_s/\hbar$.) In the Pauli notation, which we shall require in a subsequent section, χ is written as a column matrix (a/b), where a is the value of $\chi(\sigma)$ when $\sigma = 1$, and b is the corresponding value when $\sigma = -1$.

There are two basic spin wave functions that are given a special notation. These functions occur when we imagine an external magnetic field applied to define a particular direction. The electron spin can be either parallel or antiparallel to the field so that its magnitude along the specified direction—conventionally the z axis of Cartesian coordinates—can be $\pm\hbar/2$

---

[*] Others were close behind; for example, E. C. Stoner, an English physicist whose work on the magnetic properties of atoms had stimulated Pauli's researches, apparently arrived at about the same time at what is now known as the Pauli exclusion principle, although by a different route. (L. F. Bates, *Contemp. Phy.* **13**, 601, 1972.)

(Sect. 2.3). when it is $\hbar/2$, the spin function is denoted by χ_α and can have the values.

$$\chi_\alpha(1) = 1 \quad \text{and} \quad \chi_\alpha(-1) = 0 \tag{3.3}$$

Alternatively, when the value of the spin along the z axis is $-\hbar/2$, then the spin function is denoted by χ_β and can have the values

$$\chi_\beta(1) = 0 \quad \text{and} \quad \chi_\beta(-1) = 1 \tag{3.4}$$

Written in terms of the column matrices in the Pauli formulation, Eq. (3.3) and Eq. (3.4) become

$$\chi_\alpha = \begin{pmatrix} 1 \\ 0 \end{pmatrix} \quad \text{and} \quad \chi_\beta = \begin{pmatrix} 0 \\ 1 \end{pmatrix}. \tag{3.5}$$

Returning to the form Eq. (3.2), this is frequently called a one-electron wave function (or spin orbital) to distinguish it from the orbital (or positional coordinate) wave function ψ_{n,l,m_l} of Eq. (3.1). The latter as we saw, is a function of only the positional coordinate of the electron $\psi_{n,l,m_l} = \psi(\mathbf{r})$; however the one-electron wave function is a function of both positional and spin coordinates $\phi_{n,l,m_l,m_s} = \phi(\mathbf{r},\sigma)$. It is $\phi(\mathbf{r},\sigma)$ which specifies a one-particle state, such as the one implied in the opening paragraph of this section, which we shall see can accommodate only one electron; the orbital state specified by $\psi(\mathbf{r})$ can accommodate two, providing they are of opposite spin. For the one-electron system we are considering, the one-electron wave functions can be formed by assigning χ_α and χ_β successively to the orbital part. Thus

$$\begin{aligned} \phi_u &= \psi_u(\mathbf{r})\chi_\alpha(\sigma) \\ \phi_v &= \psi_v(\mathbf{r})\chi_\beta(\sigma) \end{aligned} \tag{3.6}$$

are the one-electron functions corresponding to the orbital wave function $\psi(\mathbf{r})$. This is illustrated in Fig. 3.2a: two electrons in a central field of force can occupy the same atomic orbital, but we shall see that they must have opposite spins. Because of the spin-orbit coupling effect (introduced below) the orbital motion of the electron is, in fact, dependent upon its direction of spin. Therefore, the one-electron wave function ϕ_u and ϕ_v actually correspond to two slightly different distributions in space, even if there is no external magnetic field. The corresponding energy difference between the states represented by ϕ_u and ϕ_v is rather small, however, and generally is negligible for all but the heaviest atoms. But the important principle illustrated by all of this is that one cannot usually change the direction of an electron's spin without altering slightly the corresponding spatial charge distribution.

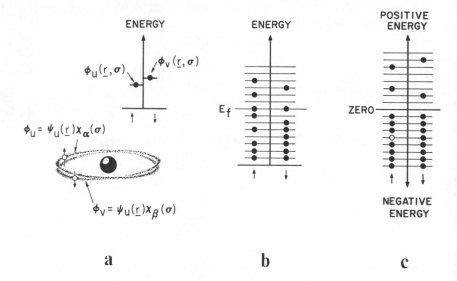

Fig. 3.2 Illustrating the operation of the Pauli principle for electrons in different situations. The electron has two possible internal alternatives available to it (known as spin directions), and in the atomic context (a) the principle dictates that the electron's probability distribution in space (described by the positional coordinate ψ_u) is slightly different for the two alternatives. There is thus a small but generally negligible energy difference between the two spin orbitals ϕ_u and ϕ_v. In a metal (b) the delocalized electrons occupy a series of closely spaced energy levels. The principle permits a limit of two per level, providing they have opposed spin directions. In the ground state (that is, at absolute zero) the levels up to some maximum E_f (called the Fermi level) are all filled; above E_f they are all empty. It is the availability of these easily accessible vacant energy levels that permits the electrons to be itinerant and characterizes the metallic state. Finally, we show Dirac's view (c) that a vacuum is not just an empty space but a region filled by unobservable particles (the extraordinary or antiparticles) having negative mass and energy. Ordinary particles, of which our universe is thought to consist primarily, are those prevented by the Pauli principle from falling into the extraordinary negative energy states of the antiparticles—of which the bulk are believed to be in any case physically separated from our universe. Only when there is a hole in the extraordinary energy distribution—which is just the corresponding antiparticle—can the ordinary particle enter an extraordinary state and so be annihilated from our universe. The result of such an annihilation is clearly a pulse of energy. Since the figure applies to electrons, the hole represents a positron (the electron's antiparticle) and the interaction is described as an electron–positron annihilation (Fig. 4.8d). On a grander scale, it has recently been suggested that at least one antiparticle meteorite has collided with the earth (*Nature,* **248**, 396, 1974).

We should note that the separation made in Eq. (3.2) can only be valid if the relativistic effect known as intrinsic spin-orbit coupling* can be overlooked. The reason for this is that, even though the electron in the one-electron atom sees only the electric field of the nucleus together with some average field induced by the other electrons, the relativistic transformation between electric and magnetic fields causes the electron to be subjected to a magnetic field because of the apparent motion about it of the nucleus (if we consider the electron to be at rest). (If E is the total electric field strength at the electron, the resulting magnetic field it experiences is of the order† Ev/c, where v is the electron's velocity and c is the velocity of light. In practice the coupling amounts to energies of about 2×10^{-21} J ($\sim 10^{-2}$eV), which is a small quantity relative to the energy of the electron; see Table 1.) There is, therefore, a force acting upon the electron mainly because it possesses an intrinsic magnetic moment and is an orbital about an electrically charged nucleus in circumstances where the nonrelativistic requirement is not satisfied.‡ This is the spin-orbit coupling referred to above and neglected when writing Eq. (3.2). Physically, this neglect corresponds to the assumption that reversing the spin of an electron would not change the form of its orbital state. This is generally a reasonable assumption, since the coupling is usually a small effect in atoms of low atomic number, but it becomes increasingly questionable as the atomic number increases. Some of its consequences will be encountered in subsequent chapters.

With the introduction of s (and hence m_s), Pauli showed that his deductions could be summarized in the statement: No two electrons can simultaneously have the same full set of quantum numbers n, l, m_l, and m_s, and, therefore, cannot occupy the same state. This is the most elementary form of the exclusion principle. It has been most useful and leads at once to an understanding of the periodic table and the main facts of chemical valence, spectroscopy, and magnetism. But it is now a rather dated form and reminiscent of the older quantum theory. For a more sophisticated version, we turn from the one-electron system to one containing N quantum particles.

3.2.4 The Pauli Principle in the N-Particle System.

An alternative statement of the Pauli principle can be formed, and some important conse-

---

* This is distinguished from extrinsic spin-orbit coupling in Sect. 3.4.6.

† Actually, the Lorentz transformation applied to the field E gives rise to a magnetic field $H = (E \times v)/(c^2 - v^2)^{1/2}$ acting perpendicular to the orbital plane of the electron.

‡ The general nonrelativistic requirement for a particle is that its energy shall be much less than the energy equivalent of its rest mass: $E \ll m_0c^2$.

quences of its existence be introduced, by emphasizing the indistinguishability of particles in a quantum system of identical but noninteracting constituents. In a system of N electrons, for example, the possibility that one or another of them could be labeled in some way and distinguished from the rest is excluded in quantum theory.* In other words, the physical interchange of any two of the particles in the system must be possible without an observer's knowledge; no manifestation or possibility of detecting the switch can exist outside the system. As we shall see, this is connected with certain restrictions upon the total wave function that describes a quantum system of identical particles, and it is part of a universal separation in nature between two classes of such systems: The fermion and the boson classes. In the particular case of a system of fermions, which is the principal interest here, the operation of the Pauli principle modifies the separation between the system's constituents in a systematic way. The magnitude of the electrostatic force between them is correspondingly modified and, rather than refer to this modification directly, it has proved convenient to picture it as the appearance of a new force in the system. This is known as the exchange force; we should realize that it is imaginary, in that it can do no work, but this is convenient terminology to describe an effect that has many and far-reaching consequences in metals. A descriptive review of these follows in Sect. 3.3.3.

First, we must appreciate the difference between a symmetrical and an antisymmetrical total wave function for a system. To see this, consider a quantum system consisting of N identical particles. The state of each of these will be specified by a function of the particle's positional coordinates and spin alternatives $\phi(\mathbf{r}, \sigma)$—exactly like the orbital wave function of Sect. 3.2.3. (A constituent could have zero spin, for we have said that usually the observable magnitude of the spin of an elementary particle is $n\hbar/2$, where n is either zero or another small integer.) The total wave function Ψ_N will likewise be some function of the spatial and spin coordinates of all the particles in the system. Thus

$$\Psi_N = \Psi(\mathbf{r}_1, \mathbf{r}_2, \ldots, \mathbf{r}_i, \ldots, \mathbf{r}_j, \ldots, \mathbf{r}_N; \quad \sigma_1, \sigma_2, \ldots, \sigma_i, \ldots, \sigma_j, \ldots, \sigma_N) \quad (3.7)$$

---

* This aspect of quantum theory has not been unreservedly accepted by all logicians. Discussion centers around the question of whether truly indistinguishable particles can even be enumerated. According to one point of view, in order to count them they must have some intrinsic distinguishing feature to differentiate one from the other, and this is prohibited in the theory. We can, of course, sidestep all of this at the present level, but it serves to illustrate how the logical foundations of quantum theory are still being worked out. Further discussion of this point of view is given by M. Jammer in *The Conceptual Development of Quantum Mechanics*, McGraw Hill, New York, 1966, p. 344.

where r_i and σ_i are, respectively, the position and spin coordinates of the ith particle. We have said that it is the core of quantum mechanics that this function Ψ_N, after its detailed form has been established for a given system, contains all the information about the motion of the system that it is possible to know. In principle, in order to extract the information, we have only to operate on Ψ_N in the appropriate manner prescribed by the theory.

Provisionally, however, we need not be concerned with the detailed form of Eq. (3.7) for any particular system, but let us consider simply what could be the effect upon Ψ_N caused by the interchange of two particles in the system. (One such interchange is known as a permutation of the particles.) Suppose the ith and jth particles undergo the general interchange involving both spin and position coordinates: $r_i \sigma_i \longleftrightarrow r_j \sigma_j$. The new wave function is

$$\Psi(r_1, r_2, \ldots, r_j, \ldots, r_i, \ldots, r_N; \quad \sigma_1, \sigma_2, \ldots, \sigma_j, \ldots, \sigma_i, \ldots, \sigma_N). \quad (3.8)$$

This can be written as $P_{ij}\Psi_N$, where P_{ij} is an operator standing for the particular permutation we have specified.

It is an established fact that any permutation P of the spin and position coordinates in the total wave function of any quantum system yet studied has one of only two consequences: Either Ψ_N is left unchanged, or it simply undergoes a sign change with no other alteration. Put another way, all known quantum systems have a total wave function which obeys

$$P\Psi_N = \pm \Psi_N \quad (3.9)$$

If Ψ_N is unchanged, it is symmetrical with respect to the operator P; if is changed then it is antisymmetrical.

Why nature should show only one or the other of these two situations is presently not fully understood, but the obvious consequence is that all the quantum systems of identical particles must fall into one of two classes depending upon whether the assemblage is described by an antisymmetrical or symmetrical total wave function. Experiments have now determined to which of these two groups the various elementary particles belong (Table 3); it emerges that the constituents of systems requiring an antisymmetrical total wave function have half odd-integer spin values (viz., $\hbar/2$, $3\hbar/2$, etc.) and that constituents of systems requiring a symmetrical total wave function have integral spin values (viz., 0, $\hbar$, $2\hbar$, etc.). Pauli has proved[*] that the incorporation of relativity into quantum mechanics requires that particles with half-integral spin values obey Fermi–Dirac statistics (these

[*] W. Pauli, *Phys. Rev.* **58** 716 (1940).

Table 3   Fermions and Bosons

Total Wave Function of the Assembly

Symmetric	Antisymmetric
photon	electron
phonon	proton
π-meson (pion)	μ-meson (muon)
K-meson	neutrino
plasmon	hyperon
deuteron	H atom
*He^4 atom	He^3 atom
graviton	neutron
magnon	
Cooper pair	
exciton	

* The isotopes of He are a very interesting example since atoms of He^4 are bosons while those of He^3 are fermions.   This leads to a striking difference between their properties at low temperatures and is a most fortunate situation provided by nature since these are the only two substances which remain liquid down to the absolute zero of temperature.   The latest development in this field is the positive demonstration [D. Kojima et al., Phys. Rev. Lett. 32, 141 (1974)] that He^3 becomes a superfluid (§4.6) when the temperature is reduced sufficiently.   Since the He^3 atoms are fermions, there must be some kind of particle pairing mechanism which produces the required boson particles for superfluidity.   Current speculation revolves around a mechanism akin to the Cooper pairing of §§ 4.5.1 and 4.6.   J. Kendall has given a nontechnical description of this subject in New Scientist 64, 100 (1974).

are called fermions) but that those with integral spin values obey Bose–Einstein statistics (and are called bosons). The conclusion is, therefore, that systems of particles for which the total wave function is antisymmetric are made up of fermions and that those for which it is symmetric are formed from bosons.

We now turn to the connection between all of this and the Pauli exclusion principle. The first step is to look at what explicity will be the form of the functions represented by Eq. (3.7) and Eq. (3.8). It turns out that not much progress can be made without invoking what is known as the single-particle approximation: In the system of N identical particles mutual interactions are ignored as a first approximation, and the particles are treated somewhat as in an ideal gas. The total energy of the system can then be written as just the sum of the energies of the individual constituents, so the system is really just a collection of N independent–particle systems.

There are N particles in the system and each will occupy a separate and generally distinct single-particle energy state. Let these states be labeled thus: u,v,w . . . . . N. Each state will be described by a corresponding single–particle wave function of the form $\phi(\mathbf{r},\sigma)$. To start, it can be assumed that the particle with coordinates $\mathbf{r}_1,\sigma_1$ is in state u, $\mathbf{r}_2,\sigma_2$ in state v, and so forth. The product of these single–particle wave functions is then

$$\phi_u(\mathbf{r}_1,\sigma_1)\phi_v(\mathbf{r}_2,\sigma_2) \ldots \ldots \ldots \phi_N(\mathbf{r}_N,\sigma_N) \tag{3.10}$$

This might appear as a likely candidate to be the total wave function of the system, but it has the defect that it is not unique because the particles could be interchanged—and hence the explicit form of Eq. (3.10) would be changed—without changing the energy of the system. For example, if we interchange the two particles having coordinates $\mathbf{r}_1,\sigma_1$ and $\mathbf{r}_2,\sigma_2$, so that we assign the latter to the state u and the former to the state v, the total energy of the system is not changed in the independent–particle approximation, but the product function becomes

$$\phi_v(\mathbf{r}_1,\sigma_1)\phi_u(\mathbf{r}_2,\sigma_2) \ldots \ldots \ldots \phi_N(\mathbf{r}_N,\sigma_N) \tag{3.11}$$

The energy of the system is said to be degenerate with respect to these product functions. (The degeneracy of a system is defined as the number of states in it having the same energy.) To remove this degeneracy all the different product functions like Eq. (3.10) and Eq. (3.11) must be formed by making all possible permutations in the system, and then a linear combination of these products must be taken. We know from the remarks made above that the only acceptable total wave functions are those that are either symmetric or antisymmetric with respect to all possible interchanges of particles.

Table 4 may help to make this clear. It shows all the possible product functions for the two- and three-particle systems. (The corresponding situation for systems containing more particles can be obtained in the obvious way, but the algebra rapidly becomes very cumbersome.) Detailed consideration, which is outside our scope, shows that the symmetrical total wave function for a system is formed by adding all the product wave functions like Eq. (3.10) and Eq. (3.11) which result from making all the possible permutations of interchanges between the particles. Thus the symmetrical total wave functions for these two-and three-particle systems are given respectively by the following equations (taking the positive signs throughout)

$$(2!)^{1/2}\Psi_N = \phi_u(1)\phi_v(2) \pm \phi_u(2)\phi_v(1) \tag{3.12}$$

Table 4

TWO-PARTICLE SYSTEM	THREE-PARTICLE SYSTEM
States labelled u and v.	States labelled u, v, and w.
Particles labelled 1 and 2.	Particles labelled 1, 2, and 3.

POSSIBLE PRODUCT FUNCTIONS

PERMUTATION	PRODUCT FUNCTION
P_0	$\phi_u(1)\phi_v(2)$
P_1 (odd)	$\phi_u(2)\phi_v(1)$

PERMUTATION	PRODUCT FUNCTION
P_0	$\phi_u(1)\phi_v(2)\phi_w(3)$
P_1 (odd)	$\phi_u(1)\phi_v(3)\phi_w(2)$
P_2 (odd)	$\phi_u(3)\phi_v(2)\phi_w(1)$
P_3 (odd)	$\phi_u(2)\phi_v(1)\phi_w(3)$
P_4 (even)	$\phi_u(3)\phi_v(1)\phi_w(2)$
P_5 (even)	$\phi_u(2)\phi_v(3)\phi_w(1)$

and

$$(3!)^{1/2}\Psi_N = \phi_u(1)\phi_v(2)\phi_w(3) \pm \phi_u(1)\phi_v(3)\phi_w(2) \pm \phi_u(3)\phi_v(2)\phi_w(1)$$
$$\pm \phi_u(2)\phi_v(1)\phi_w(3) + \phi_u(3)\phi_v(1)\phi_w(2) + \phi_u(2)\phi_v(3)\phi_w(1) \qquad (3.13)$$

These equations use the notation of Table 4, where $\phi_u(1)$ is written for $\phi_u(\mathbf{r}_1, \sigma_1)$ and so forth and where the factorial term arises from what is known as the normalization.* (The factor is incorporated to scale the wave function so that the probability of finding the electron within the

---

* This is part of a technical requirement of orthogonality of wave functions. For two orbital wave functions $\psi_u(\mathbf{r})$ and $\psi_v(\mathbf{r})$ this requirement can be written as

$$\int \psi_u^*(\mathbf{r})\psi_v(\mathbf{r})dr = 0 \text{ if } u \neq v$$
$$= 1 \text{ if } u = v$$

where the integral is taken between whatever limits define the system. We recognize the second part as the normalization condition—it expresses the fact that the probability of finding the electron somewhere in the system is unity—but the first part is more difficult to express descriptively. Mathematically, it requires that the two wave functions ψ_u and ψ_v be truly different or independent, each capable of being separately occupied by a single electron. Graphically, it means that the two functions shall have different numbers of nodes (Sect. 2.3.3).

designated system is unity, but we need not be concerned further with its origin.) The antisymmetrical total wave function is similarly obtained except that a negative sign is assigned to any product arising from an odd number of interchanges of particles (an odd permutation) but the positive sign is retained for those arising from an even number (even permutation) or from no interchange at all. Consequently the total wave functions corresponding to the cases of Eq. (3.12) and Eq. (3.13) are obtained by taking the negative sign where indicated.

For the general case of an N-particle system, this can all be written in a more concise notation. First, the symmetrical total wave function can be written

$$\Psi_N^{\text{sym}} = (N!)^{-1/2} \sum_P \prod_{i=u}^{N} \phi_i(\mathbf{r}_j, \sigma_j) \tag{3.14}$$

where the first factor on the right-hand side is again for normalization, $\sum_P$ means the sum over all permutations, and j is the subscript obtained for the $\mathbf{r}$ and σ coordinates after the permutation P.

The antisymmetrical total wave function can be written in the same notation

$$\Psi_N^{\text{antisym}} = (N!)^{-1/2} \sum_P (-1)^P \prod_{i=u}^{N} \phi_i(\mathbf{r}_j, \sigma_j) \tag{3.15}$$

where P in the factor $(-1)^P$ is the number of permutations made to obtain the particular term in the product function. We cannot leave Eq. (3.15) without referring to its alternative and familiar determinant form

$$\Psi_N^{\text{antisym}} = (N!)^{-1/2} \begin{vmatrix} \phi_u(\mathbf{r}_1, \sigma_1) & \phi_v(\mathbf{r}_1, \sigma_1) \ldots \ldots \phi_N(\mathbf{r}_1, \sigma_1) \\ \phi_u(\mathbf{r}_2, \sigma_2) & \phi_v(\mathbf{r}_2, \sigma_2) \ldots \ldots \phi_N(\mathbf{r}_2, \sigma_2) \\ \cdot \quad \cdot \quad \cdot \quad \cdot \quad \cdot \quad \cdot \quad \cdot \quad \cdot \quad \cdot \quad \cdot \\ \phi_u(\mathbf{r}_N, \sigma_N) & \phi_v(\mathbf{r}_N, \sigma_N) \ldots \phi_N(\mathbf{r}_N, \sigma_N) \end{vmatrix} \tag{3.16}$$

The form of Eq. (3.16) serves to emphasize the final point, since it is a basic property that a determinant is zero if any two rows or columns are identical. The point is that consideration of Eq. (3.14) and Eq. (3.15) shows that, when the total wave function Ψ_N is symmetrical, it does not matter if some of the single-particle wave functions are the same. In other words, more than one particle can have the same wave function $\phi(\mathbf{r}, \sigma)$ and so more than one particle can occupy the same state. But in the case where the total wave function Ψ_N is antisymmetrical such a situation is not permissible; for example, if the wave function ϕ_u was the same as ϕ_v, the antisymmetrical Ψ_N could not be formed because two of the columns in Eq. (3.16) would be identical.

This is another instance of the operation of the Pauli exclusion principle: In a system of noninteracting fermions (which automatically has an antisymmetrical total wave function) no two particles can occupy the same single-particle state corresponding to $\phi(\mathbf{r},\sigma)$, although they can occupy the same single-particle positional coordinate (or spatial) state corresponding to $\psi(\mathbf{r})$—providing they have opposite spins. No such restriction exists, of course, for a system of bosons, for which the total wave function is symmetric.

Although we have not proved it, Eq. (3.15) and Eq. (3.16) are, in fact, the only antisymmetrical combination of a given set of N one-particle functions. Consequently, a system which has an antisymmetrical total wave function will automatically satisfy the Pauli exclusion principle when the constituents are noninteracting.* The principle can be restated from this point of view: To satisfy the Pauli exclusion principle, the total wave function Ψ_N of a quantum system of noninteracting particles must be antisymmetrical with respect to the interchange of both spatial and spin coordinates of any two constituents.

The above statement implies that Pauli's principle is not just another empirical theorem in physics but is an important general requirement of nature. It is generally believed to be a fundamental precept, rather like relativity to which it is frequently compared. Both regulate the formulation of physical laws through an exclusion principle: Relativity requires its form of (Lorentz) invariance, and Pauli's principle requires invariance under the permutations outlined above.† The operation of the Pauli principle in some different contexts is illustrated schematically in Fig. 3.2. Fig. 3.2a shows two electrons in the same atomic orbital $\psi_u(\mathbf{r})$, which is the situation encountered in Sect. 2.3.1. We have seen that the principle requires the electrons to have opposite spins in this situation; as a consequence, there is a slight modification of their spatial distribution that can be regarded as a splitting of the orbital. Although it is said that the electrons occupy the same orbital, it is preferable to consider rather that they occupy separate spin orbitals, ϕ_u and ϕ_v, which are slightly separated in energy by the weak magnetic interactions between them.

---

* We emphasize this specific requirement since it is often glossed over or ignored in textbooks.

† Pauli spent a considerable amount of his research life trying to derive his principle from a more fundamental starting point. But as M. Sachs has pointed out (*The Field Concept in Contemporary Science,* Charles Thomas, Springfield, Mass. 1973), no one has yet successfully derived the principle in the most general case.

Figure 3.2*b* shows a system consisting of N fermions, such as an electron fluid. In the atomic case, such as considered in Fig. 3.2*a*, there are in general only a few hundred or so spin orbitals into which an electron in the system can be excited. But in an electron fluid, such as that formed by the itinerant electrons in a metal, there are literally billions of such quantum levels available. Because the energy spacing between these levels is extremely small (Table 1) the system gives the appearance of having a continuous distribution of possible energies, but nevertheless only the discrete quantum levels are permitted. As Fig. 3.2*b* suggests, the wave functions corresponding to the lowest energy levels are occupied preferably to capacity, this being limited by the exclusion principle to two electrons with opposite spins per wave function; in the ground state of the system the corresponding energy states are occupied up to the level E_f. The application of the exclusion principle obviously increases the range of energies of the system's constituents compared with an otherwise equivalent system in which multiple occupancy of the energy levels would be permitted.

Finally, Fig. 3.2*c* illustrates an application of the exclusion principle which, although not strictly germane to the present context, is worth a digression since it emphasizes the generality of this principle which underlies all of modern physics, and is not just restricted to the theory of solids. In what is now a famous example of a pure mathematician's confidence in the infallibility of his well-founded mathematics—no matter how fantastic the results*—Dirac proposed that each fundamental particle in nature can exist in two states: Ordinary and extraordinary, having positive and negative total energies, respectively. Furthermore, the quantum levels accomodating the extraordinary particles were postulated to be normally completely filled to the capacity permitted by the exclusion principle, as indicated in the figure, so that the ordinary quantum particles of everyday experience are prevented by the principle from giving up energy and falling into the negative energy levels of the extraordinary state. The occasional vacancy in these extraordinary states is perceived as the antiparticle of the coresponding ordinary case; thus, if Fig. 3.2*c* refers to electrons, the vacancy in the range of negative energies corresponds to a positron of

---

* Dirac, now in his seventies, has recently (*Proc. Roy. Soc. Ser.* **A338**, 439, 1974) suggested yet another upset of conventional physics, this time in the field of cosmology. Again, his mathematics have lead to an unconventional idea—that the gravitational constant is decreasing appreciably with time—but he has expressed "great confidence" in all of its radical implications. Of course, it is too soon to know whether this confidence is well placed. A short, nontechnical news report of this development has appeared in *Nature,* **250**, 460, 1974 and a more advanced but still very readable description by Dirac himself can be found in: *The Physicist's Conception of Nature,* Ed. J. Mehra, Reidel Pub. Co., Boston, 1973, p. 45.

appropriate energy. The annihilation of an electron-positron pair (such as the process represented in Fig. 4.8d) corresponds to the transition from an occupied state in the ordinary range of energies to the vacancy in the extraordinary one.*

Returning to the main argument, we have reached the statement of Pauli's principle mentioned above that is relevant to an N-particle system. We can look in two directions from this position. First, it is seen that the one-electron atom of Sect. 3.2.3 is a special case of an N-particle system containing but one fermion and that tightly bound to a parent nucleus. The single-particle state described by $\phi(\mathbf{r},\sigma)$ of the N-particle system is then just the spin orbital function of Eq. (3.2) in which the state corresponding to the coordinate $\mathbf{r}$ is specified by the three quantum numbers n, l, and m_l. So the statement of Pauli's principle given at the end of Sect. 3.2.3 is now obtained directly from the remarks following Eq. (3.16).

Secondly, we can turn to the system of itinerant electrons in a metal and, as a first approximation, look upon them also as a special case of an N-particle system of noninteracting fermions. The electrons behave more as if free than as if bound to particular atoms; thus it is inappropriate to call these single-particle functions $\phi(\mathbf{r},\sigma)$ the spin orbital functions; spin positional functions is better. The distinguishing features in this case are the electron's linear momentum $\mathbf{p}$ and spin orientation σ; as a result, the appropriate quantum numbers to specify an itinerant electron's state are just those of linear momentum and spin. In this case of noninteracting particles the appropriate Schrödinger equation has the well-known solutions $\psi(\mathbf{r})$ $\equiv \psi_k = \exp i\mathbf{k}\cdot\mathbf{r}$ corresponding to Eq. (3.1). These are traveling plane waves with a wave vector $\mathbf{k}$ ($= \mathbf{p}/\hbar$), and this is the quantum number which is used to specify the electron's linear momentum. Thus each spin positional function is specified by a pair $(\mathbf{k},\sigma)$ and it can be either empty or occupied by at most one electron. Referring to Fig. 3.1a, we see that if two itinerant electrons have the same spatial wave function $\psi(\mathbf{r})$ they must have the same momenutm ($\lambda=h/p$). It follows immediately, from the remarks made just after Eq. (3.16), that two itinerant electrons with the same momentum must, therefore, have opposite spin alternatives. Thus Pauli's principle can be stated in this context: no two itinerant and non-

---

* This is all closely related to the origin and usefulness of the Feynman diagram discussed in Sect. 4.3.3, for Feynman showed that the representation of a particle propagating forwards in time is the same as that of the antiparticle propagating backwards in time. This gives an important and wider application of the diagram than is obvious from our consideration of it, but such details are outside our present scope. However, this is the explanation of the fact that the positron of Fig. 4.8d is shown propagating backwards in time.

interacting electrons in a metal can simultaneously occupy states having the same linear momentum and spin.

Referring to Fig. 3.1b, we see that if two electrons are localized upon the same point **r** [so that they have the same spatial wave function $\psi(\mathbf{r})$] they must again have opposite spin alternatives. Electrons with like spin are thus prevented from coming close to each other in a way that electrons of unlike spin never experience. This is another manifestation of the Pauli principle; it has become known as the exchange interaction. It forms the subject of Sect. 3.3.

3.2.5 Electrons in Ions: Spin and Orbital Angular Momentum.

The manner in which an atomic electron's orbital is characterized is discussed in Sect. 2.3.1. Those remarks relate to the so-called one-electron case; that is, the electron is regarded as being alone in the isolated atom and not subjected to interactions with the other residents. But in subsequent sections we will examine the more realistic case of an ion—particularly one in a metal—in which an electron is not alone but shares the nucleus with several others. We find in that case that the mutual interactions cannot be ignored since they have important consequences for the net orbital and spin angular momenta of the electrons in the ion. This is a subject which has been well aired, since it is discussed in almost any textbook of solid state physics; nevertheless some readers will probably find it useful to have an outline of its salient features, and that is the purpose of this section.

Our focus here is how the individual orbital and spin angular momenta of the one-electron states are combined to give a net orbital and spin momentum to the many-electron ion. The observed behavior reflects the interplay of four major forces in the ion (Fig. 3.3). The two principal ones are: The Coulomb repulsion between the electrons, which makes them stay out of each other's way to the maximum extent possible and is equivalent to a coupling between their individual angular momenta, and the exchange force, which is just (Sect. 3.3) the modification of the Coulomb force produce by the exclusion principle and is tantamount to a coupling between the electrons' spin angular momenta. The remaining forces—which are generally of secondary importance, at least in the Fe group and the rare earths (Fig. 3.3)—are the intrinsic spin-orbit coupling (Sect. 3.2.3) between an electron's angular and spin momenta and forces arising from externally applied fields. Examples of the latter two forces are the electrostatic fields arising from neighboring ions in the lattice, and magnetic fields purposely applied in the course of an experiment.

Hund's rule for atoms is a descriptive expression of the influence of the two principal forces; it states that, if a given shell is filled by successively

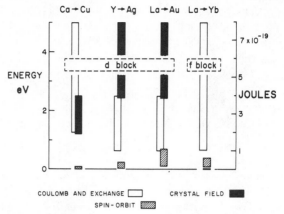

Fig. 3.3 Showing the approximate ranges of energy and relative magnitudes of the principal contributions to the energy of an ion from any of the four most important transition groups in the periodic table. The Coulomb term, with its associated modification produced by the exclusion principle, acts first to align the electron's spins to the maximum extent possible; it acts second to keep the electrons out of each other's way to the maximum extent possible. These two criteria, which together make up Hund's first rule for atoms, determine the ion's ground state and, hence, its L and S values. The energy required to reach the first excited state by altering this configuration lies in the range represented in the figure by the open rectangle (the Coulomb and exchange contribution). In the Fe series of the d block and the rare earth series of the f block, this is generally by far the most important term, but in the other transition series of the d block the energy arising from the electrostatic fields of surrounding ions (the crystal field contribution) is comparable in magnitude.

adding electrons, the occupancy of the various combinations of l and s is always such that **S** (the shell's resultant spin angular momentum) has the largest possible value consistent with the exclusion principle, and that **L** (the shell's resultant orbital angular momentum) has the largest value that can be associated with this maximum **S**. This is another way of saying that the exchange forces dominate the straightforward Coulomb repulsion between electrons: Given the choice between sharing a state with an electron of opposite spin and occupying an otherwise empty state—albeit of higher angular momentum—the electron will accept the second alternative. A specific example will perhaps help make this clear.

Consider the $4f$ shell of a rare earth ion. For this $l = 3$ and so there are $2l + 1 = 7$ distinct orbital states (Sect. 2.3.1) corresponding to the seven different components $m_l\hbar$ of angular momentum along the z-direction. In each of these there are two alternatives corresponding to the two $m_s\hbar$ components of spin angular momentum along the z-direction. Suppose three electrons occupy this shell. What choice among these fourteen alternatives gives the lowest total energy (that is, the ground state) for this shell? And what are the values of **S** and **L** for this ground state? Following Hund's rule, the answers are that, firstly, the electrons prefer to occupy separately, and with spins parallel, three different m_l states; in this way the value of $S = \Sigma m_s$ can be its maximum of $1/2 + 1/2 + 1/2 = 3/2$. Secondly, the electrons choose those m_l states which maximize $L = \Sigma m_l$. This occurs when $m_l = +3, +2$, and $+1$, are the occupied states, and, as a result, the ground state of these three electrons has $\mathbf{S} = 3/2$, $\mathbf{L} = 6$. The permissible configurations of the ion's total orbital and spin momenta are given by expressions similar to those of the one-electron case (Sect. 2.3.1). The z components of the orbital and spin momenta are, respectively, $M_L\hbar$ and $M_S\hbar$, where* M_L can have the values $\pm L$, $\pm(L-1)$, $\pm(L-2)$, $\ldots.\, 0$, and M_S can be likewise $\pm S$, $\pm(S-1)$, $\pm(S-2),\ldots.\, 0$. This approach, where m_l and m_s are summed separately, is known as Russel-Saunders coupling

Values of L and S corresponding to the ground state can be obtained by similar arguments for whatever shell and electron population is being examined. Figure 3.4 shows how they vary in the two important series that produce magnetism in metals: The iron group of the d block and the rare earth group of the f block. These graphs illustrate two important points. First, when all possible states in a shell are filled both the total orbital momentum and the total spin momentum of the shell are zero. This obviously is caused by the exact cancellation of the contributions from electrons in the different $\pm m_l$ and $\pm m_s$ states. An alternative manifestation of this is described in Appendix A where it is pointed out that combining all the different charge cloud pictures for a given n, l shell gives a spherically symmetric distribution, which, of course, can show no net angular momentum. Second, when a shell is exactly half filled, its net angular momentum is zero (because all the $\pm m_l$ orbital states are then singly occupied, and again their individual contributions exactly cancel); as a result, the shell's magnetism is caused solely by the combination of the individual spin angular momenta. Gd and Mn ions are examples of this, providing the two minor forces in the quartet described above are ignored.

The last remark brings us to the next refinement in the discussion, and that is to consider the two influences—spin-orbit coupling and externally applied fields—which we have said are of minor importance in comparison

---

* M_L and M_S are sometimes known as the total momentum and spin projection quantum number, respectively, since they define the projection along the z-axis of the angular and spin momenta of the ion.

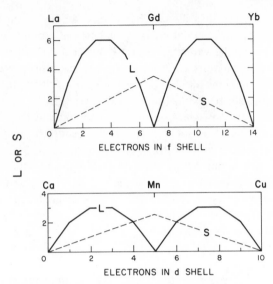

Fig. 3.4 Showing the variation of L and S for the ground states of ions where the Coulomb and exchange contribution to the total energy is dominant (Fig. 3.3). (*Upper*) The f block series La → Yb. (*Lower*) The d block series Ca → Cu. Quite a different behavior is found in a series for which the crystal field effects are comparable with the Coulomb and exchange contribution. (After D. H. Martin, 1967, *Magnetism in Solids,* Iliffe Ltd., London.)

with the Coulomb and exchange forces. First, let us consider the intrinsic spin-orbit coupling (Sect. 3.2.3). When this is significant the angular and spin momenta of an individual electron in an ion can no longer be treated as independent of each other, as they are in the Russel-Saunders coupling mentioned above. For example, if the orientation of an electron's spin momentum is shifted (say through the application of a magnetic field) the orbital momentum will be dragged along with it to an extent depending on the strength of the coupling, so that some component of the angular momentum will exist along the spin momentum's direction. If the spin-orbit interaction favors parallel alignment of these vectors, then the effective magnetic moment associated with the electron's spin will obviously be greater than for the spin alone; in other words, the effective g value (of Sect. 3.2.3) will appear to be greater than 2. If intrinsic spin-orbit coupling is very large and completely swamps the exchange coupling between the electrons' spins (which is the case for only a few metals of very large atomic number) then the resultant angular momentum **j**, which is the vector sum of **l** and **s** for each electron, has to be formed for each orbital.

These individual j's have then to be added vectorially to give the total angular momentum **J** for the ion. This is known as j-j coupling and is, of course, the opposite extreme of the Russel-Saunders coupling. As Fig. 3.3 indicates, the conditions under which it is applicable are not generally encountered in the commonly available transition elements.

Usually, the exchange interactions that couple the intrinsic angular momenta s_i of the electrons in the incomplete shell are dominant, and the spin-orbit coupling is of secondary importance. Russel-Saunders coupling is, therefore, commonly the appropriate description: The individual orbital and spin angular momenta of the shell are combined separately to give the vectors **L** and **S**, and these combine in turn to give a vector sum* **J**, but because of the spin-orbit coupling the total energy of the configuration depends upon the relative orientations of **L** and **S**. This energy takes the form $\lambda \mathbf{L \cdot S}$, where λ, which varies from ion to ion of different species, is known as the spin-orbit coupling constant.

We should recall that this coupling between **L** and **S** arises basically from the magnetic field produced by one of them acting on the magnetic moment possessed by the other. The effect of the coupling can be depicted in a familiar vector diagram (Fig. 3.5) as causing the precession of both **L** and **S** about **J**; the rate of precession being determined by the magnitude of λ. Since the magnetic dipole moments associated with the angular momenta **L** and **S** are colinear with their respective precursor vectors, their resultant $\boldsymbol{\mu}$ will also precess about **J**. This means that $\boldsymbol{\mu}$ is resolved into two components: A fixed one, which is the projection onto **J**, and a precessing one, which is normal to it. In vector notation $\boldsymbol{\mu}$ is given by $\mu_B(\mathbf{L}+2\mathbf{S})/\hbar$, where the factor 2 appears because of the anomalous g-factor of the electron's intrinsic angular momentum (see Sect. 2.3.1); its projection along **J** is $\beta\mu_B(\mathbf{L}+2\mathbf{S})/\hbar$. The magnitude of this fixed component along **J** is, therefore, $\beta\mu_B[J(J+1)]^{1/2}$. It is this component that is of interest, rather than the precessing one, because it interacts with any externally applied magnetic field, as is discussed subsequently in this section.

Next we turn to the effects upon the ion of fields that arise externally. Two kinds of fields can be distinguished: Magnetic fields, which are applied voluntarily and which couple with the magnetic dipole moments in the ion,

---

* We must remember that in all of this the **L, S, J** are vectors representing angular momenta of magnitude $\hbar[L(L + 1)]^{1/2}$, $\hbar[S(S + 1)]^{1/2}$, and $\hbar[J(J + 1)]^{1/2}$, respectively. The corresponding magnetic moments associated with the momenta are $\mu_B[L(L + 1)]^{1/2}$, $2\mu_B[S(S + 1)]^{1/2}$ and $\beta\mu_B[J(J + 1)]^{1/2}$, where β is the Landé factor; it is the ratio of the component of $\boldsymbol{\mu}$ along **J** (in Bohr magnetons) to the angular momentum component along **J** (in units of $\hbar$). Thus $\beta = (\hbar/\mu_B) \boldsymbol{\mu}\cdot\mathbf{J}/\mathbf{J}^2 = (\mathbf{L} + 2\mathbf{S})\cdot\mathbf{J}/\mathbf{J}^2$.

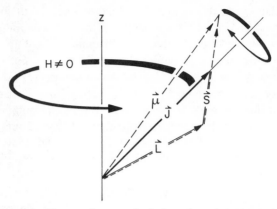

Fig. 3.5 An example of a purely metaphorical analogy between quantum processes on the microscopic scale and familiar classical processes on the macroscopic one. It shows the configuration of angular momenta (solid vectors) and their corresponding magnetic moments (dashed vectors) in the case of Russell–Saunders coupling with a small spin-orbit interaction mixed in. With no applied magnetic field the total angular momentum vector **J** is fixed in direction, but **L** and **S** precess about it because of their spin orbit coupling. The total moment **μ**, therefore, also precesses about **J** with the result that the only fixed component it possesses is its projection on **J** (say μ_J). With a magnetic field **H** applied, which is assumed small enough not to disrupt the arrangement, μ_J and, hence, **J** precess about its direction (z in the figure). Again, the only fixed components that these vectors possess are their projections on **H**. In a quantitative application of the analogy we must remember that **L**, **S**, and **J** represent momenta of magnitude $\hbar[L(L+1)]^{1/2}$, $\hbar[S(S+1)]^{1/2}$, and $\hbar[J(J+1)]^{1/2}$, respectively. The corresponding magnetic moments have magnitudes $\mu_B[L(L+1)]^{1/2}$, $2\mu_B[S(S+1)]^{1/2}$ and $\beta\mu_B[J(J+1)]^{1/2}$, respectively, where β, the Landé factor, is the projection of **μ** on **J**. The precessional motion in this model should not be taken too literally; it really expresses the fact that the vector's position on the precessional cone (Fig. 2.5) is unknown and unpredictable. Thus the corresponding vector splits into two components: one fixed and definite (along the cone's axis) and the other indefinite (normal to the cone's axis).

and electrostatic fields arising from the presence of nearby ions in the lattice. The latter are known as crystal fields, and physically they are the result of the straightforward Coulomb repulsion between the outermost electrons in a shell and the charges on neighboring ions. (They are, therefore, nonuniform electrostatic fields since they have the particular symmetry of the ion's environment.) This repulsion may be influential enough to upset the ground state dictated by Hund's rule: Some electrons may find it preferable to relocate in other m_l orbitals with shapes that will keep them further away from the repulsive ions. The crystal field can, therefore, change the electron configuration among the different orbitals and in this way change the electrons' angular momenta; this appears as a coupling between the orbital momentum and the electrostatic field, but at root its origin is the straightforward Coulomb repulsion between like charges. (Further details relating specifically to d orbitals are given in Sect. 3.4.4.) To a first approximation there is no coupling between a magnetic dipole moment and the electrostatic crystal field; the crystal field effects involve exclusively the orbital momenta of the ion's constituents. Such effects are inevitable to some degree in an incomplete shell of any ion in a crystalline environment, but their importance depends very much on whether the electrons in the shell move in the ion's outer regions (and are, therefore, exposed to the full effect of the field) or are located deep in the ion and thereby shielded by other orbitals. The latter circumstance exists for the $4f$ shells of the rare earth series (Sect. 2.3) and, consequently, crystal field effects are relatively unimportant for ions of this series (Fig. 3.3).

Before leaving the effects of the crystal field we should refer to the important phenomenon of quenching of the orbital angular momentum that it can produce. In this condition, although the total orbital momentum of an ion may have a definite nonzero magnitude, its projection on any axis is found to be zero. It is as if the crystal field makes the electrons in the ion constantly adjust their configurations to minimize the Coulomb repulsion from neighbors, with the result that the ion's total orbital momentum vector is continually changing its direction; thus the time-averaged component along any direction is zero. The magnetic dipole moment associated with this angular momentum behaves concomitantly, so that when the orbital momentum is quenched the ion has correspondingly no measurable orbital magnetic moment along any direction; this can greatly reduce its net magnetism. Whether or not quenching occurs depends upon the symmetry of the ion's environment in the crystal lattice; that is, it depends upon the symmetry of the crystal field. The rare earth metal Pr provides an interesting example of this: In its α phase it has the d-hex crystal structure (Sect. 5.2 and Fig. 5.7) in which the ions in alternate layers of the structure have environments of alternately hexagonal and cubic symmetry. In this case, if we examine the incomplete $4f$ shell, it turns out that the orbital momentum is quenched in ions occupying cubic sites and remains unquenched in those occupying the hexagonal ones. So the magnetic moments of ions of these two classes have quite different magnitudes, even though the ions are otherwise identical constituents of a pure metal. Another important case of quenching of orbital momentum occurs for the incomplete $3d$ shell of any ion in an octahedral environment, for example, the face-centered and body-centered cubic structures discussed in Sect. 3.4.4. In all cases where quenching occurs it is clear that the ion's remaining magnetism arises only from the intrinsic spins of the unpaired electrons in the incomplete shell.

Finally, we will make a few comments on the effects of a magnetic field upon an incomplete shell in an ion. As we have already pointed out, unlike the crystal field, which is inevitably present in any metal, the magnetic field is generally applied by choice as an external magnetizing force. It couples with the magnetic moments arising from the electrons' orbital and spin momenta, but in normal circumstances its contribution to the ion's total energy is quite negligible compared with the others we have considered. The energy of an ion subjected to a field of say $3T$ (30 kOe) changes by an amount typically about 1.6×10^{-22} J (10^{-3} eV), which is insignificant compared with the usual Coulomb, spin-orbit, or crystal field contributions (see Fig. 3.3). This does not imply that the application of a magnetic field has unimportant consequences (for quite the contrary is true) but it does mean that such a field rarely has a disruptive effect on an incomplete shell's configuration. The exception occurs when it becomes comparable with the spin-orbit coupling energy (this is mentioned below). With the field off, the total orbital and spin momentum vectors (L and S), which are always coupled to some extent by the intrinsic spin-orbit coupling, have a vector sum J about which they precess. But with the magnetic field H turned on, J's direction is no longer fixed in space since this vector itself now precesses about the direction of H. As long as the spin-orbit coupling between L and S is much larger than the influence of H, we can imagine L and S precessing about J at very much higher frequency than J precesses about H. Viewed from outside the system, the time-averaged result of these dynamics is that the components of J and S acting perpendicular to J appear to be zero, as do the corresponding components of the magnetic moments associated with them; the only constant nonzero component of the angular momentum is that appearing along J. It is the magnetic moment associated with this component that is acted upon by the field and which leads to J's precession about H.

Angular momentum on this microscopic scale is quantized as we have emphasized repeatedly, and is characterized by two quantum numbers: J, which specifies the magnitude of the shell's total angular momentum, $\hbar[J(J+1)]^{1/2}$ as cited above, and M_J, which specifies its projection along any given direction—say that of H. This projection has the value $M_J\hbar$, where M_J can be one of only $\pm J,\ \pm(J-1),\ \ldots\ 0$; a total of $2J + 1$ values. The corresponding magnetic dipole moment has quantized values along H of magnitude* $M_J\mu_B\beta$, but their direction is exactly opposed to

---

* It follows from remarks made in Sect. 2.3.1 that $\beta = 2$ if $L = 0$ (the spin only case) while $\beta = g = 1$ if $S = 0$ (the orbit only case). Hence, β usually has a value lying between 2 and 1.

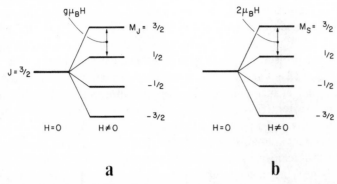

Fig. 3.6 Showing the splitting of the energy levels of an incomplete shell in an ion that is produced by an external magnetic field H. In a the ion has a chosen ground state with $J = 3/2$ (that is, $L = 1$, $S = 1/2$). The Zeeman energy separating any two adjacent levels is $\Delta M_J g \mu_B H$, which is just $g \mu_B H$ since the selection rules limit ΔM_J to ± 1. In b the orbital momentum is assumed to be entirely quenched so that $M_L = 0$. The energy steps on the Zeeman ladder are in this case $2\Delta M_S \mu_B H$, which is again limited to just $2\mu_B H$ by a selection rule. In an applied field of 1 T this energy step is about 1.6×10^{-23} J (1×10^{-4} eV).

the direction of the total angular momentum vector (Sect. 2.3.1). Any magnetic dipole moment $\mathbf{\mu}$ placed in a magnetic field $\mathbf{H}$ has a resultant potential energy of $\mathbf{\mu} \cdot \mathbf{H}$; that is, the energy depends upon the projection of $\mathbf{\mu}$ on $\mathbf{H}$. In the present case, the potential energy of the shell's dipole moment is, therefore, $(\mu_B/\hbar)(\mathbf{L} + 2\mathbf{S}) \cdot \mathbf{H}$, but since the only finite component of $\mathbf{\mu}$ lies along $\mathbf{J}$ and is $\beta\mu_B[J(J+1)]^{1/2}$ in magnitude, the potential energy can be written alternatively in the vector shorthand* as $\beta\mu_B\mathbf{J} \cdot \mathbf{H}$. However, this form does not emphasize the quantization that limits the projection of $\mathbf{J}$ on $\mathbf{H}$ to the fixed values cited above; it is easy to see that the dipole's potential energy is correspondingly limited to the values $M_J\beta\mu_B H$. Thus the $2J + 1$ different values of M_J, which give configurations of the same energy when $H = 0$, become separated in energy by an amount $\beta\mu_B H$ when the field is turned on. This energy separation is known as the Zeeman energy; an example of its magnitude is given in the caption to Fig. 3.6.

If the orbital angular momentum of the shell is quenched, so that $\mathbf{L}$ has no component along any axis, then $\mathbf{J}$ is obviously made up entirely from

---

* It is worth noting a point in passing, since it will subsequently be useful: The vectorial expression for β given in a preceding footnote leads to the relations: $\mathbf{L} = (2 - \beta)\mathbf{J}$ and $\mathbf{S} = (\beta - 1)\mathbf{J}$. Thus $\mathbf{L}$ is always parallel to $\mathbf{J}$ in the ground state, but $\mathbf{S}$ can be parallel or antiparallel to $\mathbf{L}$ depending upon whether β is greater or less than 1.

spin contributions. (This is the spin only case mentioned in the preceding footnote.) The projection of **S** on **H** is then the only quantity with significance and it too is quantized with permissible values of $M_S\hbar$, where M_S, the spin projection quantum number encountered above, can be one of only $\pm S, \pm(S-1), \ldots\ldots\ldots 0$. The corresponding magnetic moment is $M_S\beta\mu_B$ and again points in the opposite direction to the spin angular momentum vector (Sect 2.3.1). In this case the Zeeman energy between the M_S configurations permitted in the field is just $2\mu_B H$.

Under extreme circumstances of very high applied field strengths (~ 10 T), very low temperatures ($\sim 4°K$), and an ion of low atomic number, the coupling of the magnetic moments associated with the **L** and **S** vectors to the applied magnetic field may become comparable or even greater than their mutual coupling through spin-orbit interaction. Then the applied field will have a disruptive effect since under its influence the coupling of **L** and **S** to form **J** will be broken down, and **L** and **S** will precess independently about the applied field's direction. Their projections along **H** are separately quantized, in the usual fashion, and specified by projection quantum numbers M_L and M_S. The result is that the projection of the shell's net magnetic moment along **H** is changed as the field increases and overcomes the spin-orbit coupling. This leads to nonlinear dependence of the Zeeman step (Fig. 3.6) upon H; it is called the Pashen-Back effect. It is not an important effect in metals under normal circumstances.

3.3 THE EXCHANGE INTERACTION

3.3.1 Direct Exchange. At the end of a preceding section we introduced two manifestations of the Pauli principle that arise in the context of the itinerant electrons in a metal. The first is that no two such electrons can have the same linear momentum and spin, and the second is that electrons of like spin tend to stay out of each other's way to a greater extent than those of unlike spin. Each of these has important consequences leading to the distinctive features of the metallic state. The first means that a given energy state in a gas (or fluid) of electrons may not be occupied by more than two electrons, and they must have opposite spins. This obviously increases enormously the energy of such a gas over one where multiple occupancy of energy states is permitted. The following gives an idea of just how much energy this involves. Subsequent sections show that, if the itinerant electrons in a metal are bosons, at a given temperature T they will occupy energy states up to maximum of about $k_B T$, which at room temperature is about 4×10^{-21} J (0.025 eV). But in a real metal at this temperature the electrons are forced to occupy states up to an energy E_f, which is about 240 times larger than $k_B T$. The difference is due to the operation of the Pauli principle. These effects are considered further in Chapter 4; in this section we consider the consequences of the second manifestation, which is the introduction of a new, effective force between electrons having identical spin alternatives.

To illustrate this we take the conventional approach and return to the two-particle system of Table 4. Specifically, a system is considered in which two electrons are localized about the same nucleus (such as the helium atom, for example, but the reader should bear in mind that qualitatively the following remarks could apply equally to two electrons temporarily localized about the same nucleus in a metal). Note, therefore, that the following remarks relate specifically to intraatomic exchange, which is distinct from interatomic exchange where the interacting electrons are localized about different nuclei. We turn briefly to the latter at the end of the section.

In the intraatomic case two situations can be distinguished: Either the electrons are in the same orbital, say $\psi_u(\mathbf{r})$ or they are in different orbitals on the same atom, say $\psi_u(\mathbf{r})$ and $\psi_v(\mathbf{r})$. There are also two other distinct alternatives: The directions of the electrons' spin are usually either parallel or antiparallel (these are called the triplet and singlet states, respectively). (We shall ignore throughout this section the spin-orbit effects discussed in Sect. 3.2.3.)

There are actually two interactions taking place between these two electrons. The first is caused by the Coulomb force, which in this case leads to the mutual repulsion of like charges. The second, which is a purely magnetic interaction, is from the Amperian force between the magnetic dipoles that each electron possesses by virtue of its spin. In the case of two electrons localized about the same nucleus, such as we are considering, the magnetic interaction between them is quite negligible compared with the electrostatic one. Therefore, we shall henceforth be concerned solely with the Coulomb part of the total energy of the two-electron system, and we shall overlook entirely the Amperian interaction.

Concerning this Coulomb energy, we find that because of the requirement of antisymmetry imposed by Pauli's principle (since electrons are fermions) a new term appears; since it has no classical counterpart, it presents conceptual difficulties, but it can be regarded as a correction or modification of the Coulomb term produced by the Pauli principle. It is known as the direct exchange interaction, and it is peculiar to systems consisting of identical fermions in such proximity that their spheres of influence overlap so that there is a coupling between them. Its magnitude depends, in fact, upon the amount of overlap of the wave functions of the interacting particles, as well as upon the relative orientations of their spins.

Let us consider separately the two cases distinguished above in which the electrons are: (a) in the same orbital, and (b) in different orbitals on the same atom. In each of these cases we imagine for convenience that the system is set up with the mutual Coulomb interaction between the electrons turned off. The interaction in each system is then turned on, and the change it produces in the system's total energy is considered. Conventionally, this is treated as a small perturbation of the system, and the results are obtained through the application of the standard perturbation theory of quantum mechanics.* We see that in case (a) above, the system can only exist in the singlet state—as dictated by Pauli's principle. In this case switching on the interaction produces the Coulomb repulsion without embellishment; there can be no evidence of an exchange interaction here because no other distinguishable configurations are possible. However, in case (b) an energy difference appears between the singlet and triplet configurations: The triplet configuration (with spins parallel) has an energy lower than the singlet configuration (with spins opposed) by an amount that is typically about 10^{-19} J (i.e., a few eV) in magnitude, an amount that is an important quantity. All of this is summarized in Fig. 3.7.

Let us look at case (a) first: When the two electrons are in the same orbital, we know from the Pauli principle that they must have antiparallel spins. In that case the two one-particle wave functions for the system can be written [following Eq. (3.6)] as

$$\phi_u(\mathbf{r},\sigma) = \psi_u(\mathbf{r})\chi_\alpha(\sigma)$$
$$\phi_v(\mathbf{r},\sigma) = \psi_u(\mathbf{r})\chi_\beta(\sigma) \tag{3.17}$$

Here $\psi_u(\mathbf{r})$ is the atomic orbital in which the two electrons are located. The ϕ_u and ϕ_v correspond to states of identical energy, say E_u, so that the total energy of the system with the Coulomb repulsion turned off is just $2E_u$.

Equation (3.16) shows that the total wave function for the system is

$$\Psi_N = \frac{1}{\sqrt{2}} \begin{vmatrix} \phi_u(\mathbf{r}_1,\sigma_1) & \phi_v(\mathbf{r}_1,\sigma_1) \\ \phi_u(\mathbf{r}_2,\sigma_2) & \phi_v(\mathbf{r}_2,\sigma_2) \end{vmatrix} \tag{3.18}$$

---

* Considerations of space and purpose preclude a detailed coverage here of what is in any case well-trodden ground. The reader seeking more detailed coverage is referred to, for examples, S. Raimes, *The Wave Mechanics of Electrons in Metals*, (North Holland, Amsterdam, 1961) Sec. 5.7; P. W. Atkins, *Molecular Quantum Mechanics*, (Oxford University Press, Oxford, 1970) Chap. 7 and page 249; or practically any textbook on quantum mechanics.

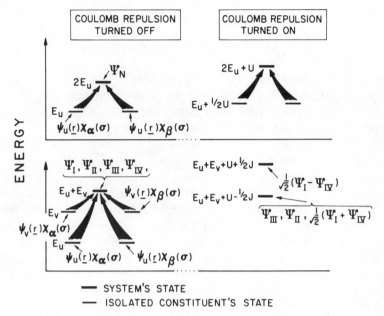

Fig. 3.7 Schematic view of the possible energy configurations of the two-particle systems discussed in the text. In the upper part the two particles occupy the same orbital in an atom; in the lower they occupy different orbitals in the same atom.

which when written out in full becomes

$$\Psi_N = \frac{1}{\sqrt{2}} \psi_u(\mathbf{r}_1)\psi_u(\mathbf{r}_2)[\chi_\alpha(\sigma_1)\chi_\beta(\sigma_2) - \chi_\alpha(\sigma_2)\chi_\beta(\sigma_1)] \qquad (3.19)$$

If we now imagine the electrostatic force between the two electrons to be turned on, with the result that the system is perturbed slightly, the methods of first-order perturbation theory can be used to give an expression for the extra energy U produced in the system by this Coulomb repulsion between the electrons. In the usual quantum fashion, this expression for U turns out to be

$$U = \iint \Psi_N^* \frac{e^2}{r_{12}} \Psi_N \, d\mathbf{r}_1 \, d\mathbf{r}_2 \qquad (3.20)$$

where $\mathbf{r}_{12}$ is the separation between the electrons, and $d\mathbf{r}_1$, $d\mathbf{r}_2$ are the elemental volumes of configuration space. Substitution of Eq. (3.19) into Eq. (3.20) gives for U

$$\frac{1}{2} \sum_{\sigma_1, \sigma_2 = \pm 1} [\chi_\alpha(\sigma_1)\chi_\beta(\sigma_2) - \chi_\alpha(\sigma_2)\chi_\beta(\sigma_1)]$$
$$\iint \psi_u^*(\mathbf{r}_1)\psi_u^*(\mathbf{r}_2) \frac{e^2}{r_{12}} \psi_u(\mathbf{r}_1)\psi_u(\mathbf{r}_2) d\mathbf{r}_1 d\mathbf{r}_2 \qquad (3.21)$$

which is just

$$U = \iint \psi_u{}^*(\mathbf{r}_1)\psi_u{}^*(\mathbf{r}_2)\,\frac{e^2}{r_{12}}\,\psi_u(\mathbf{r}_1)\psi_u(\mathbf{r}_2)\,d\mathbf{r}_1 d\mathbf{r}_2 \qquad (3.22)$$

This makes use of the fact that the bracketed spin factor in Eq. (3.21) is unity when σ_1 and σ_2 have opposite signs, and zero otherwise [see Eq. (3.3) and Eq. (3.4)]. With the Coulomb repulsion turned on the total energy of the system is just $2E_u + U$.

Next let us look at the case (b): When the two electrons are located on different orbitals on the same atom (which corresponds in the atomic context to an excited state of the atom in which one electron has been promoted to another orbital of higher energy) there is no restriction upon the electrons' spin alternatives; they can be either parallel or antiparallel. The result is that there are now four one-particle wave functions corresponding to the possible configurations which the electrons can adopt, that is $\psi_u(\mathbf{r})\chi_\alpha(\sigma)$, $\psi_u(\mathbf{r})\chi_\beta(\sigma)$, $\psi_v(\mathbf{r})\chi_\alpha(\sigma)$ and $\psi_v(\mathbf{r})\chi_\beta(\sigma)$. Correspondingly, four possible antisymmetrical total wave functions can be written down. If this quartet is denoted by Ψ_I, Ψ_II, and so forth, then the first possibility [already given in Eq. (3.18)] is

$$\Psi_\mathrm{I} = \frac{1}{\sqrt{2}} \begin{vmatrix} \psi_u(\mathbf{r}_1)\chi_\alpha(\sigma_1) & \psi_v(\mathbf{r}_1)\chi_\beta(\sigma_1) \\ \psi_u(\mathbf{r}_2)\chi_\alpha(\sigma_2) & \psi_v(\mathbf{r}_2)\chi_\beta(\sigma_2) \end{vmatrix} \qquad (3.23)$$

and the others are obtained by obvious permutations. Thus

$$\Psi_\mathrm{II} = \frac{1}{\sqrt{2}} \begin{vmatrix} \psi_u(\mathbf{r}_1)\chi_\beta(\sigma_1) & \psi_v(\mathbf{r}_1)\chi_\beta(\sigma_1) \\ \psi_u(\mathbf{r}_2)\chi_\beta(\sigma_2) & \psi_v(\mathbf{r}_2)\chi_\beta(\sigma_2) \end{vmatrix} \qquad (3.24)$$

$$\Psi_\mathrm{III} = \frac{1}{\sqrt{2}} \begin{vmatrix} \psi_u(\mathbf{r}_1)\chi_\alpha(\sigma_1) & \psi_v(\mathbf{r}_1)\chi_\alpha(\sigma_1) \\ \psi_u(\mathbf{r}_2)\chi_\alpha(\sigma_2) & \psi_v(\mathbf{r}_2)\chi_\alpha(\sigma_2) \end{vmatrix} \qquad (3.25)$$

and

$$\Psi_\mathrm{IV} = \frac{1}{\sqrt{2}} \begin{vmatrix} \psi_u(\mathbf{r}_1)\chi_\beta(\sigma_1) & \psi_v(\mathbf{r}_1)\chi_\alpha(\sigma_1) \\ \psi_u(\mathbf{r}_2)\chi_\beta(\sigma_2) & \psi_v(\mathbf{r}_2)\chi_\alpha(\sigma_2) \end{vmatrix} \qquad (3.26)$$

With the electrostatic force between the electrons turned off these represent four possible states of the system, but they all have the same energy, which is $E_u + E_v$. When the force is turned on, with the assumed result that the system is again only slightly perturbed, theory requires that the resulting perturbed states be described by wave functions formed from linear combinations of the starting functions Ψ_I to Ψ_IV. There are, in fact, an infinite number of ways in which a set of four independent combinations of these starting functions can be formed, and it requires detailed application of perturbation theory to decide what coefficients in the linear combinations give the best approximations to the perturbed states. Here the results can only be quoted, but they turn out to be relatively simple; the best linear combinations to represent the perturbed states are the quantities

$$\Psi_{III}, \Psi_{II}, \frac{1}{\sqrt{2}} (\Psi_I + \Psi_{IV}) \quad \text{and} \quad \frac{1}{\sqrt{2}} (\Psi_I - \Psi_{IV}) \quad (3.27)$$

Written out in their expanded forms these are

$$\Psi_{III} = \frac{1}{\sqrt{2}} [\psi_u(\mathbf{r}_1)\psi_v(\mathbf{r}_2) - \psi_u(\mathbf{r}_2)\psi_v(\mathbf{r}_1)]\chi_\alpha(\sigma_1)\chi_\alpha(\sigma_2) \quad (3.28)$$

$$\Psi_{II} = \frac{1}{\sqrt{2}} [\psi_u(\mathbf{r}_1)\psi_v(\mathbf{r}_2) - \psi_u(\mathbf{r}_2)\psi_v(\mathbf{r}_1)]\chi_\beta(\sigma_1)\chi_\beta(\sigma_2) \quad (3.29)$$

$$\frac{1}{\sqrt{2}} (\Psi_I + \Psi_{IV}) = \frac{1}{\sqrt{2}} [\psi_u(\mathbf{r}_1)\psi_v(\mathbf{r}_2) - \psi_u(\mathbf{r}_2)\psi_v(\mathbf{r}_1)]$$
$$[\chi_\alpha(\sigma_1)\chi_\beta(\sigma_2) + \chi_\alpha(\sigma_2)\chi_\beta(\sigma_1)] \quad (3.30)$$

and

$$\frac{1}{\sqrt{2}} (\Psi_I - \Psi_{IV}) = \frac{1}{\sqrt{2}} [\psi_u(\mathbf{r}_1)\psi_v(\mathbf{r}_2) + \psi_u(\mathbf{r}_2)\psi_v(\mathbf{r}_1)]$$
$$[\chi_\alpha(\sigma_1)\chi_\beta(\sigma_2) - \chi_\alpha(\sigma_2)\chi_\beta(\sigma_1)] \quad (3.31)$$

The point of writing out of these expansions is to note that Ψ_{III}, Ψ_{II} and $1/\sqrt{2} (\Psi_I + \Psi_{IV})$ all have the same antisymmetric orbital factor $[\psi_u(\mathbf{r}_1)\psi_v(\mathbf{r}_2) - \psi_u(\mathbf{r}_2)\psi_v(\mathbf{r}_1)]$ but that the spin factor in each of these cases is symmetric. The opposite situation holds for $1/\sqrt{2} (\Psi_I - \Psi_{IV})$; its spin factor is antisymmetric, but its orbital factor is symmetric. However, each of the four combinations represented by Eq. (3.28) to Eq. (3.31) is evidently antisymmetric under a general transformation of both spin and positional coordinates, as of course it has to be to satisfy the Pauli principle.

Having stated what are the best total wave functions to describe the system when the Coulomb repulsion between the electrons is switched on, we now turn to the corresponding energy states for the system in this condition. As in case (a) above, we are interested in the change in the energy of the system produced by switching on this interaction. We find that this energy change can be written

$$\iint \psi_u{}^*(\mathbf{r}_1)\psi_v{}^*(\mathbf{r}_2) \frac{e^2}{r_{12}} \psi_u(\mathbf{r}_1)\psi_v(\mathbf{r}_2) \, d\mathbf{r}_1 \, d\mathbf{r}_2$$
$$\pm \iint \psi_u{}^*(\mathbf{r}_1)\psi_v{}^*(\mathbf{r}_2) \frac{e^2}{r_{12}} \psi_u(\mathbf{r}_2)\psi_v(\mathbf{r}_1) \, d\mathbf{r}_1 \, d\mathbf{r}_2 \quad (3.32)$$

and has, therefore, two values.

The first term in this expression is just the Coulomb energy U: In case (a) this is just the energy of the electrostatic repulsion between two electrons in the same orbital and thus with antiparallel spins, but here it is the same quantity for two electrons in different orbitals regardless of spin orientations. This term is the only energy term that arises in case (a) when the electrostatic interaction is switched on. In case (b), however, there is also the second term of Eq. (3.32). This is known as the exchange integral and is conventionally represented by J. It represents half the difference between the energies of two electrons in states $\psi_u(\mathbf{r})$ and $\psi_v(\mathbf{r})$ on the same atom if their spins change from parallel to antiparallel. The total energy of the system with the interaction switched on is thus $E_u + E_v + U \pm J$. The lower energy state, obtained by taking the negative sign with J, corresponds to the case when the electrons' spins are parallel. It is triply degenerate; that is, three wave functions in Eq. (3.28) to Eq. (3.30) correspond to the same total energy of the system (in the absence of any external magnetic field). The state of higher energy, given by taking the positive sign with J, corresponds to the case when the electrons' spins are opposed, and it has the total wave function given in Eq. (3.31).

It is frequently emphasized that this splitting of the energy level resulting from the overlap of the electrons' wave functions is just part of a wider result in physics: Namely that whenever there is coupling between identical systems their energy levels are split in some way. The usual example cited is the case of coupled, resonant electric circuits. When they are uncoupled (that is, well separated) they may each have the tuned resonance frequency ω, but as they are brought closer together, so that their coupling is increased, the resonant frequencies become split and can then be written as $\omega \pm a$ (where a is a constant)—just as the Coulomb energy between the electrons above became $U \pm J$ when the coupling was switched on. Furthermore, it is instructive to recall that the coupling of such a system results in its energy flowing back and forth with regular periodicity from one coupled component to the other.* In the case of coupled quantum systems it is not the energy of the system which so oscillates, but rather the probability of a given configuration possessing the energy; somewhat as if a physical entity—in this case the electron—were being continuously transferred from one coupled state to the other.

In fact, in the quantum mechanical formulation the integral form of J shown in Eq. (3.32) represents the initial rate at which two labeled elec-

---

* In the analogous case of two weakly coupled identical pendulums, this can be demonstrated very graphically in a home experiment. (See Experiment 1.8 of *Waves*, Berkley Physics Course, V3, McGraw-Hill, New York, 1968, p. 38.)

trons, placed simultaneously in the state $\psi_u(\mathbf{r}_1)\psi_v(\mathbf{r}_2)$ exchange by quantum mechanical tunneling between states u and v to give the permuted state $\psi_u(\mathbf{r}_2)\psi_v(\mathbf{r}_1)$. In fact, J is known as an exchange integral, and the process is known as direct exchange because the interacting electrons are pictured as trading places directly through the overlap of their wave functions. (This contrasts with an indirect exchange, or superexchange, as discussed in Sec. 3.3.3 in which an intermediary is involved in forming the interaction between the coupled electrons.) The name "exchange integral" may be misinterpreted, however, for in spite of the above remarks, it is wrong to think of products like $\psi_u(\mathbf{r}_1)\psi_v(\mathbf{r}_2)$, and so forth as having physical significance or to think of an exchange of resonance between them as if it had real significance.

It is better to regard the integral J as a correction to the Coulomb energy U resulting from a coupling between spins of contiguous electrons. Nowadays it is called the exchange coupling (in earlier literature it used to be known as the interchange interaction force); we have remarked previously that it is an imaginary force, in that it can do no work, and is simply a convenient representation of the fact that the Coulomb interaction depends upon the relative orientations of the electrons' spins. Above all, it should be realized that the electrostatic (Coulomb) forces between the electrons are directly responsible for the exchange interaction and that the magnetic (Amperian) forces between the electrons' magnetic moments are not. As was emphasized at the outset, these magnetic dipole-dipole interactions can be ignored even in a first approximation.

To look further at this viewpoint, which regards J as a correction to U, consider what happens as the two electrons approach very close to each other so that $(\mathbf{r}_2 - \mathbf{r}_1) \rightarrow 0$. As this happens each of Eq. (3.28) to Eq. (3.30), which correspond to the triplet state, tend to zero since their common orbital factor is zero when $\mathbf{r}_1 = \mathbf{r}_2$. (This contrasts with the singlet state given by Eq. (3.31), which does not tend to zero in such circumstances.) Obviously this means that no two electrons of like spin may occupy the same position in space—as deduced in the preceding section— but this can be made even more clear by imagining one electron to pass the other in space so that $\mathbf{r}_2 - \mathbf{r}_1$ goes through zero and becomes negative. Since any wave function must be continuous and have a continuous first derivative (in other words, it can have no sharply or discontinuously changing sections) each of Eqs. (3.28) to Eq. (3.30) also passes smoothly through zero and changes sign during this maneuver as indicated in Fig. 3.8). $|\psi(\mathbf{r}_1,\mathbf{r}_2)|^2$ is proportional to the probability of finding one electron at $\mathbf{r}_1$ and the other at $\mathbf{r}_2$ [where $\psi(\mathbf{r}_1,\mathbf{r}_2)$ here represents the orbital factor in any one of the functions Eq. (3.28) to Eq. (3.30)]; its form will consequently

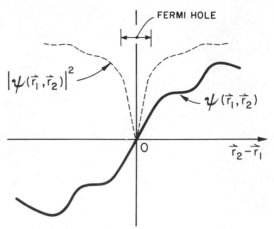

Fig. 3.8 Illustrating the origin of the Fermi hole. $\psi(r_1,r_2)$ represents the orbital factor in the wave function of the state having spins parallel (the triplet state). The probability of finding two such electrons, one at r_1 and the other at r_2, goes through a sharp minimum of zero at $r_2 = r_1$. This means that immediately around an electron there is a region deficient in electrons of like spin. This is known as the Fermi (or exchange) hole. (After P. W. Atkins, 1970, *Molecular Quantum Mechanics*, Clarendon Press, Oxford, p. 252.)

be as indicated in Fig. 3.8, having a sharp minimum of zero at $r_1 = r_2$. In other words, an electron has around it a region in which electrons of parallel spin cannot penetrate; this is described as a negative correlation in their motion since they try to stay out of each other's way. Around any electron there is thus a region of depleted parallel spin (and hence a deficiency of charge) arising from the consequence of the Pauli principle. This region is known as a Fermi or exchange hole, and it reduces the electrostatic repulsion term between electrons of like spin since it keeps them apart. Hence J is substracted from U when forming the energy of the triplet state.

The opposite effect is observed for the singlet state. In this case $|\psi(r_1,r_2)|^2$ has a slight maximum around $r_2 - r_1 = 0$ [where $\psi(r_1,r_2)$ is the orbital part of Eq. (3.31)] and electrons of opposite spin are slightly attracted to each other (all other considerations being equal) tending to pile up in each other's vicinity. The electrostatic repulsion term between electrons of unlike spin is, therefore, increased by all of this, since the electrons are drawn slightly together, and J is added to U when forming the energy of the singlet state.

To summarize we can say that when the electrostatic interaction between two electrons localized about the same nucleus is switched on, an extra energy term is added to the system due to the mutual repulsion between the electrons. Its value is $U + J$ if the electrons' spins are antiparallel and

$U - J$ if they are parallel. The electrostatic contribution, therefore, depends upon the spins' orientations in an indirect but inextricable tie, and it is convenient to look upon this whole contribution as a form of effective coupling between the spins rather than as a modification of the electrostatic interaction between charges. Historically, the idea of a spin-spin interaction was introduced by Heisenberg and by Dirac, in the earliest days, soon after Pauli had introduced his exclusion principle; in the following section we shall digress slightly to introduce the Pauli formalism that is so frequently encountered in the description of exchange effects in solids.

Before turning to the topic, however, it is important to emphasize that the preceding remarks relate entirely to the exchange interaction appearing between two electrons localized about the same nucleus (intraatomic exchange). Although we have not previously stressed the point, this is a rather special environment. Specifically, the orbitals $\psi_u(\mathbf{r})$ and $\psi_v(\mathbf{r})$ corresponding to different energy states—but associated with the same nucleus—are automatically orthogonal, and some of the specific conclusions reached in the preceding are, in fact, generally only valid in such circumstances. For example, we found above that the energy of the triplet state lies below that of the singlet one. It can be shown that this is a direct consequence of the orthogonality of the electrons' wave functions and that it does not automatically follow in the corresponding case when the electrons' wave functions are not exactly orthogonal to each other. In such a case, when the electrons are localized about separate nuclei, as in the hydrogen molecule, for example, the nonorthogonality plays a vital role. It modifies the normalization factors $1/\sqrt{2}$ appearing in the wave functions Eq. (3.27) to Eq. (3.31) for the singlet and triplet states to become a function of the internuclear separation, and furthermore the energy difference between the singlet and triplet states is no longer given by an expression like Eq. (3.32). The quantity e^2/r_{12}, which there represents the interaction between the electrons 1 and 2, is inadequate when the electrons can range over more than one nucleus, so that additional terms expressing the interactions between each electron and each nucleus have to be included. The outcome of all this is that, in the two-electron hydrogen molecule at least, the energy of the singlet state turns out to be less than that of the triplet. In a wider view, this effective attraction between antiparallel spins in nonorthogonal wave functions is at the heart of the Heitler–London scheme of chemical bonding.

3.3.2 The Formalism of Exchange. Rather than consider the Coulomb force between contiguous electrons as being dependent upon the relative orientation of their spins, it has proved preferable to imagine a new force existing between them. This imaginary exchange force is then used to account for the spin-dependent contribution to the total energy of the system. In this section we digress slightly to consider a mathematical formulation of this idea—due to Heisenberg and Dirac—which is commonly encountered throughout the theory of solids.

First, in the Pauli formalism, to which we referred in Sect. 3.2.3, the spin wave functions are represented by column matrices [see Eq. (3.3) and Eq. (3.4)], and a vector quantity **s** is introduced to characterize the spin angular momentum of an electron. When used as an operator upon the spin functions $\chi(\sigma)$, **s** gives the correct value of the spin angular momentum for the appropriate orientation. These vector operators are in the form of 2×2 matrices. To give just one example take the component s_z, which characterizes the spin state χ_α in which the spin momentum has the value $\hbar/2$ directed along the z axis. In the Pauli formalism s_z is

$$\frac{\hbar}{2} \begin{pmatrix} 1 & 0 \\ 0 & -1 \end{pmatrix} \tag{3.33}$$

and when it operates on $\chi_\alpha \equiv \begin{pmatrix} 1 \\ 0 \end{pmatrix}$ the usual rules of matrix algebra lead to

$$s_z \chi_\alpha = \frac{\hbar}{2} \chi_\alpha \tag{3.34}$$

Thus the eigenvalue of this operator equation is seen to be $\hbar/2$, which is the required result.

Returning to the main argument, the successful introduction by Pauli of his formalism showed that spin is a vector quantity. In accordance with this, we shall let the two electrons under consideration have the vector spin operators s_1 and s_2. At the same time, however, any effective interaction between two spins cannot be allowed to depend upon the direction in space along which it is chosen to lie; it has to be isotropic and a scalar quantity. This lead Dirac to propose that the basic quantity that describes a spin-spin coupling in this operator formalism should be the scalar product of the two spin vectors concerned, namely $s_1 \cdot s_2$ in this case. His work showed that this quantity is multiplied by a coefficient that is a function of the electrostatic force between the electrons; it only remains to determine this constant of proportionality. The electrostatic energy between the two electrons is $U \pm J$; therefore, if we wish to introduce a factor containing $s_1 \cdot s_2$ to replace the spin-dependent part $\pm J$, it must be suitably modified to be equivalent to the original electrostatic form. Various textbooks show* how the appropriate constants are obtained in this two-elec-

---

* A. H. Morrish, *The Physical Principles of Magnetism*, John Wiley, New York, 1965, p. 280; D. H. Martin, *Magnetism in Solids*, Iliffe Ltd., London, 1967, p. 298; D. F. G. Williams, *The Magnetic Properties of Matter*, Elsevier, New York, 1966, p. 144; M. Sachs, *Solid State Theory*, McGraw-Hill, New York, 1963, p. 134.

tron problem; here we will simply quote the result. In place of $U \pm J$ the total electrostatic energy is written.*

$$U - \tfrac{1}{2}J\left[1 + \frac{4\,\mathbf{s}_1 \cdot \mathbf{s}_2}{\hbar^2}\right] \tag{3.35}$$

The second term in this expression is the energy arising from the exchange effect and thus is known as the exchange term. The spin-dependent part originally contained in $\pm J$ is now all contained in the factor $\mathbf{s}_1 \cdot \mathbf{s}_2$, which expresses an effective coupling between the spins. It resembles the interaction between two dipoles—which contains a term of this form—but we emphasize again that it is a metaphorical description and is not at all connected with the real (but here negligible) magnetic interaction between the electrons' magnetic dipole moments. Here the sign of J has been divested of any dependence upon the relative orientation of the spins; it is now a property of the particular context. For example, in the case of electrons coupled by intraatomic exchange (that is, the electrons are localized about the same free atom) J is conventionally taken to be positive, and the spins align in an atom according to Hund's rule. In the case of interatomic exchange, on the other hand, J tends to be of the opposite sign—as the preceding remarks in Sect. 3.3.1 indicate—corresponding to a negative J and to a bonding between electrons having antiparallel spins. When the formalism is extended to the case of electrons in solids, we find that the sign of J can be either positive or negative, depending upon the problem. In fact, in the case of many solids it is often difficult to determine from experiment exactly what is the correct sign of J, but that is a separate problem that cannot be pursued here.

Expression (3.35) can be extended to a system containing more than two electrons. Suppose there are N electrons in the system. This will give $N(N-1)/2$ different pairs. Expression (3.35) holds for any one pair that is formed, say from the ith and jth electrons; thus the exchange term for the whole system becomes

$$-\tfrac{1}{2}\sum_{i<j=1}^{N} J_{ij}\left[1 + \frac{4\,\mathbf{s}_i \cdot \mathbf{s}_j}{\hbar^2}\right] \tag{3.36}$$

in which the spin-dependent term is just

$$-\frac{2}{\hbar^2}\sum J_{ij}\,(\mathbf{s}_i \cdot \mathbf{s}_j) \tag{3.37}$$

---

* Frequently in textbooks $\mathbf{s}$ is defined in multiples of $\hbar$ so that this awkward factor does not appear in such equations as Eq. (3.35) and those derived from it.

again with the summation taken over all $N(N-1)/2$ pairs. In these expressions J_{ij} is the two-electron integral connecting the ith and jth electrons [as in Eq. (3.22)]. In terms of the spin variable $\sigma = 2s/\hbar$ (Sect. 3.2.3), the expression (3.36) simplifies to

$$-\tfrac{1}{2} \sum_{i<j=1}^{N} J_{ij}(1 + \mathbf{\sigma}_i \cdot \mathbf{\sigma}_j) \tag{3.38}$$

in which the spin-dependent part is just

$$-\tfrac{1}{2} \sum J_{ij}(\mathbf{\sigma}_i \cdot \mathbf{\sigma}_j) \tag{3.39}$$

Expression (3.36) is known as the Dirac Hamiltonian, and its origin has been explained within the context of the interaction between individual pairs of electrons. Frequently, however, the context is rather the interaction between pairs of ions in which each constituent has a resultant spin. This case can still be worked within the above framework except that it is now concerned with a coupling between the resultant of all the electron spins in the ion. Suppose there are two ions labeled A and B, each having a number N of unpaired electron spins. Their total spin operators are $\mathbf{S}_A = \sum_A \mathbf{s}_i$ and $\mathbf{S}_B = \sum_B \mathbf{s}_j$, respectively, where the i and j refer to the different atoms A and B. Assuming that pairs of electrons with one constituent on each ion have the same exchange integral, say J_{AB}, then the expression (3.36) gives the exchange term for the pair of ions as

$$-\tfrac{1}{2} J_{AB} \left[N + \frac{4 \sum_A \mathbf{s}_i \cdot \sum_B \mathbf{s}_j}{\hbar^2} \right]$$

which is

$$-\tfrac{1}{2} J_{AB} \left[N + \frac{4 \mathbf{S}_A \cdot \mathbf{S}_B}{\hbar^2} \right]$$

This is the Heisenberg Hamiltonian, and its spin–dependent term is evidently $- 2/\hbar^2 J_{AB} \, \mathbf{S}_A \cdot \mathbf{S}_B$.

The preceding relates to an ion for which the individual spins of the electrons in a shell can be summed straightforwardly to give the total $\mathbf{S}(= \Sigma \mathbf{s}_i)$, and this implies (see the discussion in Sect. 3.2.5) that either the ion's total angular momentum is quenched or the spin-orbit coupling of the electrons in question is insignificant compared with their exchange interactions. However, in the opposite case when the exchange is smaller than the spin-orbit coupling—as occurs for some rare earth metals, for example Fig. 3.3—then the total spin $\mathbf{S}$ of an ion's shell can be projected onto its total angular momentum vector* $\mathbf{J}$ through the relation $\mathbf{S} = (\beta-1)\mathbf{J}$ cited in Sect. 3.2.5. Assuming that the Landé factor is the same for the two ions A and B and remembering that $\mathbf{J}^2$ is the magnitude $\hbar^2[J(J+1)]$ of the shell's total angular momentum, the spin-dependent term of the Heisenberg Hamiltonian is clearly proportional to

$$(\beta - 1)^2 (J(J + 1))\mathbf{J}_A \cdot \mathbf{J}_B \qquad (3.40)$$

The term $(\beta-1)^2[J(J+1)]$ is known as the de Gennes' factor,† and it is frequently incorporated into the exchange constant J_{AB} to cover the case where spin-orbit coupling is important.

Although the above relates to the exchange interaction between the total spins of two ions (that is, to the exchange interaction between electrons localized in an incomplete shell within each ion) the same formalism can be applied to the case of an exchange coupling between the spin of an itinerant electron and the total spin of an ion in the metal. (Such an encounter produces the exchange scattering of the conduction electron that is referred to in subsequent sections.) Suppose the itinerant electron's spin is $\mathbf{s}$ and that of an electron localized in the ion's incomplete shell is $\mathbf{s}_i$, then the spin-dependent term in the interaction is of the form $\Gamma \sum_i \mathbf{s} \cdot \mathbf{s}_i$, where Γ is a coupling constant. But $\sum_i \mathbf{s}_i = \mathbf{S}$, the total spin vector of the shell, and again this can be related to the shell's total angular momentum vector by the relation $\mathbf{S} = (\beta-1)\mathbf{J}$ of Sect. 3.2.5. So the interaction energy of this exchange coupling between the itinerant electron and the ion can be put in the form

$$\Gamma(\beta - 1)\mathbf{s} \cdot \mathbf{J} \qquad (3.41)$$

when the spin-orbit coupling in the ion is large.

3.3.3 Other Types of Exchange Interaction.
The preceding sections have illustrated how the concept of coupling between two interacting spins has been introduced to describe that consequence of the Pauli exclusion principle which makes the spatial distributions of the electrons' orbitals dependent upon the relative orientations of their spins. We discussed specifically the case of atomic electrons confined to different but specified orbitals, which interaction originally became known as a direct exchange

---

* The use of J in several closely related contexts is apt to be confusing, but we make a point of everywhere following the commonly used symbolism. J or J_{AB}, the exchange constant, should not be confused with $\mathbf{J}$, the total angular momentum vector, or with J which is also used to represent both a quantum number and the abbreviation for the Joule. The intention is always clear from the context.

† P-J de Gennes, *Compt. Rend. Acad. Sci. (Paris)* **247**, 1836, 1958.

interaction because of its analogy with the possibility of direct exchange of the interacting electrons through quantum mechanical tunneling.* (It is also frequently known as the Heisenberg interaction because it was originally introduced by him to account for the heuristic Weiss Field of the early theories of ferromagnetism.) Nowadays the term direct interaction refers rather to a situation in which the electrons having their spins coupled are the only ones involved in the interaction, and it is no longer restricted to an interaction between electrons localized in particular atoms. This contrasts with an indirect exchange interaction in which the coupling involves the spin of some intermediate electron. Depending upon the circumstances, this intermediary can either be a member of the itinerant class or be localized in a nearby ion. To distinguish these interactions descriptively, a direct process can loosely be considered as the direct interchange of places of the two electrons involved, and an indirect process as a spin-polarization effect operating over relatively longer distances.

Section 3.3.1 considers illustrative examples which show that, all other things being equal, the parallel alignment of spins (known as the ferromagnetic alignment) is energetically preferred over the antiparallel (antiferromagnetic) one between electrons localized about the same nucleus. But when the electrons are localized about different nuclei, it is conversely the antiparallel arrangement that has the lower energy. Although these conclusions were reached using the simplest of examples, experiment shows that they are generally true for more complicated atoms or molecules.

However, when these arguments are applied to solids—and particularly to nonmetals—we immediately find that something is wrong. In the class of magnetic insulators, for example, where the interacting spins have the localized atomic nature that, at first sight, makes the direct interaction cited above a seemingly appropriate description, the experimental evidence is that an antiferromagnetic alignment of the electron's spins on different atoms is found almost universally. Furthermore, in these examples the separation of the ions possessing a localized magnetic moment is found to be so great that, upon second sight, it is difficult to understand how such strong coupling could exist betwen them if their only interaction is that introduced by Heisenberg.

It was fairly soon realized that the Heisenberg interaction is, in fact, almost never the sole cause of ferromagnetic alignment of spins in solids;

---

* The probability that a classical particle can pass through a potential barrier having an energy greater than the particle's kinetic energy is zero. But for a quantum particle on the microscope scale the probability is finite and nonzero. This is described as quantum mechanical tunneling. It is another specious macroscopic analogy of an effect that exists only on the quantum mechanical scale.

other considerations have to be included. In the case of metals, one of the inadequacies of the simple interaction discussed above is immediately obvious: The Heisenberg interaction relates to a coupling between electrons localized about some atomic nucleus, but we know that in a metal this describes only one class of electrons; the other class is itinerant and its members move in the periodic potential produced by the ionic lattice. Consequently, when discussing the exchange interaction in a metal a new degree of freedom—the itinerancy of one class of electrons—which is not present in the atomic or molecular examples, has to be allowed for when applying these ideas to the metallic state.

However, the concept of an exchange interaction is obviously too fundamental to be abandoned on such accounts. Instead, the idea has been developed along different directions with embellishments being made to accomodate the experimental facts pertinent to the particular circumstances. The result is a somewhat bewildering array of descriptions in the literature of interactions that can couple the spins of two electrons in different conditions—although they all, of course, stem from the operation of the Pauli principle. Figure 3.9 shows in a very pictorial way a classification of the various interactions that have been distinguished and that are relevant here. In following paragraphs we give a description of each type in turn, but since Types 1 and 2 are the subject of Sect. 3.3.1, it is unnecessary to give more than a brief mention of them here.

TYPE 1. DIRECT INTRAATOMIC EXCHANGE COUPLING. Discovered by Heisenberg and Dirac, this is a coupling between two electrons belonging to the same atom. As a result of the Pauli principle, electrons having spins parallel to some preferred direction stay out of each other's way to a greater extent than those having antiparallel ones. The Coulomb repulsion between electrons with parallel spins is, therefore, reduced in comparison with the antiparallel case. Although this reduction in the electron's potential energy is at the cost of a certain increase in their total kinetic energy (as a result of the redistribution of the orbitals), the net result when the two electrons are in the same atom is to make the parallel alignment of spins energetically more favorable than its alternative. Hence, the spins of electrons in a free atom tend to align to give the maximum total spin permitted by the available states in the shell (this is known as Hund's rule) and this can be regarded as an effective coupling between the spins.

TYPE 2. DIRECT INTERATOMIC EXCHANGE COUPLING. Technically, whether the parallel or antiparallel spin alignment is energetically more favorable depends largely upon the degree of orthogonality of the wave

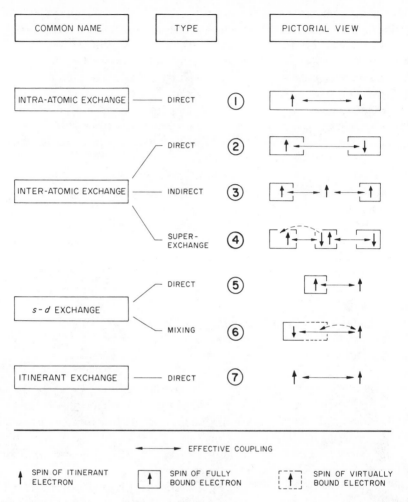

Fig. 3.9 A pictorial classification of several circumstances in which the application of the Pauli exclusion principle leads to an effective coupling between the spins of a pair of electrons. A fully bound electron can be regarded as one localized about a particular nucleus in the solid (or molecule); a virtually bound electron is merely the temporary captive of an ion (as in the resonant scattering illustrated in Fig. 3.10). A description of each interaction is given in the text.

functions* describing the two interacting electrons. The truly orthogonal case occurs when the two electrons are in the same atom and always leads to the preference for parallel alignment described in the above paragraph. But as a degree of nonorthogonality is allowed into the problem (as in the case of electrons on different atoms, for example) the end result depends more delicately upon the balance between certain terms. Calculation shows that in the majority of cases where the coupling between two electrons belonging to different atoms can be evaluated, the antiparallel (antiferromagnetic) alignment of their spins is preferred. (The hydrogen molecule is the standard example.) This result is, of course, in direct contrast to the intraatomic case considered above.

A more physical view of the origin of this difference emerges from the realization that, since in the molecular case the electrons are distributed around more than one ion core and are, therefore, inevitably more separated than in a comparable single-core case, their mutual interaction becomes relatively less important. In addition, since more than one nucleus is involved in this case, the relative importance of the interactions of the electrons with the ion cores is correspondingly increased. Both these influences tend to increase the importance of the interaction between the electrons and the ion cores. It is, therefore, perhaps not surprising that the increase in the electrons' kinetic energy required by a rearrangement of the orbitals to give a parallel alignment of spins frequently outweighs the gain from the reduced Coulomb force between the electrons. Consequently, the antiparallel alignment is often found to be energetically favorable.

This type of exchange coupling underlies the Heitler–London model of the hydrogen molecule and, hence, leads to the widely used concept of the covalent bond.

TYPE 3. INDIRECT INTERATOMIC EXCHANGE COUPLING. In spite of the importance of the Type 2 interaction—which is really the basis of modern structural chemistry—it is known from experiment that many metals and alloys show ferromagnetic (parallel) alignment of their net ionic spins. Another type of interaction in addition to the Type 2 must, therefore, exist, a type that in some way can couple the spins of electrons on different ions and into the parallel configuration; this mechanism presumably is the dominant one in those circumstances where ferromagnetic behavior is observed. The indirect interatomic interaction is of this type. The spins of two electrons localized in separate ions can be coupled over relatively large distances by means of an exchange coupling that acts between each

* Previously referred to in a footnote following Eq. (3.13).

localized spin and that of some intermediate electron. In this particular interaction this intermediary is a member of the itinerant class, so that the direct *s–d* exchange interaction (Type 5, to be considered below) is the basis of the whole mechanism. As originally conceived, this interaction could only lead to ferromagnetic alignment of the spins located on separated ions (as illustrated in Fig. 3.9), but as we point out below (in the discussion of Type 5) later refinements completely modified these earlier views. In its modern conception, either ferromagnetic or antiferromagnetic coupling can be produced by this interaction between the ionic spins, depending upon their separation.

TYPE 4. INTERATOMIC SUPEREXCHANGE COUPLING. The Type 3 interaction is not the only one that can couple the spins of electrons on well-separated ions by means of an intermediary. Another, known as superexchange, was historically the first of the two to be introduced. This occurred when it became clear that in many nonmetallic solids there are strongly coupled spins localized on magnetic ions* of a given species that are evidently very well separated from each other by ions of another species which is normally nonmagnetic. (Examples of such compounds are MnSe, MnTe, or MnO in which the spins of the Mn ions are coupled antiferromagnetically, even though well separated by the Se, Te or O ions.) Because of this separation of the magnetic species, the possibility of a direct mechanism being operable seemed unlikely.

The superexchange mechanism is still the subject of much contemporary research, and there is still not agreement among all theoreticians about the details of its origins. However, it seems clear from experiment that the nonmagnetic ions are intermediaries in this coupling; their properties are modified to such an extent by the presence of their magnetic neighbors that they are able to enter into the magnetic interaction between them. Anderson† has suggested that this interaction involves specifically an electron transfer from the normally nonmagnetic ion to a vacancy in a shell of the normally magnetic one. This would leave the previously nonmagnetic ion with an unbalanced spin, so that it would be paramagnetic and thus able to couple its spin with that of another nearby magnetic ion through the direct interatomic coupling (Type 2). We can see qualitatively that this mechanism can lead to an antiferromagnetic alignment of the spins on the magnetic ions.

---

* We mean by this an ion that has a permanent vacancy in an inner shell and, therefore, has a net magnetic dipole moment due to the spins of the uncompensated electrons.

† P. W. Anderson, *Magnetism,* loc. cit. Vol. I, p. 25.

This is illustrated in Fig. 3.9. Initially, all the shells of the intermediary nonmagnetic ion are filled; any orbital, therefore, contains two electrons having opposite spins (as in the diagram). Suppose one electron from such an orbital is transferred to the single vacancy in a shell of an adjacent ion of the magnetic species. This can only occur if the spin of the transferred electron is opposed to that of the electron resident in the state, so that there has to be essentially an antiferromagnetic alignment between the spin of the resident and of the electron to be transferred. (In the diagram the resident is shown with up spin, and the one to be transferred with down spin.) After the transfer the unpaired spin on the intermediary (up in this case) can couple with that of another adjacent magnetic ion through the direct interatomic mechanism (Type 2). Since this mechanism gives an antiparallel alignment of the two spins involved, a down spin in the second magnetic ion is coupled into the system. Hence, the overall coupling between the spins localized on the two well-separated magnetic ions is an antiferromagnetic one, and it acts through the medium of the normally nonmagnetic constituent.

Other possible electron transfer configurations can be envisaged and, indeed, have been proposed as important in particular contexts, but the one outlined above is the simplest. Whether or not it is also the most important in all cases is, however, still an open question. Finally, it must be emphasized that the superexchange mechanism relates to nonmetallic systems: Although there is frequently a very strong interaction between neighboring ions of different species in such systems—strong enough to make the permanent transfer of an electron a likely event—this is a chemical binding between them and not a metallic one involving delocalized electrons.

TYPE 5. DIRECT S-D EXCHANGE COUPLING. The need to introduce an interatomic coupling that gives a parallel alignment of spins on ions in metals was pointed out above in connection with the Type 3 coupling. Zener and Vonsovskii* independently proposed for this purpose a coupling between the spin of an itinerant electron and that of one localized in the incomplete shell of an ion. We have already seen a cooperative consequence of this in the Type 3 coupling, but here we turn to a description of the underlying mechanism.

It has become known as the *s-d* exchange interaction (or *s-f* interaction in the appropriate context) since the itinerant and localized electrons have

---

* S. V. Vonsovskii, *Soviet Phys. JETP* (Engl. trans.) **10**, 468, 1946. C. Zener, *Phys. Rev.* **81**, 440, 1951; *ibid.* **83**, 299, 1951. C. Zener and R. R. Heikes, *Rev. Mod. Phys.* **25**, 191, 1953.

generally s and d shell atomic origins, respectively. It is an extension of the direct intraatomic coupling (Type 1) considered above. As Zener pointed out, this Type 1 coupling was conceived to exist between the spins of two electrons localized in the same shell of an atom—say the d shell. But it also exists between the spins of two electrons localized in different shells of the same atom—say one from a d shell and one from an s shell. In the simple view of a metal the s electrons are itinerant, but Zener argued that the same Type 1 coupling will nontheless exist between the total unbalanced spin of an ion (due to electrons in incomplete shells) and that of any nearby member of the itinerant class. However, the electron fluid has an important property: It is easily polarized by some local perturbation such as an excess charge or (as in this case) a local magnetic moment. Consequently, an ion possessing a net spin will be able in a metal to spin polarize the surrounding electron fluid by means of this direct s-d exchange coupling. This region of spin polarized electrons can likewise couple to the spin of any other nearby ion, and thus a cooperative interaction results between the ionic spins in a metallic ferromagnet using the itinerant class as an intermediate medium. This is the indirect interatomic exchange coupling (Type 3) already referred to above.

We have said that the direct s-d coupling under consideration is an extension of the Type 1 concept. Such extension is necessary because in the present case one member of the coupled pair is itinerant, as opposed to being localized in a specific atom or ion, and this introduces a degree of dynamic freedom into the problem. Since the itinerant electrons are in a state of constant translational motion, it is seen that different members of this class are constantly coming under the influence of any given localized spin. They couple with it for a time—through the direct s-d interaction in this case—and then move on to be subsequently replaced by others. This temporary coupling, of course, usually modifies the itinerant electron's translational motion; it is, therefore, appropriate in this context to regard the coupling as an electron scattering event. This term provides a picture of how the extra degree of freedom of the itinerant electron enters the problem: It can be said that when such an electron is involved in an exchange interaction it is able afterwards to scatter into various orbital states in a manner not encountered in the purely intraatomic prototype. As an intermediary in the cooperative interaction between ions referred to above, the itinerant electron carries spin information from one scattering event (that is, a coupling with a localized spin) to another. The electron's scattering by the first ion senses the configuration of the local spin there, and this is carried to a second ion where the scattering again depends upon the local spin configuration. As a result, the two ionic spins are able to interact cooperatively. This view of exchange coupling as a scattering effect is an important one which will be elaborated in Sect. 3.4.

We should note that both this type of interaction and the concomitant Type 3 were conceived as an explanation of ferromagnetism in metals. We have seen that the mechanism applies to the case where every ion in the metal is alike and possesses a net spin arising from the unbalanced spins of electrons localized in incomplete d or f shells. Since the d electrons in several metals from the first transition series ($Sc \rightarrow Cu$) are now known to be appreciably itinerant, it is doubtful that the direct s-d interaction is a valid description of the origin of ferromagnetism in these cases. But it is certainly applicable to the ferromagnetic rare earth metals, in which the $4f$ electrons are located deep in the ion cores and are, therefore, more able to retain their atomic (that is, localized) character in the condensed state.

But examining the rare earth metals, we are struck by an important experimental fact: Although ferromagnetic ordering of the ionic spins is a common magnetic state in this group, it is not the only one; spiral magnetic ordering of various complexities is frequently observed (this is also known as heliomagnetism). Although such combinations can be accounted for in terms of a competition between ferromagnetic and antiferromagnetic interactions between adjacent spins and those further separated, respectively, it is not clear how all of this can be understood in terms of the Type 5 interaction—which we have seen is purely ferromagnetic. There is a clear suggestion here that this interaction is inadequate in some way. This feeling is strengthened by another feature, which is not obvious from this descriptive treatment but which emerges in a more quantitative one and which was emphasized not long after Zener's publications appeared; namely that the magnitude of his interaction between the indirectly coupled ionic spins turns out to be independent of their separation. This is physically unreasonable, since the interaction cannot have an infinite extent, and various theoreticians observed that further refinements in the quantitative treatment of Zener's mechanism were obviously necessary. Technically, higher-order terms needed to be included.

At about this time, Ruderman and Kittel* made a calculation (including these higher order terms) of the analogous coupling that exists between nuclear spins in metals. (This too operates through the medium of the itinerant electrons and, in this context, is called the nuclear hyperfine spin interaction.) Similar calculations for the specific case of interionic spin coupling were published soon afterwards by Kasuya and Yosida,† and the

---

* M. A. Ruderman and C. Kittel, *Phys. Rev.* **96**, 99, 1954.

† T. Kasuya, *Prog. Theoret. Phys.* (*Kyoto*) **16**, 45, 1956; K. Yosida, *Phys. Rev.* **106**, 893, (1957).

result of all this work—which is now known as the RKKY interaction—was a complete modification of the Zener picture.

In the Zener model the electron fluid in the neighborhood of the magnetic ion is imagined to be spin polarized with a parallel alignment to the ionic spin but with no particular refinement in the distribution of this polarization about the ion. The RKKY results showed, on the other hand, that this polarization alternates in sign with increasing distance from the ion and they further showed that its amplitude decreases roughly as the cube of this distance.* The causes of this rather unexpected oscillatory distribution in the polarization are discussed more fully in Sect. 3.4.3 but here we should note that, when acting as an intermediary, such a distribution can couple ionic spins in either the ferromagnetic or antiferromagnetic configuration, depending upon their separation. It is thus potentially capable of accounting for various kinds of spiral ordering. Furthermore, the quite rapid decrease in the polarization's amplitude with distance from the ion means that its influence is effectively localized close to the ion, and this leads to the physically more reasonable result that the interionic spin coupling falls off as the separation of the ions is increased.

In summary, we can say that the Type 3 mechanism, as outlined above, is now obsolete. The idea that interionic spin coupling can operate via the conduction electrons is retained, but the RKKY refinements show that the itinerant electrons' spins are not coupled to the ionic one in an exclusively ferromagnetic configuration. The specifically ferromagnetic alignment of the Zener mechanism (shown in Type 5 of Fig. 3.9) is, therefore, only part of the modern picture, since, depending upon the separation of the spins, their coupling may be either ferromagnetic or antiferromagnetic.

TYPE 6. COUPLING THROUGH S-D MIXING. We now turn to another exchange interaction that operates in a metal between the spin of an electron localized in an incomplete d or f shell and that of a neighboring itinerant electron. This is known as the s-d (or s-f) mixing exchange interaction, and, in contrast to the direct s-d coupling (Type 5) described above, its net effect is to couple the spins into an antiparallel configuration.

To understand how it operates we must use the concept of resonant scattering of an itinerant electron by an ion. This is the subject of Sect. 3.4.1, but to anticipate briefly its results we can say that an incident itinerant electron arriving in the ion's vicinity in opportune circumstances can be temporarily captured in pseudoatomic states around the ion. These are formed from the mixing (or hybridization) of the outer orbitals of the

---

* When the limitation of the electron's mean free path due to incidental scattering (by impurities, for example) is included, then an even more rapid fall-off is found.

ion with the delocalized orbitals containing the itinerant electrons. Any vacancies in the ion's inner shell give a net spin to the ion that couples by means of exchange with that of the incident electron, thus making the whole mechanism spin dependent. In other words, when the ion is magnetic, the probability of an electron in given circumstances being temporarily captured by it before quantum tunneling out again into the itinerant states depends upon the spin configurations of the incident electron and those localized in the ion's inner shells. Because this exchange interaction arises from the mixing of the two classes of states in the metal, it is generally known as the exchange mixing interaction. We should always remember, however, that basically it describes a resonance process.

The relevance of all this to the s-d mixing interaction (Type 6 of Fig. 3.9) arises from the important classes of metals and alloys in which virtually bound states around the ions (or the solute atoms) are formed from what were d or f orbitals in the isolated atoms. Furthermore, vacancies that exist in these levels in the isolated case can survive the condensation process (or the alloying process in the case of a solute atom) to give a localized net magnetic moment to the ion. The picture of the s-d mixing interaction (and it is to be understood throughout that the same general arguments apply to the s-f interaction) is, therefore, of a temporary capture of an itinerant electron by a virtual d orbital through the mechanism of resonant scattering outlined above. Since the Pauli principle allows a maximum of two electrons in any orbital, and then only if they have opposite spin directions, it is clear that the resonance capture of an itinerant electron into a virtual d state can only occur if the electron's spin is opposed to that of any resident electron in the state (as illustrated in Fig. 3.9). Consequently, the coupling of s and d spins produced by this mixing interaction leads to the antiparallel configuration, as is emphasized in the opening paragraph.

Finally, it is instructive to note two points in connection with the s-d mixing interaction. First, since the temporary capture of an itinerant electron fills an otherwise empty place in a d orbital, it will also cause a fluctuation in the net value of the ion's uncompensated spin. This is known as a localized spin fluctuation (LSF); this concept is encountered again in Sect. 3.4.5. Second, since the Pauli principle dictates the allowed mixing configurations, the s-d resonance scattering process depends upon the spin orientations of both the itinerant electron and any resident electron in the d state. The scattering is thus spin dependent; this has far-reaching consequences that are considered in more detail in Sect. 4.4.

TYPE 7. DIRECT ITINERANT EXCHANGE. The preceding six types of exchange interaction have one feature in common: In each case at least one of the coupled spins is that of an electron localized about a nucleus. We now turn to an exchange that is distinguished from the rest by the fact that no localized electrons are involved. It is the direct exchange coupling that occurs between itinerant electrons in a metal and that generally is called direct itinerant exchange.

It occurred to early workers looking for an explanation of ferromagnetism in pure metals* that just as the Type 1 effective coupling exists between the spins of electrons within an atom, so there should be an analogous coupling between the spins of contiguous itinerant electrons in a metal. This would tend to give a parallel alignment to all the spins of the itinerant electrons—and so make the electron fluid ferromagnetic—but they realized that such alignment could only be achieved at the expense of an increased kinetic energy for the system. (The lowest state of kinetic energy for the itinerant system is obviously when half of the spins are up and the other half down, so that each itinerant energy state is doubly occupied. To alter this ratio requires the promotion of some electrons into higher energy states.) It was argued, however, that in a system of sufficiently low itinerant electron concentration the exchange interaction is dominant and leads to ferromagetic alignment of the spins.

This explanation of ferromagnetism in terms of direct itinerant exchange soon encountered many objections and was the subject of various developments and modifications in the decades following its introduction. None of this is really pertinent to our present purposes, but it is worth noting that nowadays it is believed that a fluid of itinerant electrons cannot be ferromagnetically aligned by this mechanism when its density is in the range encountered in typical ferromagnetic metals.† This does not, of course, rule out the existence of the coupling under discussion; it simply reflects upon its importance in the origin of ferromagnetism.

3.4 LOCAL DISTURBANCES IN THE ELECTRON LIQUID

The fluid-like plasma formed by the itinerant electrons in a metal has a very important property: It is easily polarized by some local perturbation. For example, a foreign ion having a different net charge from the rest produces a local charge polarization in the fluid as the itinerant electrons redistribute themselves in the ion's neighborhood to screen and cancel out

---

* Notably J. Frenkel, *Z. Phys.* **49**, 31, 1928 and F. Bloch, ibid. **57**, 545, 1929.

† This subject has been reviewed at an advanced level by D. Pines, *Solid State Phys.* **1**, 1955, p. 427 and D. H. Martin, *Magnetism in Solids,* Iliffe Ltd., London, 1967, Sect. 6.3.1.

the charge difference and thereby preserve electrical neutrality locally. Similarly, an ion having a net magnetic moment can spin-polarize the fluid as the electrons in its neighborhood sympathetically align their spins under the influence of the exchange interactions discussed in Sect. 3.3.

A point first made in the preceding section under the heading of the Type 5 interaction has to be borne in mind: These polarizations are dynamic processes. It is wrong to imagine that a particular group of itinerant electrons is permanently localized in the foreign ion's vicinity and responsible for the polarization effects. Rather, all itinerant electrons arriving in the ion's neighborhood are potentially capable of contributing to the polarization during their temporary presence. We point out below that they do not all contribute equally, however, for particular conditions have to be satisfied in each circumstance, but the point to be emphasized here is that an electron's contribution to the polarization is essentially a scattering event. As a result of the electron coming under the ion's influence—through either Coulomb or exchange forces—and thereby contributing to the polarization, its translational motion is disturbed and differs from that it would have shown in the ion's absence. In short, the electron is scattered by this local perturbation in the otherwise periodic potential.

In the following paragraphs this point is pursued further; we give more detailed descriptions of several important examples of the effect of such perturbations. Firstly, we remind the reader of the salient features shown by resonant scattering in this context (Sect. 3.4.1), and then we consider the simpler case when the ion causing the scattering is nonmagnetic (Sect. 3.4.2), before moving to the magnetic case (Sect. 3.4.3) and its ramifications (Sect. 3.4.4 and following).

3.4.1 Resonant Scattering of Itinerant Electrons.

An ion in a metal, or a solute ion in an alloy, when considered to be separate from the electron fluid, possesses a net electric charge. As a result of the Coulomb force, the itinerant electrons move to neutralize this charge in whatever circumstances prevail. To fix the argument, suppose we take the case of a nonmagnetic solute ion having an excess positive charge (as is produced, for example, when an atom of Zn is dissolved in a Cu matrix). It is then necessary for extra negative charge to be built up around the ion so as to equal and, therefore, screen the excess charge on the ion compared to the host. (This is the Friedel sum rule.) This screening involves the participation of itinerant electrons of all momenta, from zero up to roughly the Fermi energy, when the metal is at a moderate temperature; however, not all itinerant electrons are involved to the same degree. The fact that the momentum spectrum of the itinerant electrons is sharply cut off at its upper end has an important consequence for the detailed distribution of the screening charge that is formed around the ion. This is described in Sect. 3.4.2, but here we need not be concerned with it.

This is illustrated in Fig. 3.10. Each itinerant electron passing in the vicinity of the ion to be shielded spends a slightly longer time there than it would if the ion were not there. But the very fastest electrons arriving in the ion's environment are affected little by its presence; they pass quickly through the region and do not spend sufficient time there either to make any great contribution to the screening or to feel significantly the ion's disturbing influence. For a different reason, the very slowest electrons

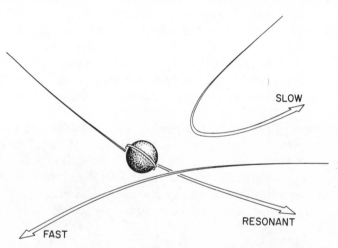

Fig. 3.10 When an itinerant electron in a metal is resonantly scattered by an ion—whether foreign or native—it can be regarded as a temporary resident of an atomic–like orbital around the ion. Such virtual atomic orbitals are formed through the mixing (hybridization) of the ion's outer atomic levels with those delocalized states of the metal that are suitably disposed. As in any resonance phenomenon, the energies of the resonating systems must be comparable; in this context this means that the incident electron must not be either too fast or too slow. Electrons temporarily captured into such virtual states form the charge cloud around the ion that is required to electrostatically screen any net charge the ion may possess either because of vacancies in its inner shells or, in the case of the foreign ion, because of its different nuclear charge. Two qualitative features are worth noting in this sketch. First, an electron that is not resonantly scattered has, nevertheless, a temporary orbital component introduced into its otherwise linear trajectory. Second, a resonantly scattered electron suffers a time delay in its general translational motion—it takes longer to traverse the ion's vicinity than would be the case in the ion's absence. Consequences of both effects are discussed in Sect. 3.4.6.

similarly do not make a significant contribution to the screening charge; their motion close to the ion is dominated by the strongly repulsive potential of the electrons localized in the outer shells of the ion. Being, therefore, strongly scattered by these outer shells (a process known as potential scattering that contrasts with the spin-dependent, or exchange, variety encountered in the following paragraph) and having low energies, these slower itinerant electrons are unable either to penetrate close to the ion or to stay long in its vicinity.

The screening charge around the ion is consequently provided almost entirely by those itinerant electrons having intermediate energies that are comparable with those of certain pseudo-atomic orbitals that can exist around the ion. These orbitals can be thought of as the remnants of the outer valence orbitals of the isolated ion that have become modified through overlap with itinerant electron states of comparable energy and suitable symmetry. (This can be looked upon as an example of the hybridization of orbitals, which is introduced in Sect. 2.3.3 and is a frequently encountered concept in the study of chemical bonding.* Also, such resonance energy levels are quite familiar in problems of particle scattering in other contexts—in atomic or nuclear physics, for example.) In the present context such resonances are known as virtually bound states, and their existence depends upon the balance of two forces that act upon an electron in an orbital about an ion: A centrifugal force tending to exclude the electron from the inner core of the ion† and the attractive Coulomb force arising from the ion's nuclear charge. In opportune circumstances, when for a particular angular momentum the combination of these two influences produces a minimum in the electron's potential energy at some radius outside the ion, an electron can be temporarily localized at this radius, thus forming a virtually bound state.

The occupant of such a state can be regarded as an electron that, having arrived in the vicinity of the ion with an appropriate energy, is temporarily captured in the virtual state. It will spend an appreciable time localized about the ion—thus contributing to the screening charge—before escaping by quantum tunneling to another delocalized energy state, and so regaining membership in the itinerant class. Looking at all of this from the itinerant electron's point of view, we see that its capture and subsequent release into

---

* F. Seel's: *Atomic Structure and Chemical Bonding*, Methuen and Co., London, 1963 remains one of the most lucid and readable introductions to this subject.

† In the isolated ion this is expressed in the term $l(l + 1)/r^2$, which appears in the usual radial Schrödinger equation that is derived in any standard textbook of solid state physics. This term is equivalent to a repulsive potential that produces a centrifugal force upon the electron.

a different itinerant state can be regarded as a scattering event, as pointed out in the preceding paragraph.

This extremely simple picture implies that an individual electron can be labeled in some way and that its progress can be traced through various energy states. This is, of course, incorrect since quantum theory explicity forbids such a procedure. However, it does allow the concepts of spontaneous annihilation and of creation of resident particles in energy states. Therefore, it is more appropriate—and, in fact, approximates closer to the quantitative approach of quantum theory—to look upon the scattering as a two-step process. First, an itinerant electron in some delocalized state is annihilated while simultaneously one is created in a localized state in the ion. Second, an electron is created in a delocalized state while one in a localized state is simultaneously annihilated.

If this scattering event involves no change in energy between the itinerant electron's initial and final states, then it is said to be elastic; otherwise it is said to be inelastic. If it changes the electron's spin orientation between the initial and final states, then it is called a spin-flip event; otherwise it is called a nonspin-flip event. And if the event leads to an unequal probability of the electron being scattered to one side of the ion rather than to the other, then it is said to be skew (or antisymmetric); otherwise it is said to be normal (or symmetric). All of these different possibilities will be encountered in subsequent sections.

It is clear why the states containing the temporarily localized electrons are referred to as quasi–atomic or virtually bound. In the true atomic states—which exist in the isolated atom or effectively in the innermost shells of the ion—the unexcited electron is for all practical purposes truly bound and unable to escape. The other extreme is, of course, the case where the outer levels of contiguous atoms hybridize to such an extent during the condensation to form a metal that the electrons in them become delocalized and completely free to range over the resultant spread of energy levels. The virtually bound state can be regarded as an intermediate situation: It is atomic-like in that it is localized about a particular ion, and yet it is sufficiently interactive with the itinerant states of comparable energy that an exchange of electrons between them is quite a likely event. A schematic view of a virtual level is shown in the left-hand side of Fig. 3.11.

Another view of the resonance process is shown in Fig. 3.12. Two cases are differentiated in this depending upon whether the ion is magnetic (that is, has a net spin resulting from vacancies in, say, its inner d or f shells that have survived the condensation process from the atomic state) or nonmagnetic (when, of course, all the inner shells are filled). In the nonmagnetic

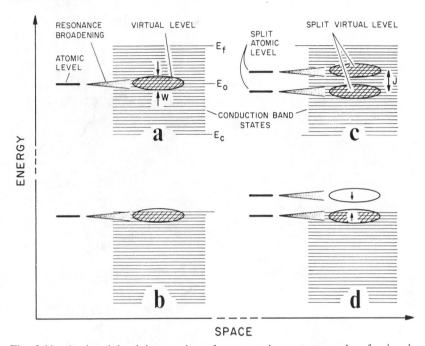

Fig. 3.11 A virtual level is a region of space and energy around a foreign ion in a metal in which itinerant electrons linger and temporarily assume to a marked degree the atomic features of the precursor atomic state. Physically, the virtual level is formed from a resonance (hybridization) between the energy level of an outer orbital of the foreign atom and those levels in the host's conduction band that have comparable energy. The virtual level is represented schematically by the cross-hatched area in this space–energy diagram. In a and b a single atomic level gives rise to a virtual level at E_0 of energy width W; in c and d an atomic level is split in energy by an interaction described in the text and gives rise correspondingly to the split virtual level.

case the example shows a virtual atomic orbital (labeled q) which is formed from the mixing of the itinerant electron states and the s levels of the foreign atom. In the magnetic case it is similarly formed from a mixing with the outer d (or f) levels of the ion. In either case, an itinerant electron incident in a delocalized state labeled k (and assumed to have an appropriate momentum) can be temporarily captured into the virtual state q. Subsequently, it returns by tunneling to another itinerant state (labeled k') and thus completes its resonant scattering from $k \rightarrow k'$.

There is, however, an important difference between the nonmagnetic and magnetic cases. In the former, the fully occupied inner shells take no direct part in the scattering process; an itinerant electron of suitable energy

I. NONMAGNETIC SOLUTE ION

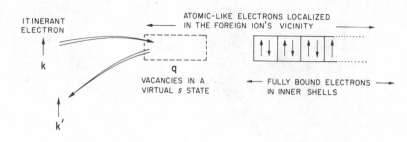

2. MAGNETIC SOLUTE ION

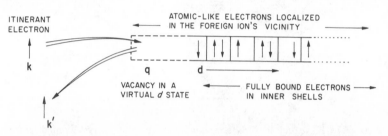

Fig. 3.12 Showing an alternative view of the resonant scattering process illustrated in Fig. 3.10. Here we differentiate between two cases depending upon whether the ion is magnetic (that is, has a net spin resulting from vacancies in its inner d or f shells that have survived the alloying process) or nonmagnetic (when, of course, all the inner shells are filled). In the nonmagnetic case the example shows a virual atomic orbital (labeled q) formed from the hybridization of the itinerant electron states in the metal and the s levels of the foreign atom; in the magnetic case it is formed by resonance with the outer d (or f) levels of the ion. In either case, an itinerant electron incident in a state labeled k and having appropriate momentum can be temporarily captured into the virtual state q before subsequently tunneling to another state (say k') of the itinerant class; the whole process is the resonant scattering of the electron from $k \to k'$. The text explains that the important difference between the two cases is that in the second the spin configuration of the incident electron enters as a parameter in the scattering.

can occupy the virtual state without regard to its spin or any coupling between its spin and those of electrons in the inner shells of the ion. (This is known as potential scattering, as indicated above.) However, in the magnetic case the occupancy of the inner shells is all-important since the net spin of the ion couples with that of the incident electron and introduces the incident electron's spin orientation as a parameter into the scattering. (This is thus the spin-dependent scattering referred to above.) The key to understanding the spin-dependent case is the recognition that the incident electron and the scattering ion must be considered as a single quantum mechanical system. Pauli's principle dictates that the spin of an incident electron has to be opposed to that of any resident in the virtual state before it can occupy the vacancy. If a spin reversal is required to achieve this, it has to be balanced (to conserve spin) by an opposite spin-flip produced somewhere among the fully bound d (or f) electrons. A consequence of this is that alternative routes for scattering an electron from $k \rightarrow k'$ in the spin-dependent case are generally not equally probable; this has several interesting manifestations which are referred to in Sect. 4.4.

3.4.2 Halos of Screening Charge.

We showed in the preceding section how the resonant scattering of itinerant electrons by a foreign ion (or other type of crystallographic imperfection) builds up a screen to neutralize whatever charge difference would otherwise exist between the ion and that of the host it replaces. We can guess that this extra itinerant electron density may or may not be uniformly distributed about the foreign ion—it depends upon whether or not the charge to be screened has spherical symmetry—but whatever the case, it seems intuitively reasonable to imagine that the screening charge will decrease uniformly as the distance from the foreign ion increases. The surprising result is that this is not so; in fact, the extra charge density shows marked oscillations in space as the distance from the ion is changed, and these decrease in amplitude as an inverse power of this distance. Moving out from the ion, we thus find successive regions of increased and decreased electron density compared to the density that existed before the foreign atom was introduced. The screening is, therefore, in the form of halos of charge about the ion (Fig. 3.13). (It is important to note in passing that these are fixed oscillations in the spatial distribution of the electron density: They do not vary with time. Temporal oscillations in the density at any point in space do occur—and are known as plasma oscillations—but they are a separate subject which arises again in Sect. 4.2.5.)

These fixed oscillations first appeared in calculations made by J. Friedel,* although their importance was not fully appreciated until some years later; they are now generally known as the Friedel oscillations. The sign and amplitude of the extra charge density at near-neighbor or next-near-neighbor sites to the shielded imperfection can produce electrostatic attraction or repulsion for other imperfections; as a result, the oscillations play an important role in the atomic theory of metals and alloys. (Although our

---

* J. Friedel, *Phil. Mag.* **43**, 153, 1952.

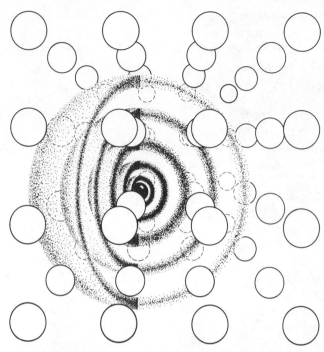

Fig. 3.13 A schematic cutaway view of the screening charge around a non-magnetic foreign ion in a lattice. The screen is built up of itinerant electrons that linger in the ion's vicinity longer than would be the case in the ion's absence. The surprising result is that the extra charge density produced around the ion is not uniformly distributed but consists of concentric regions in which the electron density is alternately increased and decreased from its normal value. (See also Fig. 3.14.)

description is in terms of the screening of a foreign ion or other charged imperfection, the oscillations also occur in the screening charge around individual atoms in a perfect metallic crystal.) The importance of this becomes still more pronounced when the foreign ion has a net magnetic moment (this case will be considered in Sect. 3.4.3). There are two approaches that can be made to a qualitative description of the origin of these oscillations. The first stresses that they are a mathematical inevitability of the sharp cutoff in the itinerant electrons' momentum distribution; the second is more pictorial in that it pictures the scattering of the itinerant electrons by the foreign ion during which the interference between the incoming and outgoing electrons produces the oscillatory screening charge. We shall look at both of these approaches. First, the least pictorial one.

We know that the screening of the foreign ion is done by the itinerant electrons, and we said in earlier chapters that at absolute zero these occupy states in the conduction band up to a maximum level E_f (Fig. 2.9). There is, therefore, a sharp cutoff in their momentum distribution. Similarly, there is a sharp cutoff in the range of the electrons' wavelengths: At absolute zero there are no wavelengths less than $\lambda_f{}^*$, which corresponds to the to the most energetic electrons (namely, those with kinetic energy E_f).

If we look on the screening of the foreign ion's excess charge as a matching between these itinerant electron wavelengths and corresponding components of the excess electrostatic potential produced by the ion, we see that it is very difficult to screen out any components of this potential having wavelengths less than λ_f. To do this matching mathematically it is necessary to resolve the ion's potential into Fourier components and to match each of these with a corresponding component obtained from the range of electron wavelengths. Since, as we have pointed out, this range is limited by a sharp cutoff, there is only a limited number of Fourier components available to construct the charge density needed to match and thus to screen the potential of the ion.

This necessarily leads to Friedel oscillations in the manner illustrated in Fig. 3.14a. This takes the simple example of a step function

$$f(\theta) = \begin{cases} +1 & 0 < \theta < \pi \\ -1 & \pi < \theta < 2\pi \end{cases} \tag{3.42}$$

which can be supposed to represent in this case the potential to be screened. We know from elementary Fourier theory† that this function can be produced (matched in this context) by the sum of the series:

$$\frac{4}{\pi}\left(\sin\theta + \frac{1}{3}\sin3\theta + \frac{1}{5}\sin5\theta + \ldots\right) \tag{3.43}$$

in which we imagine that each term arises from electrons of a particular wavelength. Wherever we decide to cut off this series, the result is an oscillatory function of wavelength comparable to that of the last term taken. (For example, Fig. 3.14a shows the effect, for the range $0 < \theta < 1/2\pi$, of taking 3, 5, or 10 terms.) Of course, the more terms taken the closer the series Eq. (3.43) approximates the function Eq. (3.42), and the more perfectly matched is the ion's potential at any given point in space, but inevitably the sum of all the terms will be an oscillatory function of the distance from the ion. This is exactly the origin of the Friedel oscillations: The electrons' wavelength distribution is sharply cut off, and this is equiva-

---

* The reader is reminded that the de Broglie relationship between a particle's wavelength λ and momentum p is $\lambda p = h$, where h is Planck's constant.

† For example, J. Mathews and R. L. Walker, *Mathematical Methods of Physics*, Benjamin, New York, 1965, p. 93.

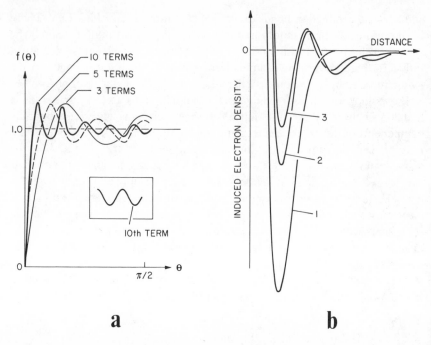

Fig. 3.14 The extra itinerant electron density induced to stay in a foreign ion's vicinity to screen electrostatically its excess nuclear charge is not uniformly distributed. Moving out from the ion, it shows marked oscillations of diminishing amplitude (Fig. 3.13). These are a consequence of the sharp cut-off in the itinerant electrons' momentum distribution at absolute zero. (*a*) an analogous situation, which shows the effect of trying to fit a step function (described in the text) by means of a Fourier series. Whenever the series is limited, whether at the third, fifth, tenth, or nth term, the function generated shows oscillations with a wavelength comparable to that of the last term included. (*b*) The effect of smoothing out the sharp cut–off in the electrons' energy distribution: As it is progressively smoothed $(3 \to 2 \to 1)$ the amplitudes of the oscillations decrease. (After M. E. Rensink loc. cit.)

lent to limiting the number of terms in an expansion like Eq. (3.43). Consequently, the excess charge distribution will be oscillatory with a wavelength comparable to that of the last term taken, which is here of energy E_f and which turns out to have a wavelength roughly of the order of a typical ionic diameter.

If the sharp cutoff in the itinerant electrons' momentum distribution did not exist, Friedel oscillations would not be expected to occur. This is indeed confirmed by Fig. 3.14*b* which shows the results of calculations* in

* M. E. Rensink, *Phys. Rev.* **174**. 744. 1968.

which the sharpness of the cutoff is artificially adjusted. Curve 1 is the case when there is no cutoff at all; all possible electron momenta or wavelengths are present to some extent and the charge density shows no oscillatory behavior and falls off rapidly as the distance from the ion increases (thus giving the behavior anticipated in the opening paragraph). But as the cutoff is made successively sharper, the behavior changes to that shown by curves 2 and 3, which do have the Friedel oscillations.

By the same token, a modification of the oscillations is expected whenever the sharp cutoff in the momentum distribution is altered. For example, when the temperature is raised above absolute zero, so that some itinerant electrons are promoted to energies above E_f and thus smooth out the distribution's cutoff, the oscillations are expected to diminish. Although experiments have been performed that indicate a temperature dependent amplitude for the oscillations, the effect is rather small, and it is difficult to eliminate other possible explanations of its origin. However, it is clear that even at temperatures above the melting point the itinerant electrons' momentum distribution is still sufficiently sharply cutoff to give Friedel oscillations of appreciable amplitude; these oscillations play an important role in the atomic properties of metals and alloys.

Turning to the other view of the origin of these oscillations, we said in preceding sections that the motion of any electron that contributes to the screening of a crystallographic imperfection is inevitably altered as a result. In fact, the screening can be looked upon as entirely a consequence of this electron scattering effect. The de Broglie wavelengths of the itinerant electrons turn out to be roughly equal to the diameter of an imperfection (whether it be a foreign ion or an unoccupied lattice site, for example), and so the problem is an example of one familiar in physics where waves are scattered by an obstacle of size comparable to the wavelength. It is well known that in the optical aspect of this problem, where an incident beam of light is scattered by an obstacle, the latter is surrounded by an interference pattern of which the shadow is a principal feature; the interference is, of course, between the incident and the transmitted beams. In the present context the difference is that the waves are approaching and leaving the obstacle from all directions (since the electrons contributing to the screening arrive from, and are scattered to, all directions), and in such a case the interference pattern has correspondingly to be averaged over all directions of incidence. Consequently, for a spherically symmetrical scatterer, any angular features of this interference pattern are removed by the averaging, but radial variations in amplitude remain. This radial modulation of the electron wave's amplitude, with its concomitant modulation of the electron density, is the origin of the Friedel oscillations.

This scattering description is usually formulated mathematically in terms of a partial-wave analysis common in formal scattering theory.* It pivots around the fact that, if the outgoing wave is resolved into partial components, the phase of each of these relative to the incident wave is shifted to some degree. These phase shifts relate directly to all that is of interest in the problem. For example, in the present case the extra time that an itinerant electron spends in the ion's vicinity, and thus the extra screening charge density associated with it, can be expressed directly in terms of the phase shift of the appropriate partial wave; waves corresponding to electrons that are resonantly scattered have by far the largest phase shifts. Further discussion of this aspect is outside the present purpose, but it is important to be aware of the partial-wave formulation because within it Friedel introduced and developed his description of the oscillatory screening charge. It has since become a well-established construction in the theory of dilute alloys, with expressions for various experimental parameters (such as the electrical resistance, the thermoelectric power, the Knight shift, as well as the screening charge distribution) having been formulated in terms of these phase shifts.

Finally, we should emphasize that the existence of the oscillations in the screening charge density cannot be confirmed by direct experiment; there is no way that we can make direct observations inside a metal. Their presence has to be inferred from experiments that reflect the consequences of their existence. It would again take us too far afield to discuss any of this in detail, but particularly striking confirmation of the oscillatory nature of the screening charge has come from studies of the atomic arrangement in liquid metals (which reflects the oscillatory dependence of the interatomic interaction), from studies of the electric field caused by a foreign ion at the nuclei of surrounding host ions (the Knight shift), and from other experiments† relating to the equilibrium separations between foreign ions in a metal.

3.4.3 Halos of Spin Polarization. We now turn to the extension of the above description to the case where a foreign ion in a metal has a net magnetic moment (arising, for example, from the uncompensated spins of electrons in an incomplete d or f inner shell). We have seen in the non-

---

* N. F. Mott and H. S. W. Massey, *Theory of Atomic Collisions*, Oxford University Press, Oxford, 1949, is the standard treatise on this subject. A more elementary treatment given in the context of metals can be found in C. P. Flynn, *Point Defects and Diffusion*, Oxford University Press, London, 1972, Sec. 14.2.

† Recent lists of such have been given by A. Blandin in section C of his review: *Alloying Behavior and Effects in Concentrated Solid Solutions*, Ed. T. B. Massalski, Gordon and Breach, New York, 1965 and by F. Gautier in, *Propriétés Electroniques des Métaux et Alliages*, Mason et Cie, Paris, 1973, Chap. 4.

magnetic case that the itinerant electrons incident upon the ion and contributing to its screening are scattered by it. Furthermore, there is an interference between the incident and scattered electrons which gives rise to a distinctive standing distribution of electron density about the ion. When the ion has a net magnetic moment there is an exchange interaction between the itinerant electron's spin and that of the unpaired electrons localized on the ion. Because of this—no matter which type of interaction dominates (Sect. 3.3.3)—it is clear that the itinerant electrons contributing to the screening are scattered differently depending upon the orientation of their spin relative to that on the ion. Two configurations are possible: The itinerant electron's spin can be either parallel or antiparallel to that on the ion, and it is convenient to look upon these two classes of itinerant electron as forming separate groups. The Friedel oscillations exist in each group, for exactly the reasons described in the preceding section, but in space these two distributions of electron density are mismatched somewhat because of the exchange interaction with the ion's spin. (When the foreign ion has no net moment, the mismatch is, of course, eliminated and the distributions for the two groups then superimpose.)

The net effect is clearly an oscillation in space for both the number density and the spin density of the itinerant electrons temporarily in the vicinity of a magnetic ion. The former is the Friedel oscillations which are part of the electrostatic screen, as in the case of a nonmagnetic ion, and the latter is effectively an oscillating magnetization of the electron fluid around the ion and is a byproduct of the exchange force acting between the spins of the itinerants and the ion. Moving out from the ion in this case, we find regions in which the spins of the itinerant electrons are polarized successively in the up and the down configurations with respect to the spin on the ion. Since this polarization arises basically from the Friedel oscillations in the electron density, it also simultaneously decreases as we move further from the ion.

The importance of the spin-dependent scattering of the itinerant electrons by the magnetic ion, and their concomitant oscillatory spin polarization about the ion, has already been referred to in Sect. 3.3.3 in connection with s-d exchange coupling. When acting as intermediaries, such itinerant electrons can couple ionic spins in either the ferromagnetic or antiferromagnetic configuration, depending upon the ionic separation. Such is the RKKY interaction (Sect. 3.3.3) which will be encountered frequently in subsequent chapters.

3.4.4 Virtually Bound States and the Magnetism of a Foreign Ion.
The preceding leads naturally to a subject that has a number of interesting consequences and that has been the source of much recent research. It is

known generally as the localized magnetic impurity problem since it concerns the case of a foreign ion from a d or f block in the periodic table that is dissolved in a simpler, nonmagnetic host metal (such as Cu or Al, for example); its implications are much wider than this, however, since the subject has led to a reconsideration of some fundamental aspects of magnetism in metals.

The interest in the problem lies in the fact, which was introduced above, that the partial filling of the d or f shells in the foreign atom can in some cases survive the alloying process to give the ion a net magnetic moment because of its uncompensated electron spins. We should note that a contribution to the ion's total magnetic moment can of course arise from the electrons' orbital momenta. This is potentially a big contribution, but frequently the net orbital momentum of an electron in an orbit is found to be zero, so that its corresponding orbital magnetism is nonexistent. This is quenching of the orbital angular momentum (Sect. 3.2.5) and it is caused by the effects of electrostatic fields produced in the ion's environment by the surrounding ions of the lattice. It is as if the total angular momentum vector for an orbital (or shell) is caused by the electrostatic fields to tumble wildly through all possible orientations so that its time-averaged effective value in any given direction is zero. In such a case the ion's magnetic moment arises solely from its uncompensated electron spins; this is our exclusive concern in this section. The $3d$ electron shell of an ion in its normal (octahedral) crystalline environment has quenched angular momentum—although this can be undone somewhat by the spin-orbit effect described in Sect. 3.2.5. Ions such as Mn or Fe dissolved in a Cu or Au matrix exemplify the case under consideration.

Ideally, the problem is concerned with a single foreign ion sufficiently distant from others of its kind in the alloy to be unaffected by them; in practice this means the foreigner's concentration cannot amount to more than about ten or a hundred for every million ions of the host metal. A couple of decades ago, when interest in this subject began, that was typically the concentration of the metallic impurities present in the highest purity metals available, and so the problem became known by the title mentioned above. Nowadays, however, it is more appropriate to call it the localized magnetic solute problem, for improvements have made it possible to prepare alloys with controlled solute concentrations approaching one part per million.

To recapitulate briefly the results of preceding sections, we recall that, when a foreign ion replaces a host in a metal, itinerant electrons can become temporarily localized about the ion as part of its screening charge. They are contained in virtually bound states that are formed from resonances between the energy levels of the ion's outer electrons (which in this context will be of d or f shells) and the conduction band states of the host. This resonance (or hybridization) can only take place efficiently between quantum states satisfying the three criteria cited at the beginning of Sect. 2.3.3. Of these, the requirement we stress in the present context is that the states must have comparable energies. As a result, for each atomic level of the foreign atom which falls within the total energy range of the host's conduction band ($E_c \rightarrow E_f$ of Fig. 3.11) we can imagine a corresponding virtually bound state to exist in the alloy, a state which occupies a limited region of space about the ion. This is illustrated in the space/energy diagram of Fig. 3.11a. The cross-hatched area represents schematically the extent in space and energy of the virtually bound state arising from the atomic level shown; it is a range of the parameters in which the itinerant electrons temporarily assume to some degree the atomic characteristics of the precursor state. The resonance broadening of this state through hybridization with the conduction band states (that is, through the s-d mixing of Sect. 3.3.3) is a description of the efficiency of the interaction; for a d or f level it gives a virtually bound state of width typically up to about 0.8 to 1.6×10^{-19} J. It turns out that for s or p states this broadening is so great that they merge into the metal's conduction band and lose any specific identity so that the virtually bound concept is not appropriate.

There are a total of 10 and 14 spin orbitals forming the d and f shells of an atom, respectively. If a virtually bound state such as that in Fig. 3.11a arises from a d (or f) shell, it, therefore, has room for 10 (or 14) electrons divided equally between the two possible spin alternatives. Since it is entirely within the range $E_c \rightarrow E_f$ spanned by the conduction band states, the bound state has its full complement of electrons and, hence, has no net magnetic moment arising from uncompensated electron spins (assuming, of course, that it is under observation for a time sufficiently long compared with the lifetime of a localized spin fluctuation; see Sect. 3.3.3). Even if the circumstances bring E_o, the bound state's mean energy level, very close (or even above) the Fermi level E_f, so that the latter cuts through the level's energy range (as in Fig. 3.11b) the state remains nonmagnetic since it empties to the appropriate level but without regard to the resident electrons' spin direction.

We should note that in this configuration the electrons that hybridize to form the virtually bound state are those with energies close to E_f and are thus the ones easily excited to vacant energy levels by an outside influence. For example, at any temperature T only the electrons within $k_B T$ of E_f contribute to the electrical conduction. Consequently, when the virtual state enters this range, it can be expected to figure prominently in the electrical resistivity. That is not a concern here, but the subject arises again in Sect 3.4.5. [$k_B T = \sim 10^{-21}$ J ($\sim 10^{-3}$ eV) when $T = 12°$K.]

In order to produce a magnetic state, what has to be accounted for, and what has so far been ignored in preceding sections, is the fact that the energies of the individual electron states forming the d or f shells (that is, the 10 and 14 spin orbitals, respectively) may be split and separated by various influences that arise either within the foreign atom itself or in the ion. as a result of its substitution into the alloy. Whatever the causes, and some of them will be discussed below, the effect of splitting the d (or f) levels is potentially to create an equivalent number of separated, virtual levels in the conduction band of the host. This is illustrated in Fig. 3.11c, which represents an atomic level split into two by some presently unspecified interaction. (This means that in the atom the electrons occupying this level have a choice between two alternatives of slightly different energy.) When the atom is dissolved into a metal, each component of the split level gives rise to a virtually bound level around the ion in the manner described above, and the whole is referred to as a split virtual level. This splitting can have important consequences, as will be seen below, but to be effective its magnitude J (Fig. 3.11) must obviously be greater than the typical width W, otherwise the effect of the splitting is swamped by the resonance broadening.

We now turn to consider four potential causes of virtual band splitting. The first is known as crystal field splitting and, as is implied, it results from the electrostatic field acting on the foreign ion substituted into the lattice of the host metal. To fix the argument, suppose an outer d shell in the foreign ion forms a virtual level in the metal's conduction band and suppose further that this host's lattice has cubic symmetry. (Chapter 5 shows that this covers about half the elemental metals at room temperature.) The foreign ion, when at the origin of a set of rectangular axes, has in this cubic case six equivalent nearest neighbor ions of the host—one located along each direction of each axis. (This is known as an octahedral coordination for the foreign ion.) These nearest neighbors are the main reason that the crystal field acts at the foreign ion, which here is just the electrostatic repulsion between the foreign ion's d electrons and those electrons in the outer shells of the host's ions. It is a standard result* that the five atomic

---

* See, for example, M. J. Sienko and R. A. Plane, *Physical Inorganic Chemistry*, Benjamin, New York, 1963, p. 55. A comparison of the effect upon d orbitals of crystal fields of different symmetries is shown in Fig. 13.17 of W. J. Moore, *Physical Chemistry*, Prentice-Hall, Englewood, N.J., 1962, p. 550. Crystal splitting of f orbitals is discussed by .H. G. Friedman et al., *J. Chem. Educ.* **41**, 354, 358, 1964.

d orbitals, which, of course, are degenerate in energy in the field-free case, are split in energy into two groups by such field:* One group consists of the d_{z^2} and $d_{x^2-y^2}$ orbitals (Appendix A) which are raised in energy an equal amount with respect to the field-free case, and the other consists of the d_{xy}, d_{yz}, and d_{zx} orbitals, which are likewise lowered equally in energy by the crystal field. In the particular symmetry notation of the subject, these are known as the e_g and t_{2g} groups, respectively. The total capacity of the e_g and t_{2g} groups is clearly four and six electrons, respectively; the energy separation between these groups turns out to be typically about 1.6×10^{-19} J. Since the width W of a virtually bound state (Fig. 3.11) is usually as big as this, any crystal field splitting effect is, therefore, often swamped by the resonance broadening and tends not to be very important in the localized magnetic impurity problem in metals. It is, however, an important effect for such ions in many magnetic salts or other insulators.†

The second possible cause of splitting a virtually bound level is the intraatomic Coulomb repulsion between electrons within the atomic shell producing the virtual level. Since electrons have like charges, they repel each other and stay out of each other's way. If this Coulomb correlation of their motion is strong enough within a given shell, its effect is to stabilize states containing integral numbers of electrons and to separate them in energy. [This can be seen from the description given in Sect. 3.3.1; in case (a) for example, an electron of spin up experiences the Coulomb repulsion U from that of spin down in the same orbital. If U is large enough, these electrons occupy effectively separate energy states.] Thus the d and f shells can be split by this effect into ten and fourteen well-separated energy states, respectively. The virtually bound state must then be thought of as arising from the hybridization of each of these split states with those of the host's conduction band. Thus, in the way illustrated in Fig. 3.11c, an atomic shell containing n electrons gives rise to n distinct components in the split virtual level. The experimental evidence is that this happens with the f shells of the rare earth lanthanide series; it is often significant when one of these is a localized magnetic solute, but it is probably far less important in the case of d shells of solute atoms from the first transition series. This difference is believed to be due partly to the more extended nature of the d orbitals (Sect. 2.3)—which of course reduces their

---

* Knowing the shapes of these orbitals (Appendix A), this is not difficult to understand qualitatively. The electrical repulsion will raise the energy of a d orbital directed along any axis and will lower that of any directed between axes. This segregates the orbitals into the two groups since the d_{z^2} and $d_{x^2-y^2}$ are of the former type, and the d_{xy}, d_{yz}, and d_{zx} are of the latter.

† J. B. Goodenough's *Magnetism and the Chemical Bond*, Interscience, New York–London, 1963 is the standard description of this aspect.

Coulomb correlations—and partly to their generally more efficient hybridization with the conduction band states, which in turn tends to swamp the correlation effects through resonance broadening and subsequent overlap of the virtual level's components. However, such Coulomb correlations are the main pivot in Anderson's model of an ion, which is described later in this section.

In fact, for d electrons forming a virtually bound level, the most important cause of the virtual level's splitting is probably the intraatomic exchange correlation; the third cause considered and that to which we now turn. The direct intraatomic exchange coupling between electrons in a given shell (Type 1 of Sect. 3.3.3) can split an atomic level into two spin-polarized states separated by the exchange energy J of Sect. 3.3.1. For example, split d and f states can accomodate up to five and seven electrons of fixed spin alternative, respectively; the quantized spin direction conventionally called up is the lower energy configuration. As described in Sect. 3.2.5, this is the origin of Hund's rule for atoms: As a shell is filled, the exchange correlation favors parallel alignment of the spins whenever possible.

Friedel proposed that if such an atom is dissolved into a metal this exchange correlation may still be expected to be an important quantity in the ion. Thus an atomic level split by this interaction as shown in Fig. 3.11c can give rise in this view to a split virtually bound state in the host's conduction band. Once again, the first important feature is the magnitude of the splitting J with respect to the width W of a component; for, if the splitting is relatively small, the resonance broadening swamps it completely, as was already pointed out. But when the splitting is of intermediate strength, the situation can be like that in Fig. 3.11c; the virtual state then has two distinct components, the lower being populated by up-spin electrons and the upper one by down-spin electrons. In this intermediate case, both components lie entirely within the range of energies $E_c \to E_f$ of the conduction band, and both—again, when observed over a sufficiently long period of time—have their full complement of electrons. Consequently, the virtual level (and hence the ion it surrounds) has no net magnetic moment arising from uncompensated spins.

But if the splitting is large enough and if the position of the atomic levels with respect to E_f is appropriate, it is possible for one or both of the components of the virtual state to be shifted partially above E_f (even entirely above it in the case of the higher energy component). The virtual level must, of course, empty to the appropriate energy level (as in Fig. 3.11d) but in this case the down-spin component empties before the up-spin and the virtually bound state is populated by electrons of a preferred spin direction. Consequently, the virtually bound states (and hence the ion they surround) have a net magnetic moment because of these uncompensated spins.

This approach by Friedel through resonant scattering and the virtually bound state was the first attempt to account for a magnetic moment on an isolated ion in an alloy. It was followed soon afterwards by Anderson's* work, which essentially put Friedel's qualitative approach onto a more quantitative footing. There are strong similarities between these two approaches (indeed, we often see reference to the Friedel–Anderson model) but there are also important differences, and this subject should not be left without at least a brief reference to them. First, there is a difference in emphasis: Friedel's picture is concerned with the origin of a localized magnetic moment in a metal, and Anderson's starts from the assumption that such a moment exists and considers rather how it can survive in the circumstances. (As it turns out, subsequent developments have favored the latter emphasis because it fits better with the more dynamic view already encountered in Sect. 3.4.1 and to be discussed in the following section.) However, a more important difference exists between these approaches concerning the origin of the virtual state's splitting. Friedel, as described above, attributes the splitting to the intraatomic exchange (expressed as J in Sect. 3.3.1); Anderson, on the other hand, attributes it to the electrostatic Coulomb repulsion between two electrons having antiparallel spins and in the same orbital (U in Sect. 3.3.1).

To appreciate Anderson's view, we must imagine a single atomic orbital (say a d orbital) with an energy (say E_o) which lies within the conduction band of the host metal as shown in Fig. 3.11a (in this case the energy level is discrete, however, for we have not yet let it broaden through mixing with the conduction band states). Suppose the level is occupied by a single electron of spin up. This leaves a single vacancy in the level, but the Pauli principle forbids another spin-up electron to occupy it. However, a spin-down electron can be added to make up the full complement, but it experiences the Coulombic repulsion U. In fact, if it does join the state, the total energy of the electrons is $2E_o + U$, just as in the case (a) of Sect. 3.3.1. The spin-down state is, therefore, situated at an energy U above the spin-up one and, since the whole is magnetic by assumption, the spin-down state must lie above E_f and thus be empty. The energy broadening of both of these states is again assumed to arise from the s–d mixing interaction (Type 6 of Sect. 3.3.1). This single orbital picture can be extended to the case where the full five orbitals of a d shell are present, but

* P. W. Anderson, *Phys. Rev.* **124**, 41, 1961.

then an extra Coulomb energy term enters because electrons with parallel spins can now exist on different orbitals and their repulsion amounts to $U-J$ for each pair (Sect. 3.3.1). In fact, at any given instant the total Coulomb energy of the electrons trapped in the virtual state is the sum, taken over all the orbitals in the state, of: $U \times$ (product of the numbers of spin-up and spin-down electrons present) and $(U-J) \times$ (product of the numbers of spin parallel electrons present).

It is this energy that separates the two spin components of the virtual state and gives the configuration its stability. The $s-d$ mixing upsets this stability, however; not only does it broaden the upper and lower energy levels through resonance with the conduction band states, but it also brings them closer together in energy. (This aspect is described in the following section.) Too much $s-d$ mixing and the virtual state's components coalesce to give a nonmagnetic level.

But the final picture of Anderson's magnetic state is not unlike Friedel's phenomenological model illustrated in Fig. 3.11, except for the important difference that the virtual band's splitting arises from the intraatomic Coulomb correlation in Anderson's case. Whether the Anderson or Friedel view is more reasonable is still a controversial point; probably neither is universally applicable—each being appropriate for particular combinations of solute and solvent metals. We may think it is possible to decide between the two approaches in any given case by comparing their parameters (such as W, U, J, or E_f-E_o) with experiment, but unfortunately the critical parameters for both models are almost inaccessible by direct measurement, and in any case there are often grounds for explaining away "unreasonable" values of say U or J in terms of effects that these models ignore. This aspect irritates some experimentalists in this field. They would like to see a real theory (as opposed to descriptive models) that not only could be tested more directly but would predict from first principles whether or not a given solute ion in a given matrix will be magnetic at a given temperature—something which neither model can presently do. Nevertheless, the gross features of both are well established[*] and unlikely to be toppled in the foreseeable future—particularly when their dynamic aspects are included. A description of these is given in the following section, but first we must consider the fourth and final cause of virtual band splitting.

Since this arises from the effect of intrinsic spin-orbit coupling (Sect. 3.2.5) upon the electrons that are temporarily resident in a virtually

---

[*] Examples of the probable arrangement of this model for certain solutes in Cu or Ag have been given by H. P. Myers, L. Wallden, and A. Karlsson, *Phil. Mag.* **18**, 725, 1968 and may interest a reader looking for more quantitative details.

bound state, it will be referred to simply as spin-orbit splitting. It is a mechanism that operates in an ion (generally a foreign ion in a host matrix) and that may or may not have a net magnetic moment but that is surrounded by an occupied virtually bound state. The latter is part of the ion's electrostatic shield, but, as described above, when it is incompletely filled it can also be the source of the ion's magnetism. Whether or not an atomic level that is split by spin-orbit effects becomes incompletely filled when the atom is dissolved in the host depends upon the energies of the split components with respect to the alloy's Fermi level—exactly as in the cases of Fig. 3.11c and 3.11d.

The effects of intrinsic spin-orbit coupling in an ion's shell are described in Sect. 3.2.5. Briefly, when the coupling is neglected the orbital and intrinsic (spin) angular momenta of all the electrons in the shell can be summed independently to give a net value for each quantity. However, since it is a charged particle that is in motion, the orbital and spin momenta—whether of a single electron in an orbital or of a populated shell—each have an associated magnetic field that couples with the other's to make the orbital and spin motions interdependent. The combination of the orbital and spin momenta gives the total angular momentum of a shell, which is characterized by the quantum number J. When spin-orbit coupling is operative, it produces different internal energies for the ion for the different combinations. In the Ce ion, for example*, which has a lone electron in the $4f$ shell, the large spin-orbit coupling puts the energy of the $J = 7/2$ state ($l = 3$, $s = \frac{1}{2}$) in which the electron's orbital and spin momenta are parallel, far above the ground state $J = 5/2$ ($l = 3$, $s = -\frac{1}{2}$) in which the momenta are antiparallel. Consequently, when Ce is dissolved in an appropriate host, the conditions for resonance and the formation of a level are different for the two configurations, and a split virtual level can result (Fig. 3.11c and 3.11d).

Another example from the rare earth series is provided by the sequence Nd→Yb when dissolved in La, Y, or Ag. In this case the ion has a moment arising from its incomplete $4f$ shell, which lies relatively deep within the ion's core. It is thought† that in the alloy the ion is surrounded by a completely filled virtually bound d state that is needed to provide sufficient electrostatic screening for these generally trivalent ions. (It is presumed to be completely filled since the incomplete $4f$ shell can itself account for the observed magnetic moment.) This virtual d state is believed to be split by spin-orbit effects, thus giving a situation somewhat like that shown

---

* We recall from Sect. 3.2.5 that the rare-earth atoms have large spin-orbit effects in their f shells (Fig. 3.3).

† A. Fert, *J. Phys. F.* **3**, 2126, 1973.

in Fig. 3.11c: The $J = 3/2$ and $5/2$ states (arising from $l = 2$, $s = \pm\frac{1}{2}$) are thought to be split in energy by an amount that is of the order of W, the energy width of the virtually bound state. In fact, it is not simply this energy separation that is important but also the effect it has upon the scattering of an itinerant electron that is temporarily localized in the virtual state. It turns out that the scattering has left-right asymmetry with respect to the electron's direction of incidence; more electrons are scattered to (say) their right than to their left because of the split virtual level. This is one cause of skew scattering, which is described in Sect. 3.4.6.

To summarize all of this it can be said that whether or not a split virtual d or f state shows a net magnetism depends generally upon the energy width W of its components (which is a measure of the s–d or s–f mixing) compared with that of their splitting. Three rough categories can be distinguished (Fig. 3.15) depending upon the relative magnitudes of W and the combination of U and J. If W is greater than $U + J$, then the state has no magnetic moment. Examples of this are the virtual states around ions of the $3d$ transition series when dissolved in A1 and those around $4d$ and $5d$ ions when dissolved in Cu, Ag, or Au. If W lies roughly between $U + J$, and $U - J$, then the virtual state shows magnetism from uncompensated

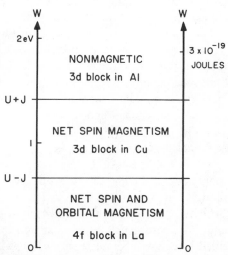

Fig. 3.15 Whether a split virtual d or f state possesses a net magnetism depends generally upon the energy width W of its components compared with their energy separation. Three rough categories can be distinguished that depend as shown upon the relative magnitudes of W and the combination of U and J (these being, respectively, the energies arising from the electrostatic Coulomb repulsion and the intraatomic exchange force between the virtually bound electrons).

electron spins only (that is, there is no orbital component). An example of this is provided by the $3d$ transition elements when dissolved in Cu, which as we pointed out above show a net spin magnetism except that the orbital component is quenched. Finally, if W is less than $U - J$ then the virtual state shows both spin and orbital components in its net magnetism. The behavior of the $4f$ state of a rare earth metal dissolved in a suitable matrix exemplifies this class. (However, there are three notable anomalies—Ce, Eu, and Yb—where the $4f$ level often lies close to the Fermi level in the alloy and thus complicates this simple categorization.)

3.4.5 The Magnetism of a Foreign Ion and the Scattering of Itinerant Electrons.

It was first emphasized in Sect. 3.3, and later in Sect. 3.4.4 in the discussion of the virtually bound state, that the polarization of the electron liquid about an ion—whether it be of the charge or spin variety—is a dynamic process; the s-d mixing interaction, which is the heart of the Friedel and Anderson models, can be looked upon as the constant capture and release of itinerant electrons by pseudo-atomic orbitals existing around the ion. Here our aim is to consider further this aspect; we are particularly concerned with the way that it affects the motion of itinerant electrons moving under the influence of externally applied forces (such as electric fields or temperature gradients) in a metal containing localized magnetic impurities. This is all by way of introduction to the electron transport properties and particularly to the Kondo effect, which is discussed in Sect. 4.4.

To fix the argument, take the now familiar case of a d shell of a foreign ion that forms a virtually bound state in a nonmagnetic metal. First, the importance of the strength of the s-d mixing (that is, the number of $s \longleftrightarrow d$ electron interchanges per unit time) should be made clear; perhaps Fig. 3.16 will help in this. It shows in a very pictorial way how to look upon the occupancy of a virtual state in an energy/space plot (such as Fig. 3.11). Figure 3.16a shows the situation when the exchange and Coulomb interactions giving an energy difference between the two spin configurations are negligible; a virtual state is formed that has, when observed for a long enough period of time, its full complement of five up-spin and five down-spin electrons. These are distributed in accordance with Pauli's principle among the five orbitals, which in this case are degenerate in energy. This virtual state has seemingly no net magnetic moment from uncompensated electron spins, but this is a time-averaged result that is not true on an infinitesimal scale, for the electrons in the virtual level are constantly interchanging with those in properly disposed delocalized states of the host's conduction band and within the compass of the virtual level.

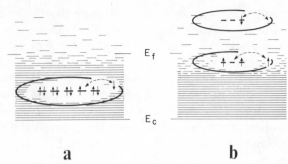

Fig. 3.16 An energy/space plot of the form of Fig. 3.11 which shows pictorially the occupancy of a virtually bound state at some temperature above absolute zero. The electrons in the virtual energy levels are constantly interchanging with those in properly disposed delocalized states of the host's conduction band (as the dashed arrows indicate). As is explained in the text, it is the time-averaged result of these interchanges that gives the virtual bound state's observed magnetic properties. (a) The case where the Coulomb and exchange effects compared to the level's energy width (W of Fig. 3.11) are insufficient to produce any appreciable energy separation of the electron's spin configurations. (b) The converse case where the splitting is pronounced.

One such event (or localized spin fluctuation) is illustrated in the figure: The virtual state has an instantaneous net moment because of the vacancy (or hole) which leaves an uncompensated spin, but this is about to be eliminated by an electron from a delocalized state that will fill the vacancy. This electron must, of course, have its spin antiparallel to that of the resident in the state. We can see that, if the frequency of these localized spin fluctuations is much less than the natural frequency* of the net moment they produce, the ion will appear nonmagnetic.

The same *s-d* mixing takes place when the virtual level is split by, for example, Coulomb and exchange effects. This is illustrated in Fig. 3.16*b* which shows the situation in a level like that of Fig. 3.11*d,* where the up-spin component is assumed to contain three localized *d* electrons. The figure shows localized spin fluctuations in both up- and down-spin compo-

---

* It would not serve our present purposes to become involved too deeply with this point, but we should realize that all observations of a magnetic moment involve its repeated transition between at least two states that are separated in energy: The Zeeman levels in a magnetic field, for example. In given circumstances, the frequency with which the moment tumbles between these energy levels because of its thermal energy is what is here called its natural frequency. The lifetime of a localized spin fluctuation is roughly $\hbar/E_f$, and only when this is long compared with the sampling time of whatever experimental probe is used will there appear to be a magnetic moment.

nents. In the former case the mixing electron from the delocalized states has to have up-spin, since down-spin electrons feel the repulsion U that puts their energy above the Fermi level E_f. At all temperatures above absolute zero, however, some states above E_f are occupied, as the drawing suggests, and electrons from these states are able to mix with the down-spin component to produce localized spin fluctuations of the type shown. The effect of these fluctuations is always to reduce the number of electrons in the up-spin component below its normal value and to increase above zero the number in the down-spin one. Because of this s-d mixing there is a reduction of the net moment, but whether or not this is observable again depends upon the relative values the moment's natural and spin fluctuation frequencies. However, it does frequently seem to be observable inasmuch as it gives an immediate explanation of the previously puzzling result that many ions show moments corresponding to nonintegral numbers of localized d electrons and, therefore, always slightly less than predicted by Hund's rule.

The s-d mixing has another important consequence: It effectively reduces the Coulomb repulsion U between opposite spin electrons in an orbital. We can see qualitatively how this comes about from the change produced in the probability of occupancy of the d states. For example, we described above how this probability for an up-spin state is reduced by s-d mixing from unity (certain occupancy) to something slightly less. A down-spin electron experiences the full repulsion U from the up state when the latter is occupied but does not experience it during the temporary vacancies. Averaged over a sufficiently long period, the net repulsion will be less than U. When the down-spin electron's probability of occupancy is likewise uncertain, the effect is compounded. As the s-d mixing is imagined to increase, the energy of the up-spin component of the virtual state, therefore, moves upwards toward E_f, and that of the down-spin moves correspondingly downwards. Since these components are also broadened in energy by the mixing, there comes a point when the state collapses to a nonmagnetic configuration.

For present purposes, the preceding is a sufficient description of the importance of s-d mixing in this context. Let us now turn to the second objective which is to outline the consequences of all this for the motion of the itinerant electrons. First, it is important to bear in mind a point already noted in passing in Sect. 3.4.4: Of all the electrons in delocalized states forming the conduction band, at a temperature T only those having a kinetic energy within roughly k_BT of E_f are able to take part in the normal electron transport that produces the everyday electronic properties of metal. This is because electrons can only be excited into states that are initially vacant. Consequently, only those electrons lying within a given energy reach of the vacant states can take part in the electron transport processes. The rest of the electrons, that is, those with energies in the range $E_c \rightarrow E_f - k_B T$, are unable to be excited into vacant delocalized states by outside influences (such as electric fields or temperature gradients) of normal magnitudes. Thus if a component of a virtual level lies deep within the conduction band (as in Fig. 3.11a) the delocalized electrons involved in the s-d mixing are not those normally contributing to the electron transport. Alternatively, when the component lies within $k_B T$ of E_f (as in Fig. 3.11b) then its presence will be felt in the transport effects because the electrons involved in the mixing are also those involved in the transport process. In the latter case it is appropriate to emphasize the scattering of these electrons by the level, for their motion can be easily changed by temporary capture into the virtual level since there are numerous alternative and vacant delocalized states available for escape. However, when the virtual level is deep in the conduction band, the surrounding delocalized states of comparable energy are generally filled and the electron's alternatives are much more restricted by the Pauli principle. This leads to a distinction in terminology: We generally speak of s-d mixing when the level is deep within the conduction band, and of s-d scattering when the mixing electrons are also those taking part in the electron transport.* The rest of this section is concerned with the latter, which is of course just the spin-dependent (or exchange) variety already encountered in Sect. 3.4.1. It contrasts with the ordinary potential variety experienced by the electrons when the virtual level is not split.

We explained above that, if the s-d scattering is not strong enough to collapse the state, a split virtual level can be described by a well-defined local magnetic moment or spin. In the usual classical analogy (Fig. 3.5) this is represented† by **S**, and it is the algebraic sum of all the unpaired electron spins in the level. We have seen that, because of the dynamic nature of the problem, **S** can be equivalent to a nonintegral number of unpaired electrons. On the microscopic scale the orientation of such a moment is quantized (Sect. 3.2.5) and is allowed only $2S + 1$ different

---

* This is illustrated by the behavior of ions of the rare earth metals (Ce→Yb) when dissolved separately in either La or Y. Although all the ions show a localized magnetic moment in the solid solution and are thus surrounded by a split virtual 4f level, only alloys containing Ce show the resistance minimum effect. This occurs because only in Ce does the lower 4f level happen to lie close enough in energy to the Fermi level of the alloy (Fig. 3.11d) for s-f scattering (as opposed to s-f mixing) to take place.

† The reader is reminded that the orbital momentum is assumed to be quenched.

alternatives or orientations with respect to some fixed direction. Conventionally this is taken as the z-direction, and this component of the moment is always the objective of calculations and theoretical discussions. It is specified by the total spin-projection quantum number M_S, which characterizes the component of the ion's spin moment projected along some specified direction. It clearly can also have only $2S + 1$ values. When the ion's moment is free from external influences, such as applied magnetic fields, these different alternatives, which each correspond to a separate Zeeman energy level (Sect. 3.2.5), are degenerate in energy. Since energy has to be conserved, an s-d scattering of an electron cannot be other than elastic in this case; in other words, there can be no transfer of energy between the itinerant electron and the ion during the former's temporary captivity. However, there can be an exchange of spin information (as was described towards end of Sect. 3.4.1). The itinerant electron's spin can be reversed during the scattering: Thus an electron from an up-spin state could be scattered to a down-spin one (of equal energy); to conserve spin in the interaction M_S would have to increase to $M_S + 1$ by reversing the spin of a d electron resident in the virtual level; correspondingly, a down-spin state could empty into an up-spin one, and M_S would decrease to $M_S - 1$. But whatever the moment's orientation among the $2S + 1$ alternatives, there can be no change in its potential energy when the Zeeman levels are degenerate, and consequently there can be no change in the itinerant electron's kinetic energy as a result of the scattering.

The above are examples of a spin-flip scattering event, where the itinerant electron's spin orientation is permanently reversed in the process. This is one type of elastic exchange scattering of itinerant electrons; the other is the nonspin-flip variety in which the electron's spin orientation (and hence the ion's M_S value) are unchanged in the event. To be specific, suppose the itinerant electron is in a delocalized conduction band state $\mathbf{k}, \sigma$ before scattering and ends up in $\mathbf{k},' \sigma$ after it. Looking at this solely from the itinerant electron's point of view there are just four possibilities (where the arrow indicates the corresponding spin orientation):

	BEFORE		AFTER
1.	$\mathbf{k}\uparrow$	$\rightarrow$	$\mathbf{k}'\uparrow$
2.	$\mathbf{k}\downarrow$	$\rightarrow$	$\mathbf{k}'\downarrow$
3.	$\mathbf{k}\uparrow$	$\rightarrow$	$\mathbf{k}'\downarrow$
4.	$\mathbf{k}\downarrow$	$\rightarrow$	$\mathbf{k}'\uparrow$

Clearly, types 1 and 2 are the nonspin-flip processes, and types 3 and 4 are the spin-flip ones; all are elastic in this case since the Zeeman levels are degenerate. Technically, these alternatives describe the scattering as a first-order transition, which is a way of saying that we look at the scattering event purely from the itinerant electron's point of view of passing simply between initial and final energy states. No consideration is given to any intermediate configuration. But from the magnetic ion's point of view (or, more properly, from that of the split virtual state around the ion), as a result of its magnetism it has an internal degree of freedom that allows different pathways between a given pair of initial and final states. In fact, as will be described in Sect. 4.4, when this extra degree of freedom is taken into account, each of the four processes listed above has itself four possible alternative routes between the same initial and final states. Technically this expanded view describes the scattering event as a second-order transition, which is now a three-step process as opposed to the two-step first-order one. J. Kondo* first took the trouble to consider the second-order effects in this localized impurity problem that everyone had previously thought was of no significant interest beyond the first-order approximation. The outcome was an explanation of many anomalous data, some of which had been in the literature for thirty years, and the discovery of an effect which is now named after him. However, this forms the subject of Sect. 4.4, and it would serve no purpose to anticipate that description here. Instead, we break off at this point and turn to the situation when the Zeeman levels of the local moment are not degenerate.

This degeneracy is removed when the moment on the ion experiences a magnetic field. This could be applied externally as part of an experimental set up, for example, or it could arise internally if other ions with local moments were brought close enough to the one in question. In either case, mutual interaction then separates in energy the Zeeman levels of each moment. We recall that the whole description here and in preceding sections has centered upon an ion assumed to be isolated from others of the same kind. This is a very difficult—if not impossible—circumstance to produce with certainty in practice, and there is no doubt that for all except the most extremely dilute alloys some interaction takes place. (When the interaction is appreciable, the splitting of the Zeeman levels in fact leads to a local maximum at low temperatures in the temperature dependence of the electrical resistivity.) With the degeneracy removed by whatever cause, an itinerant electron that is scattered by the split virtual level can change its energy as well as its spin in the process; the scattering is now inelastic, where the difference between the kinetic energies of the incident and scattered electrons will be the Zeeman energy of the localized moment;

* J. Kondo. *Progr. Theoret. Phys. (Kyoto)* **32**, 37 1964.

actually it is $g\mu_BH$, where in this spin-only case (Sect. 3.2.5) $g = 2$ is the electron's g factor, μ_B is the Bohr magneton, and H is the magnetic field strength at the virtual level.

Let us concentrate upon the effects of applying an external field H to remove the degeneracy. In zero field the Zeeman levels of the ion's moment are equally populated; because of its thermal energy, the moment tumbles at random through the various alternative orientations, but all are equally probable. As the field strength is increased however, and the degeneracy is progressively lifted, the lower Zeeman levels bcome preferentially occupied. In other words, the moment takes up those orientations with respect to the field that require the least energy. For strong fields this becomes a very pronounced effect indeed, and only the lowest Zeeman levels are filled. This has an immediate effect upon the inelastic spin-flip scattering because those processes in which the itinerant electron's kinetic energy is increased (at the cost of a corresponding reduction of the ion's Zeeman energy) become increasingly improbable; the ion is already in its lowest energy state before the event and cannot go lower. Therefore, the above type of scattering is forbidden by the energy conservation principle. (To anticipate the results of a forthcoming section, Fig. 4.10d shows that this forbidden process requires a localized d electron in the virtual level to go from the spin-down to the spin-up configuration, thus increasing the moment's M_S to $M_S + 1$.) Correspondingly, the forbidden spin-flip for the itinerant electron would be from spin-up to spin-down. Hence, as the applied magnetic field increases, the latter type of spin-flip scattering becomes increasingly improbable.

The alternative type, which scatters the itinerant electron from a spin-down to a spin-up state, also becomes increasingly forbidden as H is increased, but for a different reason. As H increases, the Zeeman levels of the ion's moment become increasingly separated in energy, and the separation becomes progressively greater than the energy width k_BT about E_f that contains all the significant itinerant states in the conduction band. Any itinerant electron scattering from a spin-down to a spin-up state (with the resulting surrender of energy $g\mu_BH$ to the ion) produces a kinetic energy at least $g\mu_BH - k_BT$ less than E_f. But this brings its energy into a range in the conduction band consisting of entirely filled states; therefore, this scattering process is forbidden by the Pauli principle.

The net effect of all this is that as an external magnetic field is applied the total amount of spin-flip inelastic scattering of itinerant electrons is reduced. Essentially, this is because the field removes (or freezes out) the virtual level's internal degree of freedom conferred by its magnetism (to which we referred above); in the jargon of the subject the consequent reduction of the electron scattering is known as the freezing out of the spin-flip processes. It provides an explanation* of the fact that the electrical resistivity of many alloys containing localized magnetic impurities can often be decreased by applying a magnetic field (an effect known as negative magnetoresistance). This had been a puzzle because all simple arguments describing the effect of a magnetic field upon the movement of itinerant electrons lead to the conclusion that invariably the field should increase the sample's electrical resistivity (as in the normal positive magnetoresistance described in Sect. 6.3).

3.4.6 Asymmetric Scattering of Itinerant Electrons. The resonant scattering of itinerant electrons by an ion possessing a net magnetic moment, which was the subject of the preceding section, has another important aspect: In certain circumstances it can have left-right asymmetry; that is, the plane defined by the ion's moment and the electron's direction of incidence separates regions of different scattering probability. In other words, the electron is preferentially scattered to one side of the ion rather than to the other. This is also known as skew scattering, and it is one form of asymmetric interaction that itinerant electrons experience with the magnetic ions in a metal. Another is the side-jump effect that is described at the end of this section.

Turning first to skew scattering, it is a microscopic phenomenon that will obviously only have macroscopic effects when a majority of the ions have their moment aligned along a common direction—by an applied magnetic field, for example. In this case, at each scattering the electrons forming a uniform current in the metal are deflected in a preferred direction and produce an internal current in the sample that, since any sample has finite dimensions, eventually builds up a corresponding internal electric field as the electrons are piled up against the sample's inside surface. The macroscopic result of this is known as the extraordinary Hall effect; other experimental circumstances make, in principle, a corresponding extraordinary contribution to the magnetoresistance. These macroscopic manifestations of skew scattering are not the concern of this section; our aim is rather to give a descriptive account of its microscopic origins.

The first point to notice is that a foreign or native ion, or other scattering center—whether or not it has a magnetic moment—can in principle cause asymmetric scattering in an alloy if the scattering center is attractive so that the itinerant electron experiences a certain orbital motion during the scattering process. It is the electron's intrinsic spin-orbit coupling which

---

* M.–T. Béal–Monod and R. W. Weiner, *Phys. Rev.* **170**, 552, 1968.

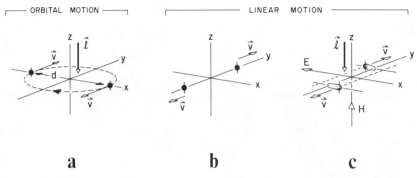

Fig. 3.17 When an itinerant electron is scattered by an attractive scattering center, such as a foreign, or even a native ion in the lattice, its normally linear motion is perturbed so that the electron experiences a temporary orbital component during the scattering. Since the electron has an intrinsic spin, there is a corresponding coupling between the spin and the temporary orbital moment. Here the effect of such intrinsic spin-orbit coupling is shown schematically. (*a*) Emphasizes a feature of pure orbital motion, which is that the particle's center of mass is displaced if its velocity **v** in the orbital plane is reversed. (*b*) Linear motion with no spin-orbit mixing. This feature, of course, does not exist; the electron's center of mass follows the same path whatever the sign of **v**. (*c*) Spin-orbit coupling mixed into the linear motion. The electron is essentially distorted along the x direction so that for $\pm$**v** its center of mass moves along different paths parallel to the y axis. The electron is thus attributed an electric dipole moment directed in this case along the x direction. This interacts with the electric field **E**, which is also along that direction and which was produced by the relativistic transformation of the magnetic field **H**. The latter exists in the electron's frame of reference because of the relative motion within it of the charged scattering center (Sect. 3.2.3). Still within that frame of reference, the electron's magnetic moment arising from its intrinsic spin is coupled to the field **H**. Thus we have alternate descriptions of the effect of spin-orbit coupling, depending upon which frame of reference is preferred. **l** is everywhere the orbital angular momentum vector associated with the temporary orbital motion.

is at the heart of the matter, and here we digress slightly to consider its consequences for an itinerant electron's translational motion as opposed to the purely orbital motion of the atomic electron considered in Sect. 3.2.3 and Sect. 3.2.5. What does it mean to say that as a result of spin-orbit coupling an itinerant electron temporarily acquires a certain orbital component in its motion? The simple answer is illustrated in Fig. 3.17. First, consider again purely orbital motion (Fig. 3.17*a*). Apart from the fact that the electron has no net displacement, the characteristic feature here is that for some instantaneous velocity along any direction in the orbital plane (say along the y direction in the figure) the electron's center of mass is displaced if this velocity is exactly reversed. (In the figure the electron is displaced by d if $\mathbf{v}$ is reversed.) This is a distinguishing feature of a particle having orbital angular momentum. Consequently, if a component of such momentum is temporarily introduced into an itinerant electron's linear motion, this feature must be present.

This is illustrated schematically in Figs. 3.17b and 3.17c. With no spin-orbit interaction present (that is, without orbital angular momentum) the center of mass of an electron moving with a velocity $\mathbf{v}$ along a y direction follows the same path whatever the sign of $\mathbf{v}$ (Fig. 3.17b). However, when some spin-orbit coupling is mixed into this situation, so that the orbital component again lies in the xy plane of the figure, the electron's center of mass is displaced along the x axis when $\mathbf{v}$ is reversed. Compared with the sitation in Fig. 3.17b, the center of mass is shifted in, say, the $-x$ direction, when $\mathbf{v}$ is along the $+y$ axis, and is shifted an eqal amount along the $+x$ direction when $\mathbf{v}$ is along the $-y$ axis. Consequently, the paths followed by the electron's center of mass are different for $\pm\mathbf{v}$. In this way one obtains the characteristic feature of angular momentum. It is as if the electron's charge distribution is smeared along the x direction by the spin-orbit coupling. In other words, we have to attribute an induced electric dipole moment to the electron during the temporary existence of the orbital component in its translational motion.* Of course, it is wrong to think of an electron as a particle that can be distorted in this manner. We recall from Sect. 3.2.3 and Fig. 3.1 that a localized electron is appropriately described by a wave packet, so that technically we would describe the effect of spin-orbit coupling as a distortion of the wave packet's envelope that produces a shift in its center of area. The result is the electric dipole moment.

But whatever view is taken, the important feature is the appearance of this electric dipole moment because it couples with an electrostatic field ($\mathbf{E}$ of Fig. 3.17c) acting in the plane of the orbital motion. This field arises from the relativistic transformation of the magnetic field $\mathbf{H}$ that in the electron's frame of reference is perpendicular to the plane of the temporary orbital motion and is produced by the apparent motion about the electron of the electrically charged ion responsible for the spin-orbit coupling (exactly as in the purely orbital case described in Sect. 3.2.3). It seems that we have, therefore, two ways of regarding the spin-orbit coupling of an itinerant electron, depending upon the frame of reference we choose to emphasize. We can regard it either as a coupling between the induced electric dipole moment and the electrostatic field $\mathbf{E}$ or as a coupling between the electron's magnetic dipole moment (caused by its

---

* J. Smit, *Physica* **21**, 877, 1955.

intrinsic spin) and the magnetic field **H**. In the following, where we discuss the asymmetric scattering of an electron subject to spin-orbit coupling, we shall emphasize the latter point of view.

If an itinerant electron's temporary orbital angular momentum is **l** and its spin momentum is **s**, then the spin-orbit interaction energy depends upon the relative orientations of these two vectors through a term **l·s**. There are two possibilities: **s** may be either parallel or antiparallel to **l**, the antiparallel being the configuration of lower energy because the electron carries a negative charge (Sect. 3.2.5). We see from a very simple picture of the scattering process (Fig. 3.18a) that an electron of given spin (up in this case) couples oppositely with the corresponding orbital angular momentum vector when it passes to the right and left of the scatterer. Since it is energetically favorable for scattered electrons to be involved in the configuration of lower internal energy, they pass preferentially to one side of the scattering center (all other considerations being assumed equal). Since the electrons of down spin, of course, have exactly the opposite asymmetry to those of the up spin class, the two asymmetries cancel out

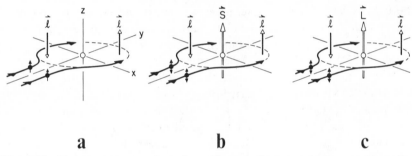

a   **b**   **c**

Fig. 3.18 Nonresonant scattering of an itinerant electron introduces a temporary orbital component into the otherwise linear trajectory (Fig. 3.10). In such an event, as might be caused by the attractive potential of a foreign ion, for example, there is the possibility of asymmetric (or skew) scattering in which the electron is deflected preferentially to one side of the ion. Here are illustrated three interactions that can in principle give skew scattering. (a) The case of any attractive potential—which may or may not have an associated net magnetic moment—where the intrinsic spin-orbit coupling of the electron in its brief orbital path gives skew scattering for incident electrons of a given spin direction. The preferred path has this spin coupled antiparallel to the electron's orbital angular momentum **l**. (b) The case of extrinsic spin-orbit coupling where the ion possesses a magnetic moment **S** that couples with the electron's temporary orbital momentum. The preferred configuration has **S** and **l** antiparallel, and the asymmetry in this case is independent of the incident electron's spin direction. (c) The case where the ion has a net nonzero orbital angular momentum **L**. This can also couple with **l**, as explained in the text, to give a preferred path with **L** and **l** parallel. The asymmetry is again independent of the incident electron's spin direction.

unless the up and down spin itinerant electron populations are assumed to be unequal. Such inequality can occur in certain circumstances but we need not pursue its causes here because calculation suggests that the simple mechanism shown in Fig. 3.18a is too small in all cases of interest to give rise to the macroscopic effects of skew scattering that are observed.

If the scattering center is an ion with a net spin magnetic moment arising from an incomplete d or f shell, another skew scattering mechanism is possible. It comes from what is called* extrinsic spin-orbit coupling. As pictured in Fig. 3.18b, an itinerant electron approaching the ion, of course, experiences direct exchange interaction (Sect. 3.3.1 and Sect. 3.3.2) between its intrinsic spin and the net spin S of the ion. This causes exchange scattering (Sect. 3.3.2) of the electron, which is an important effect and a prominent contributor to the sample's electrical resistance, but it is incidental here because it is a symmetrical scattering. The skew component arises from the effect upon the electron's temporary orbital motion of the magnetic field associated with S. The electron's temporary orbital angular momentum l has an associated magnetic moment (directed opposite to l) that couples with the field produced by S. Whether the electron passes to the right or left of the ion determines whether S and the orbital magnetic moment are parallel or antiparallel; the electron again prefers to pass to the side involving the lowest coupling energy (which is to the electron's left in Fig. 3.18b). Whereas the skew scattering mechanism introduced in the preceding paragraph involved the coupling of the electron's orbital momentum with its own intrinsic spin (intrinsic spin-orbit coupling), this one couples the electron's orbital momentum to an ion's net spin (extrinsic spin-orbit coupling). Note that the second mechanism is independent of whether the incident electron's spin is up or down, for it couples only the electron's orbital angular momentum. Consequently, this mechanism does not require unequal spin populations and is, therefore, expected to be a much more significant effect than that arising from the intrinsic spin-orbit coupling. Nevertheless, calculation suggests that it is too small (by a factor of about 20) to explain the skew scattering effects observed in real metals. However, in common with the other mechanisms described in this section, its quantitative significance in any given context is presently not certain, and the disentanglement of the various skew mechanisms that can conceivably contribute to a given observation remains an active field of research.

* This type of skew scattering was introduced by F. E. Maranzana, *Phys. Rev.* **160**, 421, 1967 to account for the extraordinary Hall effect in ferromagnetic metals. Note that throughout the discussion in this section we are ignoring spin-flip scattering events(Sect. 3.4).

Whereas extrinsic spin-orbit coupling is a magnetic effect that arises basically from the relativistic transformation between electric and magnetic fields, there is another interaction that is electrostatic in origin and that can produce asymmetric scattering. It arises when the incomplete d or f shell giving the ion its spin moment S also has a nonzero orbital angular momentum L. This might arise from the unquenching of the orbital angular momentum, as is generally the case for the f shell, or from the intrinsic spin-orbit coupling between L and S which can lift some of the quenching in the case of the d orbitals (Sect. 3.2.5). In any case, the mechanism requires some nonzero component of L to exist along the direction of S. It is found for this case that the expression for the electrostatic interaction energy between the ion and the incident electron no longer consists simply of the usual Coulomb repulsion term and an exchange term of the type $(s \cdot J)$ that expresses the spin dependence of the interaction [Eq. (3.41)]; there are now other terms, including one of the form $(L \cdot l)$, which shows that an effective coupling exists between the angular momenta. This coupling gives rise to skew scattering (as indicated in Fig. 3.18c) that is again independent of the incident electron's spin direction. It was introduced by J. Kondo* in an interpretation of electron motion in ferromagnetic metals, particularly the rare earth metals.

However, this mechanism immediately runs into a difficulty. As Kondo pointed out and as is clear from Fig. 3.4, the f shell in the Gd ion is exactly half filled. Consequently the shell has no net angular momentum and thus the above Kondo mechanism will be nonexistent. Nevertheless, experimentally it is found that the macroscopic effects that Kondo sought to explain are greatest in the case of Gd. Since this mechanism evidently cannot be the whole story, even more skew scattering mechanisms have been put forward; this time it is resonance scattering that provides the twist.

All of the above skew scattering mechanisms involve an itinerant electron that comes under the influence of the scattering center so that its trajectory is modified simply to have a temporary orbital component; none of them involves resonance scattering of the electron. When this type of scattering is introduced, the subject takes on a new aspect because the itinerant electron is now pictured as being temporarily captured into a virtually bound state around the ion. This is a pseudo-atomic state (Sect. 3.4.4) and has a legacy of atomic properties that a virtually bound electron must inherit. This is the new aspect of the problem.

It was again Kondo† who introduced the resonant scattering refinement in order to overcome the impasse with Gd. In this mechanism the energy

---

* J. Kondo, *Progr. Theoret. Phys.* **27**, 772, 1962.
† J. Kondo, *Progr. Theoret. Phys.* **28**, 846, 1962.

states of the ion's incomplete shell (an f shell in this case, but the mechanism has since been extended to ions with incomplete d shells) are assumed to satisfy the requirements for resonance with neighboring delocalized states. An incident itinerant electron is, therefore, imagined to be captured into the f shell through the s-f mixing interaction (Type 6 of Sect. 3.3.3) before tunnelling out again into a delocalized state. While it is temporarily in the shell, the electron experiences intrinsic spin-orbit coupling. This introduces an energy term proportional to $\mathbf{l}\cdot\mathbf{s}$, where these are again the temporary resident's orbital and intrinsic spin momenta. We saw above (and in Fig. 3.18a) that such coupling can lead to skew scattering when the up and down spin populations of the itinerant electrons are unequal, and it may at first sight seem that we have just returned to that first mechanism. But there is an important difference between the two situations: Here the ion has a net magnetic moment due to its incomplete shell, and, as a result, the incident electron also experiences the exchange coupling $\mathbf{S}\cdot\mathbf{s}$. The combination of these two couplings produces the skew scattering; it seems as if the exchange coupling, which depends upon whether the electron's spin is parallel or antiparallel to $\mathbf{S}$, acts to spin-polarize the incident electrons and the intrinsic spin-orbit coupling produces the asymmetry.* Thus by describing the scattering of the itinerant electron as an s-f mixing interaction (or s-d mixing as the case may be), Kondo was able to show that it could give a skew component.

This idea has since been extended to ions in other circumstances. For example, suppose the ion's incomplete shell is split in energy by a large spin-orbit coupling and, therefore, forms a split virtual level in the alloy. Ce is the example cited in Sect. 3.4.4, and it is used again here in Fig. 3.19 to illustrate how skew scattering can arise from Ce ions dissolved in a host matrix (La, for example). We recall that the Ce ion has a single electron in its $4f$ shell. The very large spin-orbit effect in this case puts the energy of the $J = 7/2$ state far above that of the $J = 5/2$ ground state. In the alloy, the latter happens to lie close in energy to the Fermi level; as a result, s-f mixing with delocalized states occurs with the consequent formation of a split virtually bound level. The resonant scattering of an electron is, therefore, only with the $J = 5/2$ state; in other words, only those

---

* As we pointed out above, to make the effects of skew scattering obvious on a macroscopic scale it is necessary to align the spins of a majority of the ions in a sample. When this is done by means of an external magnetic field, there will also be a spin polarization of the itinerant electrons. This is the Pauli paramagnetism of the electron fluid (Sect. 4.7) but with the field strengths available in a laboratory it is too small to contribute significantly to the skew scattering and is neglected in the following descriptions.

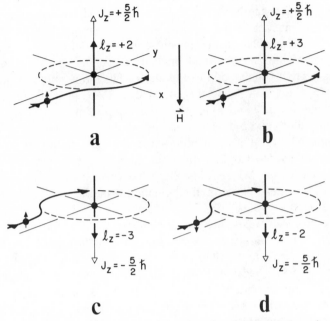

Fig. 3.19 An itinerant electron that is resonantly scattered by an ion is temporarily captured into a pseudoatomic orbital surrounding the ion (Fig 3.10). The dynamics of the temporary resident are, therefore, dictated by the atomic characteristics of its virtually bound state. When the orbital is only partially filled, it can give rise to skew scattering in an applied magnetic field. This example relates to the case of a Ce ion dissolved in such as La, for which the energy of the $4f$ state of total angular momentum $J = 5/2$ is close enough to the Fermi level to mix with the itinerant electron's delocalized states. Itinerant electrons able to mix into states characterized by $J = 5/2$, $l = 3$, $m_s = \pm 1/2$ can, therefore, be resonantly scattered. The total angular momentum of the pseudoatomic state acting along any arbitrary direction (here the z axis) can be one of $2J + 1$ values characterized by $M_J = \pm 5/2$, $\pm 3/2$, $\pm 1/2$, and the corresponding component of the captured electron's orbital angular momentum l_z is given by $J_z = l_z \pm 1/2$ and so depends upon the resident's spin direction. The diagram shows the cases for the two extremes of the range, namely $M_J = \pm 5/2$. When there are no external fields these quantum states are equally probable so that an electron that is resonantly scattered could resonantly mix into any state of permissible l_z without prejudice. But when a magnetic field **H** is applied, say along the negative z axis in the figure, an incident electron prefers to mix with a state of $+ M_J$ rather than one of $- M_J$. As the figure shows, this results in the electron passing preferentially to one side of the ion.

itinerant electrons able to mix with states characterized by $J = 5/2$, $l = 3$, $m_s = \pm\frac{1}{2}$ are resonantly scattered. (Ordinary exchange scattering resulting from the interaction between the ion's net spin and the itinerant electron's intrinsic spin also occurs, of course, but is not of concern here.) The component of the f shell's total angular momentum projected onto any direction—say the z direction in Fig. 3.19—has the permitted values $M_J \hbar$, where $M_J = \pm\frac{5}{2}$, $\pm\frac{3}{2}$, $\pm\frac{1}{2}$ in this case (Sect. 3.2.5). The figure shows the two extreme cases, $J_z = \pm\frac{5}{2}\hbar$, when the component of the total angular momentum along the z direction has its maximum magnitude. The corresponding orbital angular momentum along the z direction that an electron must have is specified by the quantum number l_z ($J_z = l_z \pm \frac{1}{2}$) and is shown in each case. We see that this varies between extreme magnitudes of 2 and 3, depending upon the orientation of the electron's spin with respect to J_z.

To see how skew scattering arises, compare the situations for the different $\pm M_J$. When M_J is positive (upper half of the figure) the captured electron has to pass to the right of the ion to produce the required sign for l_z, and when M_J is negative (lower half of the figure) the electron has to pass to the opposite side of the ion. When the ion is free of all external influences, the various M_J values have equal probability of representing the ion's angular momentum component J_z. Consequently, if we imagine a stream of electrons incident on the ion over a period of time and resonantly scattered by it, they will adopt with equal probability one of the range of permitted values of l_z during their temporary residence. It will be that value appropriate to the ion's J_z value at the instant of scattering and one of $2J + 1$ values produced as the ion's total angular momentum oscillates between its different quantum states (of which Fig. 3.19 shows only two). The scattering will, therefore, be symmetric since all possibilities between the extremes shown in Fig. 3.19 will occur with equal probability: There will be as many electrons deflected to the right of the ion as to the left. But if the probability of a given M_J state for the ion is weighted in some way, then skew scattering can occur. For example, if an external magnetic field $\mathbf{H}$ is applied (down in the figure) it couples with the shell's net magnetic moment and has two effects: First, it makes the configurations of Fig. 3.19a and 3.19b ($M_J = + 5/2$) the lowest Zeeman levels and, therefore, energetically more favorable than Fig. 3.19c and 3.19d ($M_J = - 5/2$), and second, it defines a specific direction in the sample (the z direction) that is now common to all ions. The net result of this influence is, therefore, to make the scattering by each ion skew, since whatever the incident electron's spin there is a preferential scattering to the right of the ion (as in the upper half of the figure); these microscopic events are at the same time combined additively throughout the whole sample to give a net macroscopic effect. This is the origin of skew scattering from an incomplete virtually bound state in an applied magnetic field.

It is also believed that a completely filled virtual bound state that is split by spin-orbit coupling (as in Fig. 3.11c) can give skew scattering when it is part of the electrostatic shield of a magnetic ion. To see how this comes about, it is convenient to refer again to the processes of Fig. 3.19, but now from the point of view of the incident electron's spin. Imagine, for example, a stream of spin up itinerant electrons to be resonantly scattered by the $J = 5/2$ component over a period of time. Again, when the ion is free of external influences, each scattered electron will adopt an orbital momentum l_z arising at random from the range corresponding to the $2J + 1$ quantum states of the ion's total angular momentum. The extremes of these l_z values are shown in Fig. 3.19a and 3.19c, and they have unequal magnitudes. This means that over the period of time the mean orbital momentum for a temporary resident will be nonzero; in this case more electrons will have passed to the left of the ion (in the sense of Fig. 3.19c) than to its right (Fig. 3.19a). Thus the scattering is skew: From the point of view of a single up spin electron, it has a greater probability of being scattered to the left of the ion than to the right. Only if the mean value of l_z was zero will the scattering be symmetric. Exactly the opposite asymmetry exists for incident electrons of down spin, as Figs. 3.19b and 3.19d show; on balance there will be no net skew effect from an ion free of external influences when the up and down spin populations of the incident electrons are equal. However, the case is just like that described above when Kondo first introduced resonant scattering into the problem: Although the virtual level here is filled and has no net magnetic moment, we have said that it is part of an ion possessing a magnetic moment from some other incomplete shell. As a result, the incident electrons, although they resonate with states of a completely filled level, do experience the ordinary exchange scattering because of the ion's net magnetic moment. Once again, it is the combination of the exchange and resonance scattering that produces the skew component.

To give a specific example, we again fall back on the rare earth ions because they have large spin-orbit effects, and consider what is thought to be their screening configuration when dissolved in a noble metal such as Ag. This is described in Sect. 3.4.4, and here we recall that a completely filled but split virtual d state is thought to surround the ion and that a net magnetic moment arises from an incomplete f shell located deep within the ion. Resonant scattering of itinerant electrons takes place in this case by s-d mixing with the filled virtual levels: Those itinerant electrons able to mix with states characterized by $J = 3/2$, $l = 2$, $s = \pm\frac{1}{2}$ and $J = 5/2$, $l = 2$, $s = \pm\frac{1}{2}$ (corresponding to the two components of the split level) are so scattered. The higher energy component of the level ($J = 5/2$) is sixfold degenerate corresponding to $M_J = \pm 5/2$, $\pm 3/2$, $\pm\frac{1}{2}$, and the lower component ($J = 3/2$) is fourfold degenerate since $M_J = \pm 3/2$, $\pm\frac{1}{2}$. However, to simplify the illustration we again consider only the extreme cases of $M_J = 5/2$ and $M_J = 3/2$ for each component. The corresponding possibilities for the resonant scattering of an incident up spin electron are sketched in Fig. 3.20, which shows the l_z component (z again defines an arbitrary direction) of the resident electron's orbital momentum that is required in each case. The argument follows closely the case of scattering by an unfilled virtual state described above: Comparison of Figs. 3.20a to 3.20d shows that scattering of the up spin electrons by $J = 5/2$ and $J = 3/2$ components has a mean l_z that is nonzero (it is positive).

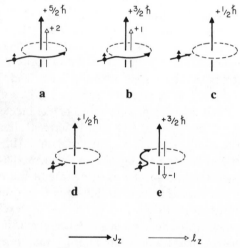

Fig. 3.20 Illustrating different mixing configurations in the resonant scattering of an itinerant electron of up spin by a split virtual d level about a foreign ion in an alloy. J_z is the instantaneous component of the total angular momentum of the state along some arbitrary direction; l_z is the corresponding component of the captured electron's orbital angular momentum. Possible resonances with the higher energy component of the split level ($J = 5/2$), are shown by a, b, and c, and those with the lower energy one ($J = 3/2$) by d and e. The choice of l_z is therefore limited to the range 2,1,0,−1, which in two cases out of three gives a scattering to the right of the ion. Exactly opposite asymmetry exists for a down spin electron but, as the text explains, a net asymmetry can result when the virtual level is part of a magnetic ion.

Consequently, an up spin electron is more likely to be scattered to the right of the ion (in the sense of Fig. 3.20) than to its left; the opposite asymmetry exists, of course, for the down spin case. There is, therefore, no skew scattering by this virtual d state when the up and down spin populations of the incident electrons are equal. But again, account has to be taken of the influence of the direct exchange between the incidental electron's spin and the total spin of the ion, which in this case comes from an incomplete f shell, and when combined with the resonant scattering of the d level this leads to skew scattering if an external magnetic field is applied to align the ion's moments.

Finally, we turn to the asymmetric electron scattering interaction which is known as the side-jump effect.* This arises from the delay introduced in the electron's motion by the scattering process. To see this, it is helpful to consider first what may by analogy be called the forward (or backward) jump effect. This is a phenomenon implicit in the resonant scattering of any particle by a transparent scattering center (that is, one which allows the particle to pass through on the same trajectory) although it is generally of no experimental consequence. Suppose Fig. 3.21 represents plots in space of the electron's paths before and after a scattering. These are assumed to be straight lines. The region in which the scattering takes place is represented symbolically by the dotted circle. It is not necessary to have any detailed knowledge of what happens to the trajectory during the scattering process, and this is left blank. Figures 3.21a and 3.21b represent the case of a resonant scattering via a virtual state—of a potential well, for example—that allows the electron to pass through on the same trajectory. The effect of the resonance is to cause a delay in the electron's motion: The time taken to traverse the scattering region is different from that which would be required to traverse the same region in the absence of the well. Depending upon the nature of the scattering center, the electron may be slower to pass through the region than otherwise, spending a considerable time trapped in the virtual state (positive delay), or it can alternatively be accelerated through the well so passing faster than normal through the region (negative delay). (On a grander scale, the positive delay suffered by photons passing through the gravitational field of a massive body like the sun is a measurable effect.†) If the scattering event can be said to start at a definite time $t = 0$ and to end at $t = \Delta t$, then in Fig. 3.21 the electron's trajectory will end at the point reached at $t = 0$ (chosen to be the origin of the space coordinates) and will start afresh at some

---

* Introduced by L. Berger, *Phys. Rev.* **B2**, 4559, 1970; ibid **B5**, 1862, 1972.

† C. M. Will, *Sci. Amer.*, November 1974, p. 25 gives a nontechnical account.

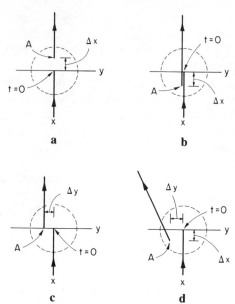

Fig. 3.21 An itinerant electron that is resonantly scattered suffers, in addition to the temporary orbital motion acquired by its trajectory, a time delay in its translational motion (Fig. 3.10): The time taken to pursue its trajectory is different from that taken for the same trajectory in the absence of the scattering center. In principle, this delay can be positive or negative, and it is equivalent to a displacement Δx of the electron's trajectory after scattering. The two upper parts of the illustration show cases of symmetric scattering in which the electron's trajectory is unchanged. (a) The electron's accelerated through the scattering region by the scattering potential. (b) The electron retarded. (c) The situation when spin-orbit effects are included in the scattering, so that a short-lived transverse component of the trajectory exists along the y direction. Different delay times can exist along the positive and negative y directions; the net effect is a side-jump of the trajectory (Δy). (d) The situation when skew scattering is included, so that the direction of the electron's trajectory is no longer preserved through the scattering. It has been estimated that the side-jump for an itinerant electron is typically about 10^{-11} m.

point A after a lapse of Δt. What happens inside the scattering center is not of immediate interest. The effect of the delay appears as a finite jump (Δx) made by the electron during the scattering process, and the cases of negative and positive delay are illustrated in Figs. 3.21a and 3.21b, respectively.

If the electron's trajectory during the scattering time Δt has a temporary orbital component—when the electron moves past an attractive ion, for example—it gives the particle a short-lived but effective velocity along the y direction during the scattering. This motion has an associated delay time just like that in the x direction discussed above, but when there are no spin-orbit effects, the velocity can be along either direction of the y axis with equal probability. Consequently, the separate delay times for the motion along these two directions are equal, and the scattering of the electron has no asymmetry; in a stream of electrons undergoing scattering, there is an equal number of positive and negative side jumps (Δy). However, when the electron's intrinsic spin-orbit coupling is included (Fig. 3.18a) and unequal populations of up and down spin itinerant electrons are assumed,* the effect is to make the delay times different for motion along the positive and negative y directions. If for example, the positive delay time along the $+y$ direction exceeds that along the $-y$ direction, the net effect is a positive delay time for motion along the $+y$ direction. By analogy with the case shown in Fig. 3.21b, we see that this is equivalent to a side-jump (Δy) made by the electrons during the scattering. This is illustrated in Fig. 3.21c. The effect of spin-orbit coupling is, therefore, to add the side-jump Δy to the backward jump Δx, which is inevitable in any resonance mixing but is ignored for clarity in Fig. 3.21c. If in addition there is also skew scattering arising from one of the above sources, the trajectory starting at A is no longer parallel to the incident section terminated by the scattering. Usually when spin-orbit effects are operative, both skew scattering and side-jump effects occur simultaneously. The result is illustrated in Fig. 3.21d.

Coherent and repeated skew scattering of an electron by a species of scattering center in a metal obviously has macroscopic effects whatever the electron's mean free path between scattering events, because the direction of the trajectory is changed at each event. But the same is not true for the side-jump effect. The direction of the trajectory is not changed, but rather the center of gravity of the electron's wave function undergoes a finite lateral displacement: The trajectory is displaced but its direction is unchanged. (Again, we should strictly think of this jump as a distortion of the corresponding wave packet's envelope rather than as a finite displacement of a particle—just as in the case of Fig. 3.17c.) The displacement amounts to about 10^{-11} m at each scattering, and whether or not this has significant macroscopic consequences depends upon the electron's mean free path between scattering events. If the mean free path is long— say of the order of 10^{-3}–10^{-4} m, as in reasonably pure metals at low

---

* The side-jump effect has been suggested in the context of electron scattering in ferromagnetic metals. In this case each ion has a net magnetic moment (taken to be along the z axis in Fig. 3.21) with a common alignment in a domain, and the spin up and down populations of itinerant electrons are—in one view of ferromagnetism, at least—taken to be unequal.

temperatures—then the side-jump effect is such a minute perturbation of the electron's trajectory that it is insignificant. However, it is thought to become significant in some ferromagnetic metals and alloys for which the mean free path at higher temperatures can be as small as 10^{-8} or 10^{-9} m. When an external magnetic field is applied to align all the ion's moments in a sample, so that as an electron is scattered its side-jumps become accumulative, these samples show anomalous effects when an electron flow is established in them. It is believed that such effects are a direct consequence of the side-jump mechanism.

3.5 CHAPTER SUMMARY

There is a rather cynical statement, which contains nevertheless an element of truth as well as some comfort to the struggling student, to the effect that the difference between an expert and a neophyte in quantum mechanics is that although the former knows how to use it neither really understands it. Electron spin is part of this enigma.

On the macroscopic scale we are used to being able to characterize the state of any body—whether it be a billiard ball or a planet—by specifying its translational and rotational motions. On the quantum scale things are different; for reasons that are presently not fully understood, quantum particles have discrete internal alternatives available and these give internal degrees of freedom that are not evident in the particle's normal translational or rotational motions. This is something outside our macroscopic experience, and it first became prominent when attempts were made to interpret the anomalous Zeeman effect in which the electron is, of course, the quantum particle involved. In this case the quantum effect is manifested as two alternative internal states for the electron, and for many purposes it is convenient to look on this as an intrinsic spin having only two permissable orientations. There are dangers in this analogy, but it is a very firmly established one.

The spin of a quantum particle is inextricably connected with a wider and equally opaque aspect of nature. This is the universal separation of conglomerations of identical quantum particles into two classes: Those for which the total wave function describing the system is symmetric and those for which it is antisymmetric. There are no intermediate situations, and there is no simple classical analogy of this aspect, although a case can be made to show that it is somehow connected with the requirement that identical quantum particles in a system are indistinguishable and cannot be labeled or traced from outside. The electron falls into the latter of the above classes—as, in fact, do all quantum particles having half-integral spin values. Such are known as fermions.

The operation of the Pauli exclusion principle in a fermion system leads to a modification of the normal Coulomb repulsion between the constituents: They have like charges and so repel each other, but they tend to stay out of each other's way to different extents depending upon their internal alternatives (or spins). For example, two electrons in the same atom tend to approach closer when their spins are antiparallel than when they are parallel. Consequently, to an outside observer the Coulomb repulsion is greater in the former than in the latter case. We can look upon this either as a modification of the Coulomb force or as the introduction of an imaginary new force that depends upon the spin configuration. This is nowadays known as the exchange force; it is a pliable concept that has been manipulated to fit numerous contexts, some of which are described in Sect. 3.3.3.

With this background established the chapter turns in Sect. 3.4 to the main argument, which is that a metal is a lattice of positive ions immersed in fluid-like plasma of itinerant electrons. This fluid is easily polarized by some local perturbation, whether it be a foreign ion with a different net ionic charge from the host's or one with vacancies in inner shells giving rise to a localized magnetic moment due to the uncompensated spins of the residents in the ion. In the former case, the electrostatic requirements of such a plasma make it adjust to screen out whatever electrostatic field the foreigner produces. This is done by a dynamic process; if there is an excess positive charge to be screened, the itinerant electrons remain slightly longer in its vicinity than would be the case in its absence. This gives a pile-up of negative charge about the ion (charge polarization), but surprisingly its radial distribution is not uniform: Moving out from the ion we find successively regions of increased and depleted electronic charge density. These are the Friedel oscillations.

Exactly analogous oscillations occur in the spin polarization that is set up about ion with a localized magnetic moment, and it is in this context that we meet the concept of the virtually bound state about an ion. This is an electron from the itinerant class which is temporarily captured into pseudo-atomic states about the ion; it is as if the electron shows characteristic atomic features typical of the foreign ion's outer shells in its isolated state, before tunneling back into the itinerant class at the end of its capture. This can also be viewed from the itinerant electron's point of view as a scattering event: Its translational motion is perturbed by the foreign ion, and thus the electron is effectively scattered between initial and final states. This scattering can be elastic, inelastic, or even dependent upon the spin orientations of the incident and resident electrons involved. All of this can be manifested in the electron transport properties of such alloys, and this subject is introduced in Sect. 3.4.5 and discussed further in Sect. 3.4.6.

RELATED READING

D. H. Martin, 1967, *Magnetism in Solids,* (Iliffe Books: London). Generally intermediate to advanced standard, this book gives a good coverage of the magnetic properties of metals.

THE VIRTUAL BOUND STATE

J. Friedel, 1963, "The Concept of the Virtual Bound Level" in *Metallic Solid Solutions* (Benjamin: New York), Ed. J. Friedel and A. Guinier. Paper XIX-I.

J. Friedel, 1958, "Metallic Alloys" in *Nuovo Cimento,* Suppl. to Vol. VII, Series X, p. 287 (In English).

J. Friedel, 1969, "Transition Metals: Electronic Structure in the d band" in *The Physics of Metals,* Ed. J. M. Ziman (Cambridge University Press: Cambridge), p. 340. Of the three reviews listed, this is the most advanced and least descriptive treatment.

HISTORICAL PERSPECTIVE OF PAULI'S PRINCIPLE

M. Jammer, 1966, *The Conceptual Developments of Quantum Mechanics* (McGraw-Hill: New York); particularly pp. 133–156.

B. L. Van der Waerden, 1960, in *Theoretical Physics in the Twentieth Century* (Interscience: New York), Ed. M. Fierz and V. F. Weiskopf; particularly pp. 199–244.

INTERACTIONS WHICH CAN COUPLE THE SPINS OF TWO ELECTRONS

There are few descriptions of this topic at an intermediate level. D H. Martin's book cited above gives in Chapters 5 and 6 a thorough discussion at the graduate's level; at a still more specialized level there is the very comprehensive treatise: [1]

H. J. Zeiger and G. W. Pratt, 1973, *Magnetic Interactions in Solids* (Oxford University Press: London). At a similar level there are various reviews in the series:

G. T. Rado and H. Suhl, Ed., *Magnetism* (Academic Press: New York), Vols. I to V. These make a useful starting point for advanced study.

4

THE ELECTRON LIQUID

II

GENERALLY MANY-BODY EFFECTS

4.1 INTRODUCTION

The preceding chapter is concerned with ideas developed generally within the framework of the independent electron model that hinges on the assumption that the mutual interactions between electrons (except the one or pair of principal interest) are weak enough to be neglected. In metals this is known to be untrue, but it is a simplification that allows some understanding of the consequences of electron spin, particularly those that relate to what could be called the energy structure of the electron liquid; that is, the way in which the liquid is built up by adding electrons one after another to the system.

But what of the response of such a liquid to outside influences such as electric fields, temperature gradients, or bombardment by energetic particles? It has been said that there are typically 10^{29} itinerant electrons in each cubic meter of a metal, each influenced through electrostatic forces by the rest of the electrons and by the positive ions. How is one to make any progress in the understanding of such an enormously compli-cated system? (We should stress that it is not simply the large number of electrons that poses the problem, for a system of any number of non-interacting particles can be treated exactly, providing sufficient time and computing power is available, but what leads to intractability is the fact that the electrons are interacting via mutual forces.) In principle in such a system, anything done to one constituent will sooner or later be felt in some degree by the others; in this way the concept of an individual constituent obviously tends to be lost. Such a collection of mutually inter-acting particles is known as a many-body system.

Insuperable difficulties are faced when trying to understand the detailed motion of any given constituent in such a system. In fact, the most compli-cated many-body problem that can be solved exactly is the two-body problem considered in Sect. 4.2.2; the three-body problem can be solved only by approximate methods (although it is true that any required degree of accuracy can be reached, providing one has the time and patience to make enough successive approximations) and to obtain even approximate solutions of more complicated systems remains beyond anyone's present capability. This seems unfortunate at first sight because almost all of the physical systems of interest, whether on the microscopic, macroscopic or cosmic scale are many-body systems.

Fortunately, there are two features of the behavior of a many-body system on a microscopic scale that can be used to advantage when ap-proaching the problem. The first is that for many purposes the details of any individual particle's motion are not of interest. For example, returning to our specific case, we do not care about the details of any individual electron's motion under an external influence; it is the response of the electron fluid as a whole that is of interest, for that determines the macro-scopic behavior of the metal leading to parameters of interest, like its electrical or thermal conductivity.

The second fortunate feature is that, after a suitable conceptual and mathematical transformation, the behavior of a many-body system of inter-acting particles can often be represented by a hypothetical system of non-interacting particles (or at least only very weakly interacting ones). For convenience this procedure can be called the reduction of a many-body system to a representative (and hypothetical) one. This subject is a sophisticated branch of theoretical physics, having immediate applications in a wide field of problems in the physical sciences with foreseeable applications in the biological sciences too. It is the subject of Sect. 4.2, but it obviously would not serve our present purposes to become involved in its intricate mathematical details. Consequently, we shall not attempt to prove that such a reduction is possible in any but the most trivial case of Sect. 4.2.2. The aim throughout is not a microscopic justification of the statements made but is rather to illustrate the basic ideas and results of the theory—which are often quite simple—and to give special emphasis to the particular many-body system of interest. In any case we give in Sect. 4.2.4 a simple physical explanation of why such a reduction can be made: It is again the electron spin that is at the heart of the matter.

Paragraph 4.3 turns to the origin of the Coulomb force. This subject introduces several concepts from many-body field theory that have been incorporated into the modern description of metals. Using these concepts, subsequent sections (Sect. 4.4 to Sect. 4.7) describe several manifestations of the many-body nature of the electron fluid.

4.2 REDUCTION OF THE MANY-BODY PROBLEM

The layout of this section is illustrated in Fig. 4.1 and follows closely the approach made by D. Ter Haar.* The first step is to introduce the concept of an elementary excitation in a many-body system (Sect. 4.2.1) for such quantities have a direct correspondence with the noninteracting particles of the hypothetical, representative system. Two classes of elementary excitation can be distinguished, and they correspond to rather opposed points of view in the theory, although the end results of both approaches are somewhat similar. Subsequently, the specific system of interest is examined progressively in Sect. 4.2.2 and the following sections. First, Sect. 4.2.3 considers the trivial case of the free electron fluid; second, in Sect. 4.2.4, the electron-electron interactions are finally allowed so as to give a many-body system of fermions. Sect. 4.2.5 describes several other types of elementary excitation encountered in solids. The fact that in a real metal this electron fluid is not simply an isolated system of fermions but a system that envelopes a lattice of positive ions (and, therefore, superimposes upon a regular periodic potential) is inevitably encountered from time to time in this chapter, but its detailed consideration is reserved for Chapter 6.

4.2.1 Elementary Excitations.† We pointed out above that in trying to understand the gross physical properties of metals, such as the thermal, electrical, magnetic, or optical properties, we are trying to understand the response of a system of strongly interacting particles to an external influence. In a metal these particles could be the itinerant electrons, the positive ions, or the interacting spins of a ferromagnet, for example; each comprises a many-body system in which the concept of an individual

---

* D. Ter Haar, *Introduction to the Physics of Many–body Systems,* Interscience, London–New York, 1958 and *Contemporary Physics* **1**, 112, 1959–60 .
† L. D. Landau, a Russian theoretician, introduced this name in the early 1940s; it is he and his colleagues who are most closely associated with the subject's early development.

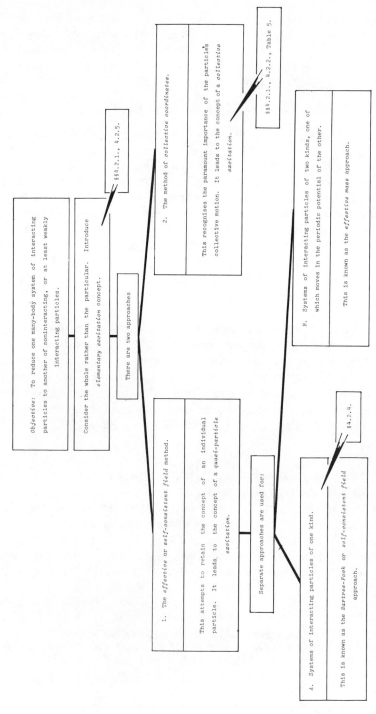

Objective: To reduce one many-body system of interacting particles to another of noninteracting, or at least weakly interacting particles.

Consider the whole rather than the particular. Introduce *elementary excitation* concept.

There are two approaches

§§4.2.1, 4.2.5.

1. The *effective* or *self-consistent field* method.

This attempts to retain the concept of an individual particle. It leads to the concept of a *quasi-particle excitation.*

Separate approaches are used for:

§4.2.4.

A. Systems of interacting particles of one kind.

This is known as the *Hartree-Fock* or *self-consistent field* approach.

2. The method of *collective coordinates.*

This recognises the paramount importance of the particles' collective motion. It leads to the concept of a *collective excitation.*

§§4.2.1, 4.2.2., Table 5.

B. Systems of interacting particles of two kinds, one of which moves in the periodic potential of the other.

This is known as the *effective mass* approach.

Fig. 4.1 Summary of the two approaches to the reduction of the many-body problem and the arrangement of Chapter 4. (After T. der Haar.)

148

constituent tends to be lost, and in which the collective behavior of the system is pronounced. What many-body theory has suggested is that rather than concentrate upon the detailed responses of the individual constituents to such an influence we should look at the system as a whole, somewhat as if the constituents formed a gigantic molecule.

The overall behavior of an assembly is always easier to predict than that of any individual constituent—this being the basis of all statistical methods—and the electron fluid is no exception. Therefore, it turns out that we should forget about the myriad electrons that constitute the fluid and look instead at the whole. This gives a system, the internal workings of which we provisionally ignore, that can be excited to higher energy states by an external influence and that relaxes to the unexcited (ground) state when the influence is removed (just as an atom or molecule can be so excited and will so relax).

At absolute zero, if there are no external influences, the system is in the state of lowest possible energy. This is the ground state and represents the total binding energy. The magnitude of this quantity, which is the value of the energy of the ground state, is usually quite unimportant when trying to understand the physical properties of metals. What are important are the magnitudes and the nature of the deviations from the ground state that are produced when the external influence is switched on. These correspond to the low-lying excited states of the molecule; accordingly, once the form of the ground state has been agreed upon, it can generally be forgotten altogether. Instead, the behavior of these excited states relative to the ground state can be looked upon as being the only quantities that are of interest in determining the metal's physical properties. This is the idea upon which the concept of an elementary excitation rests; a switch of the system from the ground state to an excited one is regarded as the creation of an elementary excitation where none existed before. Similarly, a deexcitation of the system corresponds to an annihilation of an elementary excitation.

Many-body theory, therefore, requires a shift of emphasis from the particular to the whole, but to be useful this transition has to be more than just a conceptual one; it has to lead to a mathematical simplication. It turns out that for most cases of physical interest a mathematical transformation can be invoked to show that the elementary excitations of low energy in a many-body system behave like a collection of noninteracting (or at least only weakly interacting) particles in a low pressure gas. In mathematical terminology the elementary excitations of an interacting system are said to be in a one-to-one correspondence with the excited states of a gas of noninteracting particles (Fig. 4.2); this is the origin of the use-

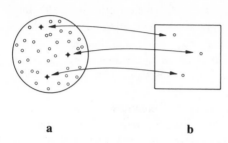

Fig. 4.2 Illustrating Landau's simplification for the many-body system of interacting fermions. The set of all particles in such a system when at a low temperature is represented in the left-hand side; only three in the sample remain excited above their ground state. Landau showed that such low energy excited states (known as elementary excitations) are in a one-to-one correspondence with the excited states of a hypothetical system composed of noninteracting fermions (represented by the right-hand set). This fact stems from the application of the Pauli principle and is given a physical explanation in Sect. 4.2.4. It enables a problem of enormous complexity to be reduced to one of manageable proportions.

a   **b**

fulness of the concept of an elementary excitation, whose discovery was Landau's initial contribution to this subject that he established in the context of the fermions which compose liquid He³.

When elementary excitations of low energy are created in a many-body system, they are relatively stable entities. Furthermore, they need not stay put but can move throughout the entire system as a quantized wave motion of the system's constituents. For example, in the case of an electron fluid, we point out in Sect. 4.2.4 that, by means of a complicated resonance process, an excited electron communicates its energy to all the others within its sphere of influence. As a result, the excitation is passed through the electron liquid. Since this occurs on an atomic scale, the laws of quantum mechanics apply and the energy transported by the excitation must be in the form of discrete quanta. These behave somewhat as free particles, having only infrequent collisions with each other or with the constituents of the system.

We make a distinction in terminology between two classes of elementary excitations: The quasi-particle excitations and the collective excitations. In a sense these represent extreme views of the excited states of a system of mutually interacting particles. The quasiparticle concept is an attempt to retain, in a system where the concept of an individual constituent tends to be lost through mutual interactions, the idea that an elementary excitation can be an individually excited particle. This approach is only appropriate when the interaction is relatively weak and when the energy of the excitation above the ground state is small. It is only in these circumstances that the lifetime of an excited particle between interactions is long enough to make the individual particle concept viable.

The collective excitation, on the other hand, is applicable in those cases where the interaction is so strong, or of such a long range, that there is little point in trying to isolate for consideration the behavior of any one constituent. In such a system, whatever excitation a constituent is given is immediately transmitted to its neighbors and quickly becomes the property of the entire collection of particles rather than of any individual member; the interaction effectively introduces a new degree of freedom into the system that is associated with the collective behavior of the constituent particles and that is in addition to the normal ones associated with the motion of the individual particles. In a solid the collective excitation has the form of waves (either propagating or standing) in the motion of frequently large numbers of constituent particles. The wavelength is dependent upon the excitation's energy and, as is noted above, the energy must be transported by the excitation in the form of individual packets or quanta. A collective excitation thus shows effectively no resemblance to the constituent particles supporting its existence (just as, for example, energy is transported by waves on water that give no clue to its molecular structure). A quasiparticle, on the other hand, being just a modified form of the real version, shows a strong resemblance to the constituent particles of the many-body system. In fact, if the interaction between the constituent particles in such a system could in some way be turned off, the quasiparticle would become just an ordinary excited particle of the system, whereas a collective excitation in such a circumstance would disappear.

Although the quasiparticle and collective excitation concepts represent extreme views, we should not assume that their existence in a system is mutually exclusive. On the contrary, electrons in metals can show a response characteristic of the simultaneous presence of both types, including interactions between them. It is just that this dichotomy is inherent to the simplification introduced by the many-body theory.

We should also caution that the above terminology is not uniform in the literature. Some authorities* refer to all elementary excitations as quasiparticles. This usage arises because the energy transported by a collective excitation is in the form of quanta and these exhibit particle-like properties, such as charge, mass, and the conservation of momentum, so that it is appropriate in this sense to refer to these quanta as quasiparticles. Con-

* Notably D. Ter Haar (loc. cit) and the Russian school: A. A. Abrikosov: *Introduction to the Theory of Normal Metals,* Academic Press, New York, 1972 or M. Ya. Azbel et al., *Sci. Amer.,* January 1973, p. 88.

Table 5   Some Elementary Excitations

Type	Name
collective excitation	total mass of a system plasmon roton phonon magnon
quasi-particle excitation	effective mass conduction electron quasi-electron quasi-hole polaron Landau particle bogolon exciton

sequently, in that terminology all elementary excitations, whether at the collective or single-particle extremes, have an associated quasiparticle that is just the quantum of energy created in the excitation. The choice of convention employed is ultimately a matter of taste, and throughout our discussion the distinction will be made between the quasiparticles and collective excitations outlined above.

Table 5 shows a list of the elementary excitations referred to in subsequent paragraphs.

4.2.2   A Two-Body Problem.   The basic ideas behind much of this description emerge from the consideration of the simple problem of the motion of two point particles that are unrestrained except that they experience a mutual attraction or repulsion that is a function of their separation. This illustrative problem will be considered in some detail. In fact, it has been of fundamental importance since the earliest times, particularly in classical dynamics and celestial mechanics, and the postquantal era has added new relevance to it. Let us look at examples of this. If the particles carry charges of opposite sign, then the problem is equivalent to the bound positron-electron pair (the positronium atom) of atomic physics or the bound electron-hole pair (the exciton), which is encountered in Sect. 4.2.5.

We have said that this two-particle problem is the only many-body problem that can be solved exactly. This is true whether the particular conditions are classical or quantal. For example, in the two cases above, which are specifically quantal, the Coulomb attraction leads to a series of bound configurations for the pair (just as in the hydrogen atom) in which the particles revolve about their common center of mass with quantized angular momenta. It turns out that such a system in free space has energy levels that are given exactly by a modified Bohr model of the hydrogen atom, the modification being simply the replacement of the free electron mass by a reduced mass for the system, which is defined below.

All of this relates to the situation where the two-particle system is considered to be at rest. But more commonly the center of mass of the system can be in motion as the result of some external influence. The total energy, of course, is that of the stationary system plus the extra kinetic energy of the moving pair. In practice such a motion occurs and is often of great importance (moving excitons, for example, are a significant method of energy transfer in certain insulating crystals and in the macromolecules of biological systems). To include this possibility of motion of the system's center of mass, the case of two interacting particles in a uniform, common gravitational field is taken for illustration. This is treated in classical circumstances, but the results carry over directly to the quantum context with appropriate modification of the details that does not interfere with the purpose.

Figure 4.3 shows two particles having masses m_1 and m_2 and located at $\mathbf{p}$ and $\mathbf{q}$, respectively. They are subjected respectively to gravitational forces of $\mathbf{F}_1$ and $\mathbf{F}_2$, and $\mathbf{F}_{21}$ and $\mathbf{F}_{12}$ are respectively the (attractive) forces exerted by particle 2 on 1 and vice versa. We assume that this is a system of conservative forces so that the force acting on any body at $\mathbf{r}(=\mathbf{i}x+\mathbf{j}y+\mathbf{k}z)$ can be expressed as the negative gradient of a corresponding scalar potential energy function; for example

$$\mathbf{F} = -\nabla_r V \tag{4.1}$$

where ∇ is the gradient operator. The use of a subscript on this operator indicates that the derivatives are to be taken with respect to the components of the subscript. Thus,

$$\nabla_r \equiv \mathbf{i}\,\frac{\partial}{\partial x} + \mathbf{j}\,\frac{\partial}{\partial y} + \mathbf{k}\,\frac{\partial}{\partial z} \tag{4.2}$$

If the potential of the interaction between the particles is written as V_{12} and if V is the gravitational potential, then it follows from Eq. (4.1) and Eq. (4.2) that

$$\begin{aligned}
\mathbf{F}_1 \,(= m_1\mathbf{g}) &= -\,\nabla_p V \\[4pt]
\mathbf{F}_2 \,(= m_2\mathbf{g}) &= -\,\nabla_q V \\[4pt]
\mathbf{F}_{12} &= -\nabla_q V_{12} \\[4pt]
\mathbf{F}_{21} &= -\nabla_p V_{12}
\end{aligned} \tag{4.3}$$

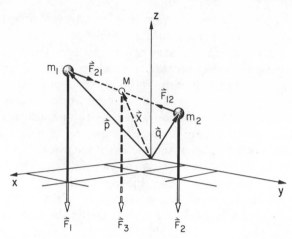

Fig. 4.3 Illustrating the arrangement of forces in the two–body problem discussed in the text. M, which is the hypothetical particle located at the system's center of mass X, is the quasiparticle of the new representative system to which the two-body system can be reduced.

Here **g** is the acceleration due to gravity and ∇_p and ∇_q indicate the gradient operators with respect to **p** and **q**, respectively. Note that as a direct consequence of Newton's third law

$$\mathbf{F}_{12} = -\mathbf{F}_{21} \qquad (4.4)$$

The equations of motion for the particles can be written down directly. For particles 1 and 2 they are, respectively, (where the double dot indicates the second differential with respect to time)

$$m_1\ddot{\mathbf{p}} = \mathbf{F}_{21} + \mathbf{F}_1$$
$$m_2\ddot{\mathbf{q}} = \mathbf{F}_{12} + \mathbf{F}_2 \qquad (4.5)$$

By straightforward substitution from Eq. (4.3) these become

$$m_1\ddot{\mathbf{p}} = -\nabla_p V_{12} - \nabla_p V \qquad (4.6)$$

$$m_2\ddot{\mathbf{q}} = -\nabla_q V_{12} - \nabla_q V \qquad (4.7)$$

When expressed in terms of their coordinates, Eq. (4.6) and Eq. (4.7) correspond separately to three differential equations. Consequently, a total of six such equations is needed to describe the motion of the system and twelve constants of integration appear in the general solution if it is necessary to obtain it. But the objective can be achieved without this effort. All that is required is to express the above equations in a new form and to demonstrate how this conceptual shift can reduce the complexity of the problem. This is done by introducing two new coordinates: First, the relative coordinate $\mathbf{x}$ ($=\mathbf{p}-\mathbf{q}$), which is the separation of the particles, and second, the position $\mathbf{X}$ of the system's center of mass (Fig. 4.3), which is given by

$$\mathbf{X} = \frac{m_1\mathbf{p} + m_2\mathbf{q}}{m_1 + m_2} \tag{4.8}$$

The equations of motion of the system can be rewritten in terms of these new coordinates. The first rewritten equation is obtained directly by the addition of Eq. (4.6) and Eq. (4.7) which, after appropriate substitution from Eq. (4.3) Eq. (4.4) and Eq. (4.8) gives

$$M\ddot{\mathbf{X}} = M\mathbf{g} \tag{4.9}$$

where $M = m_1 + m_2$. The second is obtained by considering the quantity $m_1 m_2 \ddot{\mathbf{x}}$. Since $\ddot{\mathbf{x}} = \ddot{\mathbf{p}} - \ddot{\mathbf{q}}$, we can write from Eq. (4.6) and Eq. (4.7)

$$m_1 m_2 \ddot{\mathbf{x}} = m_1(\nabla_q V_{12} + \nabla_q V) - m_2(\nabla_p V_{12} + \nabla_p V) \tag{4.10}$$

But from Eq. (4.3)

$$m_2 \nabla_p V = -m_1 m_2 \mathbf{g} = m_1 \nabla_q V$$

and from a standard result †

$$\nabla_p V_{12} = \nabla_x V_{12} = -\nabla_q V_{12}$$

we see that Eq. (4.10) becomes

$$m^* \ddot{x} = -\nabla_x V_{12} \tag{4.11}$$

where $m^* = m_1 m_2/(m_1 + m_2)$ is known as the effective mass.

Both Eq. (4.9) and Eq. (4.11) are the equation of motion of a single particle. In fact, the two-particle problem has been reduced to one of two separate and noninteracting particles. They are both hypothetical and can be thought of as being positioned in hypothetical systems. The first,

---

† To see this, consider for simplicity just the first component of $\nabla_p V_{12}$; let this be $\mathbf{i}\,\dfrac{\partial V_{12}}{\partial p_1}$.
By the rules of differentiation

$$\frac{\partial V_{12}}{\partial p_1} = \frac{\partial V}{\partial x_1}\frac{\partial x_1}{\partial p_1} + \frac{\partial V}{\partial x_2}\frac{\partial x_2}{\partial p_1} + \frac{\partial V}{\partial x_3}\frac{\partial x_3}{\partial p_1}$$

where x_1, x_2, and so forth are the coordinates of $\mathbf{x}$. We see immediately that $\partial x_1/\partial p_1$ is unity while $\partial x_2/\partial p_1$ and $\partial x_3/\partial p_1$ are both zero, so that $\partial V_{12}/\partial p_1 = \partial V_{12}/\partial x_1$. Similar arguments for the other coordinates give the ultimate result that $\nabla_p V_{12} = \nabla_x V_{12}$, and it can be shown by the same approach that $\nabla_q V_{12} = -\nabla_x V_{12}$.

described by Eq. (4.9), has mass M, is located at $\mathbf{X}$, and is subjected to the gravitational force $M\mathbf{g}$ (in other words, it moves in a potential of $\mathbf{g} \cdot \mathbf{X}$). The second has a mass m^*, is located at $\mathbf{x}$, and moves in a potential of V_{12}.

This result illustrates the two extreme approaches to the many-body system, which are referred to in Sect. 4.2.1. Equation (4.9) describes the motion of the whole system by relating to the center-of-mass motion; it represents the collective approach in which we concentrate upon the behavior of the system as a whole and overlook the nature of the constituent particles. Equation (4.11), on the other hand, represents the quasiparticle approach; we try to make the end result of the reduction retain the concept of an individual particle and, further, to make it retain as closely as possible the properties recognizable in the behavior of the free constituent of the original system. It happens that the result of the collective approach in this case is an entity having three degrees of freedom, a mass, and other particulate properties that give it a strong resemblance to the constituents of the original many-body (two-particle) system. This is particular to the present case, however, and in general the results of the collective approach do not at all resemble the constituents of the original system.

But such simplification is bought at a price. In Eq. (4.11) it is the concept of a fixed mass that has been sacrificed in order to eliminate the interaction between the particles. Thus the original pair of interacting particles with masses m_1 and m_2 has been replaced by a single free particle of mass m^*. This effective mass contains the effect of the pair's interaction. The single free particle is the hypothetical quasiparticle, and its effective mass is often quite different from the mass of a free constituent taken from the real many-body system. Under certain circumstances it can even have the opposite sign, as is described in Sect. 6.2.2.

The concepts of bare and dressed particles are often invoked in this context, but since they form the subject of Sect. 4.3.2 it is pointless to go into details here. Briefly, they make up just another metaphorical device that is useful in the attempt to retain the idea of an individual constituent in a many-body system. In this view, the hypothetical quasiparticle is just a bare constituent from the system surrounded (or dressed) by a cloud of virtual particles that constitute the mutual interaction operating beween constituents. As is described in Sect. 4.3.2, these virtual particles are short-lived deviations from nature's conservation rules that are permitted by the Heisenberg uncertainty principle. They give a picture—intelligible to our macroscopic experience—of not only the origin of the interaction in question but also the modification of the bare particle's inertial properties, which was seen above as an inevitable consequence of the problem's reduction.

Put simply, the quasiparticle's effective mass is different from that of its precursor because the former has to drag around its surrounding cloud of virtual particles. The result is the mass enhancement (or mass renormalization) effect that is described for various situations in Sect. 4.5.

4.2.3   Quasiparticles in a Free Electron Liquid.   We have in mind a system of noninteracting particles, such as the itinerant electrons in the free electron (Sommerfeld) model of a metal. In such a case it is assumed that both the electron-electron and the electron-ion interactions are unimportant. A possible development with time for such a system is shown schematically in Fig. 4.4a. When time is equal to zero the system is shown in its ground state, and, since electrons are fermions, it has each energy level up to a maximum E_f filled with two electrons of opposite spin (Sect. 3.2). Suppose now that some influence is switched on at time t and that this raises the energy of a single particle to E and above E_f. (Typical is thermal excitation which involves an energy of only $k_B T$, which at room temperature in many metals is only about $1/100$ of E_f, indicating that this particular quasiparticle is truly a weak excitation.) The excitation that is created is known as an electron-hole pair. These are well-defined, filled, and vacant energy states lying respectively above and below the ground state energy. Since there are no mutual interactions in this case, the excitation has no effect upon any of the other particles, and its lifetime is at least as long as the duration of the applied influence. If more than one elementary excitation is created, the energy of the system above its ground state is just the sum of these individual parts. Elementary excitations in such a noninteracting system are known as zero order or single particle excitations. The example we have considered is a rather trivial case because the lack of mutual interaction means that the system is not technically many-body.

Such excitations can be represented alternatively in a diagram of the occupied region of wave vector $\mathbf{k}$ space, as is shown in Fig. 4.4c. (We recall that this is equivalent to a plot in momentum $\mathbf{p}$ space since $\mathbf{p} = \hbar\mathbf{k}$.) We described in Sect. 3.2.4 how an itinerant electron is specified by the pair $(\mathbf{k}, \sigma)$, $\mathbf{k}$ being the electron's wave number and σ being its spin alternative. Consequently, if we plot the region of occupied states in $\mathbf{k}$ space, the ground state of the system is represented simply by a sphere of radius k_f (this being the $\mathbf{k}$ value corresponding to the energy E_f). Within this sphere are all points corresponding to occupied electron states $\mathbf{k}$, each of which contains two electrons of opposite spin. This boundary between occupied and unoccupied electron states, when the entire system is in its ground state (in this case, a sphere) is called the Fermi surface. A single particle

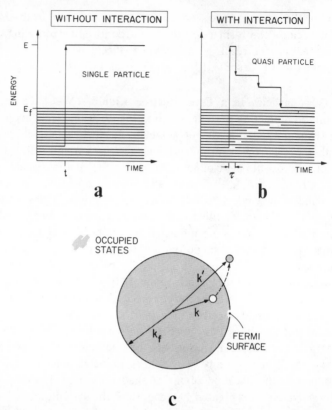

Fig. 4.4 Illustrating schematically the possible development with time of an individual elementary excitation after creation in a system of particles. (*a*) *A* system without mutual interactions that supports single particle excitations whose behavior closely resembles that expected of one of the system's constituents. (*b*) A system with mutual interactions that supports the quasiparticle type which, as a result of the mutual interactions, often has a behavior quite unlike that of any constituent. (*c*) The alternative representation in terms of the occupied region of wave vector space that is conventionally employed to describe the elementary electron-hole excitations in a system of itinerant electrons such as the delocalized electrons in a metal.

excitation is formed by lifting an electron from an occupied state (say **k**) within the Fermi surface to an unoccupied one (say **k'**) outside it. In Fig. 4.4*c* this corresponds to the creation of a vacancy or hole within the Fermi surface together with an occupied electron state outside it. If more than one elementary excitation is created simultaneously, then we can imagine electrons from occupied states say $\mathbf{k}_1$, $\mathbf{k}_2$, $\mathbf{k}_3$ . . . . being placed into unoccupied ones say $\mathbf{k}'_1$, $\mathbf{k}'_2$, $\mathbf{k}'_3$ . . . .

Since the total number of electrons in the system must be conserved, the number of energy states within the Fermi surface that are emptied when excitations are created must be equal to that of occupied states correspondingly introduced outside this surface. However, there is a danger of misconception here, for it is wrong to imagine that the particular electron excited from say k_1 is the one that occupies k'_1 and so forth. We recall from Sect. 3.2 that the operation of the Pauli exclusion principle in a system is tantamount to the requirement that its constituents be indistinguishable. Therefore, it is meaningless to inquire whether an electron excited from a given state is same as the one that eventually occupies any other state; all that can be said to characterize the excited system is that a number of electron states within the Fermi surface have been vacated k_1, k_2, k_3, ...) and an equal number of previously unoccupied states outside the surface have been filled (k'_1, k'_2, k'_3, ...). In other words, because of the operation of the Pauli principle, it is adequate to describe the excited state of the whole system simply in terms of the electron states outside the Fermi surface that are occupied and those inside it that are unoccupied. It is this feature, which stems from the operation of the Pauli principle, that makes possible Landau's approach outlined in Sect. 4.2.1. The occupied states in question can even be regarded as totally independent elementary excitations—electron and hole excitations, respectively—but as seen above their numbers must be equal for any state of the whole system.

4.2.4 Quasiparticles in an Electron Liquid with Interactions.

The case of a noninteracting system considered above is, of course, trivial. Even though there are many bodies in the system, their lack of mutual interaction leaves the concept of an individual constituent still quite appropriate, and we do not have in this case a true many-body system. An elementary excitation is then just a single electron, having inertial properties no different from a free particle, that is excited into an energy state lying outside the Fermi surface. The lifetime of such an excitation is as long as the duration of the exciting influence.

However, if we imagine some mutual interaction between the constituents to be turned on, then the energy of an elementary excitation is expected to decay because of the interactions (that is, collisions in this case) between the excited and the less excited constituents of the system. Eventually, these collisions lead to the disappearance of the elementary excitation as its energy decays to zero and is dissipated throughout the whole system.

This is illustrated schematically in Fig. 4.4b for the case of a single electron-hole excitation of initial energy E. The form of decay depends upon the strength and nature of the mutual interaction between the system's constituents, and in this case, where the Pauli principle is operative, it turns out that the lifetime of an elementary excitation decreases roughly as the square of its energy above the ground state (E_f). Therefore, the form of the decay is somewhat as indicated in the figure. It is important to realize that this is a relatively weak decay rate that in any case vanishes for states on the Fermi surface.

The important step forward made by Landau was to point out that such elementary excitations in a system of mutually interacting particles (providing their energy above the ground state is low enough to give lifetimes long enough to make the concept viable) behave like the excited states of the noninteracting system discussed in the preceding section. [It follows from Heisenberg's uncertainty principle (Sect. 4.3.2) that the minimum lifetime of an elementary excitation of energy E has to be large compared with $\hbar/E$ if the excitation is to be a well-defined independent entity.] In other words, what would be single electron excitations in the case of a free electron liquid are quasiparticle excitations when the mutual interactions in the liquid are turned on. As already said, the different intertial properties of these quantities reflects the effect of the continual interactions between the constituents in the second case. As a result, the effective mass and the Fermi velocity, v_f ($\simeq 10^6$ m/sec) of the quasiparticle both differ from their free electron values. But apart from the weak decay described above, there is otherwise no essential difference between a quasiparticle of an interacting system and the corresponding excitation of a noninteracting one.

The origin of these inertial differences is discussed in more detail in Sect. 4.5.2, but here we should note that in the interacting system the constituents are subjected to exchange and correlation effects. The result is to surround each electron by a region [roughly about 10^{-10} m (1 Å) in diameter] from which exactly one unit of electronic charge has been pushed away. This is the same as saying that because of these exchange and correlation effects the quasi-electrons stay out of each other's way. The result is that the quasiparticle appears to be electrically neutral except when it is approached within less than about 10^{-10} m, which is approximately the typical interionic spacing. This mutual avoidance, of course, exists whether or not the electron is excited. In the circumstances under consideration the unexcited electrons overwhelmingly outnumber the excited ones, and thus one of the latter will spend most of its lifetime trying to avoid the former—just as in the ground state the unexcited ones try likewise to keep away from each other. To maximize this correlation in their motion, excited electrons, by means of the mutual interactions in the system, constantly give up temporarily part of their excess energy to unexcited ones, only to reclaim it at a later time. This spreading out of the excitation energy in a controlled manner over many electrons, which can be regarded as a form of complicated resonance between contiguous electrons, is part of the dressing of the bare electron referred to in Sect. 4.2.2; it is also the essence of the quasiparticle.

We may wonder how the quasiparticle remains a stable entity long enough to be viable. After all, if this constant interaction between an excited electron and its unexcited counterparts shares out the excitation's energy to some degree, why is this not dissipated quickly throughout the entire system? How can the quasiparticle (that is, the bare electron and its dressing of interactions) be not only stable, having a relatively weak decay rate, but also able to move as an entity throughout the crystal? The answer lies in the combined influence of the Pauli principle and the principle of energy conservation. To see this we must remember that the quasiparticle description relates to the case where only excitations of sufficiently low energy are created. Consequently, when such an excited electron attempts to share out its energy by means of collisions with unexcited counterparts, there is in the overwhelming majority of cases insufficient energy transmitted to lift any such unexcited electrons into vacant energy states outside the Fermi surface. Since for low levels of excitation such as we are considering the majority of states within the Fermi surface are still filled to the maximum permitted by the Pauli principle, the energy transfer to an unexcited electron is impossible as it would violate this principle. The only time a collision between excited and unexcited electrons can occur in these circumstances is when the process fails to conserve energy, by an amount say ΔE, which is sufficient to lift the unexcited electron into the energy range of vacant states. This occurence is permitted by the Heisenberg uncertainty principle, but it demands in return that the situation can last only a very short time (Δt) before a reverse collision restores the initial circumstances ($\Delta E \, \Delta t = \hbar$). This idea of continual and temporary deviations from energy conservation in a process introduces the concept of a virtual particle, which is the subject of Sect. 4.3.2. At this point, it is sufficient to note that this is how the excited electron comes to be dressed as a quasiparticle, and this is how it maintains its relatively stable structure; basically it is all due to electron spin acting through the Pauli principle.

To summarize this qualitative picture of the behavior of an interacting electron liquid, we can say that its low energy excitations (such as are produced, for example, by moderate increases in temperature above absolute zero) can be pictured as quasiparticles (quasielectrons and corresponding quasiholes). These are relatively stable entities. At normal temperatures their number is very small; they are itinerant and, for all but the most extremely high temperatures, behave rather like the constituents of a free electron gas. Thus they keep out of each other's way to the maximum extent possible and give the appearance of being noninteracting particles. (This is why the simple Sommerfeld free electron model of a metal works so well when describing electron transport effects.) However, the inertial response of a quasiparticle to outside influences is different from that of a free electron in the same circumstances; this can be described as a change in the effective mass of the quasiparticle. This change stems, of course, from the continual interactions that the excited electron suffers as it moves. In a metal these can be collisions with other electrons or with the ions of the fixed lattice, but whatever the case their effect can be described as virtual particle interactions, and we shall elaborate upon this view in Sect. 4.5.2. Finally, we should stress for those readers having a corpuscular view of electronic motion in metals that this is the place to make a subtle modification of that mental picture. Henceforth, when the description refers to an itinerant (or conduction) electron in a metal it is understood that it refers to a quasiparticle; if it helps, we can think of this as the familiar bare corpuscular entity together with an accompanying disturbance arising from its various interactions with both itinerant electrons and those localized in the ion cores.

4.2.5 Other Elementary Excitations of Interest.

The quasi-electron (with its corresponding quasihole) introduced in the preceding section is perhaps the most frequently encountered elementary excitation in solids; its behavior is certainly prominent in the present context. But as Table 5 suggests, several other elementary excitations exist and are of either direct or indirect interest in this subject. It is convenient at this point to give brief descriptions of them. We first consider three separate quasiparticle excitations before turning to examples of collective excitations in solids.*

The polaron is a quasiparticle closely related to the Landau quasielectron of the preceding section. If the excited electron of that case had been instead in the conduction band of an insulating ionic crystal (such as rock salt), the same arguments would apply. However, in that case the electron is strongly coupled to the optical mode of the ionic lattice vibrations (which are described below) and consequently it is accompanied by a region of charge polarization arising from its constant interactions with the ions. This combination—the electron and its polarization cloud—is a

---

* Collective excitations can exist in liquids and in atomic nuclei too, although they are not of concern here. The roton, for example, is one peculiar to liquid He in which a few atoms collectively undergo a vortex motion somewhat like that of particles in a smoke ring, whereas in nuclei there are various vibrational and rotational motions that have associated collective elementary excitations.

polaron. Again, its effective mass is different from that of an electron free of interaction with the lattice and depends upon the dielectric properties of the solid.

Within the field of insulators (and, in this case, also certain semiconductors) is the exciton. This is a quasiparticle formed when an electron, after being excited across the forbidden energy band of a crystal, associates with its corresponding hole to form a bound electron-hole pair known as the exciton. Since there is an attractive force between the electron and hole, this quasiparticle has a lower energy than the separated pair of its constituents, and energetically it is located within the range of forbidden energies in the crystal. Typically, in the case of an ionic crystal, the exciton is an elementary excitation in which an electron from a negative ion is transferred to a neighboring positive ion to produce two neutral atoms. The displaced electron is to some extent mobile among the ions, and its motion is referred to as excitation migration. This is an important method of energy transport in biological systems.

There are various kinds of exciton corresponding to limiting extremes of the concept. For example, the Frenkel exciton has the electron-hole pair close enough to be essentially on the same atom, but in the Wannier exciton they have a relatively large separation. A recently discovered effect in semiconductors at very low temperatures, which is causing much excitement, arises when excitons group themselves into what we called electron-hole droplets—rather like the condensation of water droplets in a moist atmosphere. Conduction within a droplet is found to be metallic, presumably because the constituents of the condensed excitons dissociate to form an electron liquid like that in a normal metal.

The final quasiparticle to be discussed brings the description back to the context of itinerant electrons in metals; it is the bogolon (or Bogoljubov quasiparticle). It relates to the phenomenon of superconductivity, which is considered in Sect. 4.6, and to the attraction between pairs of appropriate quasielectrons which is described in Sect. 4.5.1. Here it is sufficient to note that Bogoljubov showed that just as the quasielectron/quasihole combination of Sect. 4.2.4 is an elementary excitation above the ground state of a normal metal, so a combination of two quasielectrons of opposite spins and momenta could be an elementary excitation above a new ground state for the metal. This new ground state, which in fact lies lower in energy than the Fermi energy of the normal metal, is the superconducting state. In the superconducting ground state when no bogolons have been excited, all the electrons taking part in the superconductivity are paired off in the Cooper pairs to be described in Sect. 4.5.1; no Cooper pair can have a vacancy. Thus all the electrons involved are correlated in pairs with constituents having opposite spins and momenta such that every pair's center of mass has the same velocity. When the system is excited above the ground state, however, some of these pairs are split up. It requires a minimum energy to do this, since the constituents have to be promoted across an energy gap, but once across they become analogous to the usual quasielectrons of the normal state. Two quasielectrons produced by a broken Cooper pair in these circumstances constitute a bogolon. The difference between this quasiparticle and a Cooper pair exemplifies the simplification produced by the elementary excitation concept in the reduction of the many-body problem: The original Cooper pairing idea derived from the more unwieldy formulation involving the motion of all the itinerant electrons in the metal, but Bogoljubov's treatment considers only that small number of itinerant electrons that are excited above the superconductivity ground state. Qualitatively, it is all quite analogous to the Landau simplification described in Sect. 4.2.3 and Sect. 4.2.4.

Turning to brief descriptions of certain relevant collective excitations, we recall that before a system can support such excitations a strong or long-range mutual interaction is required among its constituents. It happens that in a metal there are two many-body systems that separately are capable of supporting their individual types of collective motion: The system of positive ions and the system of the itinerant electrons. We will consider examples of both of these systems; first we turn to that composed of the positive ions.

The propagation of lattice vibrations in a solid having two dissimilar atoms per unit cell is a standard problem in intermediate solid-state theory that has instructive results. First, as in any lattice, the usual transverse and longitudinal modes can be distinguished. But in the diatomic lattice each of these is found to be split into two further types known as the optical and acoustical branches. Briefly, the optical mode represents a motion in which the two types of neighboring atom in a unit cell vibrate relative to one another; they oscillate out of phase, moving in opposition to each other about their common center of mass, which remains fixed. It is, therefore, an internal oscillation of the constituents of the unit cell of the lattice. In the acoustical mode, on the other hand, both neighboring atoms vibrate in phase; they move together in the cell in such a manner that this mode is a displacement of the pair of atoms as a unit, forming part of the general motion of the lattice (Fig. 4.5). Generally there is a marked separation between the frequency ranges of these two modes. The optical mode occupies a narrow band of high frequencies in the infrared range, and in suitable lattices it can be excited by external infrared radiation (hence its title). Because of this high frequency, the group velocity of a lattice wave in the optical branch is small, and this branch makes a

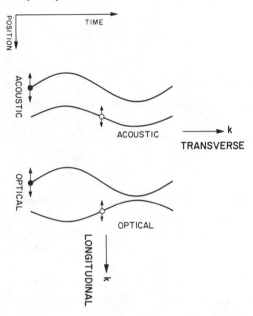

Fig. 4.5 Two representative atoms from the unit cell of a diatomic lattice with dissimilar constituents are shown oscillating about their mean positions in both the optical and acoustic modes. For simplicity the wavelengths of each disturbance are taken to be equal. The diagram shows the case of a transverse wave in which the direction of propagation is from left to right; the corresponding longitudinal case is obtained when the page is turned through 90°.

correspondingly insignificant contribution to the transport of energy in the lattice. Such transport is produced rather by the acoustic modes, which are of lower frequency, and are, in fact, in the limit of long wavelengths, just like the ordinary sound waves of the elastic medium (hence their title).

Up to this point a vibrational mode has been regarded as a transfer of energy by wave motion through a lattice. However, the amplitude of such motion is quantized, and at low temperatures, on the microscopic scale for this process (see Chapter 1) we must take into account the quantum nature of such vibrational energy. The phonon is the elementary quantum of this collective lattice excitation, and it has an energy $\hbar v$, where v is the phonon's angular frequency.* (As is pointed out in Table 3, the phonon is a boson.) A lattice wave of large amplitude, therefore, corresponds in the quantum picture to a stream of phonons of the corresponding mode.

---

* A typical phonon's energy is $k_B \Theta_D$ for most materials (i.e., the product of the Boltzmann and Debye constants). This amounts to about 4×10^{-21} J (0.025 eV) and corresponds to a phonon frequency of about 10^{13} sec.$^{-1}$.

Consequently, it is possible to encounter four forms of this collective excitation of the ionic system: Longitudinal or transverse optical phonons, and longitudinal or transverse acoustical phonons (Fig. 4.5).

The second collective excitation to be considered is one supported by the itinerant electron system. Here the elementary quantum unit of the excitation is called a plasmon, which is also a boson. For small values of its momentum, a plasmon's energy is hv_p, where v_p is the plasma frequency that can be shown from simple arguments to be given by

$$v_p = \left(\frac{4\pi ne^2}{m}\right)^{1/2}$$

Here n is the average electron density while e and m have their usual meanings. The collective excitations in question are temporal oscillations in the denstiy of the fluid of itinerant electrons; they are known* as plasma oscillations, and should not be confused with the spatial oscillations in the fluid's density that are discussed in Sect. 3.4.2. It is not difficult to see qualitatively how the plasma oscillations come about.

Their origin is rooted in the itinerant electron's high mobility compared with that of an ion and in the long range of the Coulomb interaction between them. (An ion is very sluggish compared with an itinerant electron since the latter is typically able to respond some 100 times faster than the former.) First, we recall that the itinerant electrons attempt to maintain a state of charge neutrality in the metal. For example, suppose there is some charge imbalance in a given region—it could be in say a longitudinal phonon, where the uniform distribution of positive charge is automatically upset by the displacement of the ions. The electrons rush into the positively charged region in order to restore the balance. This is screening, which was first encountered in Sect. 3.4. In the present case the electrons move in phase with the ions and are able to screen any local variations in charge density, rather like the longitudinal acoustical mode of Fig. 4.5 where the open and filled symbols represent respectively the electrons and ions. The ability of an electron to move into a region where screening is required is determined by the balance between two influences acting upon it: The electrical forces arising from the unscreened charge and the electron's intrinsic thermal motion. The former directs the electron's motion towards the unscreened charge, and the latter acts to randomize it. In other words, there is a certain range around an unbalanced charge in a metal within which the electrical forces overcome the thermal energy of the electron and are able to attract it into the unbalanced region. This range is known as the Debye length; clearly, it decreases as the temperature increases (because of the

---

* This is the name of the excitation in a fluid of charged fermions. When it arises in a fluid of neutral fermions—such as liquid He³—it is called zero sound.

larger thermal motion of the electrons) and increases with increasing electron density n (because of the increased electron potential energy).

The plasma oscillations arise from this screening mechanism. The electrons rush into the unbalanced area and acquire a certain amount of kinetic energy in the process. They consequently overshoot the target, are attracted back, overshoot again, and so it continues: A persistent, longitudinal oscillation in the electron's density is set up, but here the electrons are moving out of phase with the ions. The entire sea of itinerant electrons oscillates back and forth with respect to the ionic lattice; therefore, a plasma oscillation can be looked upon as a kind of longitudinal optical mode where the open and filled symbols of Fig. 4.5 could again represent the electrons and ions, respectively. Thus as far as the individual electron is concerned, it executes a simple harmonic motion, and it turns out that no single electron is very strongly affected. But, because of the long range of the Coulomb interaction, this individual motion is only a part of the correlated motion of all the itinerant electrons, which in the bulk are moving coherently with the frequency v_p. We find, in fact, that at metallic electron densities this plasmon frequency is quite high (about 10^{16} sec^{-1}) and it corresponds to a plasmon energy hv_p, which is very large in a metal, typically between 8 and 24×10^{-19} J (5–15 eV). Since most metals melt when the electron's thermal energy is about 0.2×10^{-19} J, it is obvious that the plasma oscillations cannot be excited thermally. Here is a distinction between the ease with which phonons and plasmons are created. If an ion is given energy, it quickly dissipates it into the ionic lattice through the creation of a large number of collective oscillations of low energy (the phonons), whereas an excited electron is unlikely to be able to excite the corresponding collective oscillations in its system because of the large energy necessary to create a plasmon. Experimental evidence for plasmons, therefore, comes from energy-loss experiments in which an incident particle is able to couple with the electron system and has enough energy to excite the plasma oscillations. For example, electrons with typical energies of $1–20 \times 10^{-16}$ J incident upon metal foils show energy-loss effects in both transmitted and reflected beams at energies corresponding to multiples of hv_p, indicating the excitation of bulk collective oscillations.*

Finally, the discussion returns to the ionic system to consider two collective excitations that it can support in the special case when interacting electric or magnetic dipole moments are localized upon each ion. As far as electric dipoles are concerned, it seems that presently there is only

---

* C. J. Powell and J. B. Swan, *Phys. Rev.* **115**, 869, 1959. We should also mention that other plasma oscillations, occurring on the metal's surface or at the interfaces between metals, can also be excited, and these have their own characteristic frequencies. But details of such complications would go beyond our present purpose.

theoretical evidence of the existence of collective oscillations. In van der Waals solids, which consist of neutral atoms mutually polarized by their neighbor's presence to become an electric dipole, it has been suggested* that a longitudinal charge density oscillation analogous to a plasma wave could be excited among the induced dipoles. However, this is contradicted by studies of the corresponding system† consisting of permanent dipoles that suggest that such a system is not able to support collective excitations.

In the case of a system consisting of localized magnetic moments, however, the existence of collective oscillations is well established. In a ferromagnetic metal, for example, when unfilled inner shells of the ion give it a resultant and localized magnetic moment, the ground state is the complete parallel alignment of these moments. It may at first appear that the lowest excited state is given by the reversal of one ion's entire spin, but it turns out that there are excited states having much lower energies than this (in fact, only slightly above that of the system's ground state) that correspond to the precession of the ionic spins about their ground state position. Such a deviation from the ground state position by one ionic spin, which could be initiated, for example, through a spin-flipping event involving a core electron contributing to the spin (as described in Sect. 4.5.4), cannot remain isolated from the rest of the ionic spins. Their mutual interaction ensures that the spin's deviation is communicated to the contiguous ions; as a result, the disturbance tends to move throughout the spin system as a collective excitation in the form of a wave (known as a spin wave).

The precession of the spins may have the same phase on all the ions (which is the uniform or coherent mode); there may be some fixed phase difference between the precession on neighboring atoms (the standing spin wave mode); or the phase may vary periodically in the direction of travel of the wave (the traveling spin wave mode). The amplitude of the spin wave is quantized, of course, and the elementary quantum unit of this excitation is called a magnon. Magnons have frequencies typically in the microwave range (about 10^{10} sec^{-1}) with corresponding energies of about 10^{-24} J.

4.3   THE COULOMB FORCE: VIRTUAL PARTICLES AND THE FEYNMAN DIAGRAM

4.3.1   The Origin of the Coulomb Force.   What is the origin of the Coulomb interaction? It may seem an unnecessary digression to ask such

* S. Lundqvist and A. Sjölander, *Arkiv. Fisik.*, **26**, 17, 1964.

† R. Lobo, *Phys. Rev. Lett.* **21**, 145, 1968.

a question in this context. After all, if the force acting between charged particles is accepted as an empirical fact—of which Coulomb's law is a quantitative statement—the study of electrons in metals can proceed without reference to the force's origin. However, this force is part of a wider problem relating to the origin in nature of what are known self-evidently as action at a distance effects, and their study is part of quantum field theory. It is useful to have an acquaintance with some of the concepts developed in this theory, since several have been incorporated into the contemporary description of a metal, and the Coulomb force provides a convenient framework for their introduction.

Chapter 1 distinguishes four fundamental interactions in nature. For convenience here the weak and strong nuclear interactions can be lumped together to leave but three, that is the nuclear, the electromagnetic, and the gravitational ones. These are the action at a distance interactions, and they formerly were distinguished from the contact interactions, that require any two interacting bodies to be in evidently very close contact before the interaction is operative. (The exchange interaction of Sect. 3.3 is an example of a contact interaction.) However, we shall see that in the modern view this distinction between the interactions is no longer maintained.

In the classical (Maxwellian) view, which is appropriate to the macroscopic scale, the action at a distance interactions are said to operate through the intermediate influence known as a field of force. This field is thought to operate throughout the whole of space around any particle that can participate in the particular interaction—the particle being seen as the source of the field, which can propagate throughout space with a velocity never exceeding that of light. Appropriate fields are defined for each context, so that there are nuclear, electrostatic or electromagnetic,* and gravitational fields. Correspondingly, there are particles that can be influenced by these fields: Nucleons, particles with charge, and particles with mass, respectively. The particles themselves are further divided into three families according to their masses and the interactions they can be coupled by. The hadrons are heavy particles that are subject to all interactions. (These are further subdivided into the baryon and meson classes which have half-integral and integral spin values, respectively.) The leptons are those particles that are only subject to the electromagnetic, gravitational, and weak interactions (the electron is a lepton). Finally, there are the mass-

---

* The electrostatic case is part of the more general electromagnetic one involving moving electric charges as opposed to the stationary ones in the electrostatic case.

less quanta (namely the photon and the graviton to be introduced below).*

Classical fields have wave-like properties and can support the transfer of energy or momentum between particles in quantities as large or as small as desired. But we shall be concerned primarily with processes that occur on the microscopic scale; thus the concepts of quantum theory dominate the discussion. When the switch is made to this scale, so that consideration has to be given to the quantization of the field, the description meets what is a typical characteristic of such a transition: The quantization gives an additional corpuscular aspect to what was an entirely wave-like nature of the field in the classical case, and it gives conversely a wave-like nature to the particles in the problem that were purely corpuscular entities under classical conditions. In other words, the wave/particle duality referred to in Chapter 1 is encountered. Taking the case of the field, the corpuscles (or field quanta) resulting from quantization are the only permissable form of energy transference in the field under quantum conditions. Depending upon the particular interaction, these field quanta may carry a charge or be neutral, have a nonzero rest mass or be massless. In the cases of the nuclear, electromagnetic and gravitational interactions, they are known respectively as the muon, the photon, and the graviton.†

The modern quantum field theory is that an action at a distance inter-action and a contact interaction are only different manifestations of the same thing. All interactions are thought to arise from the creation and annihilation of field quanta, which are the undetectable intermediaries carrying energy (that is, transmitting a force) between the interacting objects that will themselves be quantum particles. We have said that these intermediaries have wave/particle duality in the usual quantum fashion,

---

* It is currently topical to question the assumption that these quanta are massless. The possibility of a mass for the graviton, for example, has been discussed recently by M. G. Hare (*Can. J. Phys.* **51**, 431, 1973). It is part of the wider problem of whether or not the graviton exists at all, for it is still not certain that the gravitational field should be quantized. However, that issue is sidestepped here.

† Because the gravitational force is so weak (Fig. 1.1) we would expect that detectable gravitational waves would arise only from enormous bodies (such as stars or galaxies) undergoing rapid changes in speed. It has become fashionable recently to make laborious and expensive efforts to detect such waves (some fads in science seem to be virulently infectious) but the results have been inconclusive. (See, for example, the news report in *The New Scientist*, **55**, 132, 1972 or in *Physics Today*, October, 1973.) It seems fair to say at the present time that practically no one—the exception being the originator of the fashion—believes that gravitational waves have been conclusively detected.

but in a descriptive as opposed to a mathematical interpretation the common practice is to emphasize the latter aspect and to think of them as particles that can be emitted or absorbed by the interacting bodies; it is their exchange that creates the interaction. We will describe in Sect. 4.3.2 how these intermediaries are not real in the sense that real particles can be detected directly (or so one likes to think) but are extremely short-lived violations of energy conservation that are possible as a result of the Heisenberg uncertainty principle. (This idea is introduced in Sect. 4.2.4 in connection with the quasielectron.)

Turning specifically to the class of inverse-square-law interactions (which includes the electromagnetic and gravitational ones), theory shows that they are produced if the intermediary is a particle having zero rest mass. It has been said that in the case of the electromagnetic interaction it is the photon that is the intermediary and the sole mediator of the interaction. Thus when two electrons repel each other, or when an electron and a proton are attracted, it is through an exchange of photons. These are known as virtual photons, and we can imagine them to be constantly created (sent out) and annihilated (received) by an isolated electron. When two electrons at rest can exchange virtual photons, the result is the electrostatic or Coulomb interaction. A kind of balanced equilibrium can be envisaged in which the number of virtual photons emitted and absorbed by either electron in a given time is equal. But if one of the electrons is in motion, then the electromagnetic interaction proper is involved, and we expect it to upset somewhat the balance in the virtual photon exchange. Some of the photons then are not exchanged but escape from the interaction to become real in the sense that they are detectable (as described in Sect. 4.3.2), whereas those that are exchanged impart momentum between the two interacting electrons. Since the virtual photon is undetectable, the apparent effect is a force between the two electrons, and this is coupled with the emission of radiation.

Classical analogies of this type of interaction, involving ball players and the like on ice skates, can be found in elementary texts, but we should emphasize that the whole concept of exchange of virtual particles is only a metaphorical description. It is just that the mathematical description of the interaction turns out to be the same as the one that is applicable to the propagation of the free photon, and this has given rise to the common practice of describing the interaction in terms of photon exchange.

In the case of gravitation, the modern view is that this too arises from the exchange of a massless quantum—the graviton—that is the intermediary and the sole mediator of this interaction. Consequently, what are observed as gravitational effects on the macroscopic or cosmic scale can be regarded as just the accumulated effects of the graviton interactions taking place on the microscopic one.

We see now how action at a distance is reduced to a contact interaction in the microscopic view: The latter interaction is between the virtual photons and the charged particle, and it takes place at the time of creation or annihilation of the virtual photons; macroscopically, however, the interaction appears as an action at a distance. Some examples of interactions where the virtual particle terminology is commonly employed will be described in Sect. 4.3.3 in connection with Fig. 4.8.

4.3.2   Virtual Particles.   One form of Heisenberg's Uncertainty Principle is: The product of the uncertainty in the energy of a system (ΔE) and the uncertainty in the time at which the energy is measured (Δt) is a constant; $\Delta t \Delta E = \hbar$ (where $\hbar$ is Planck's constant h divided by 2π). This means, for example, that to test the conservation of energy in a particular process (which obviously would automatically need a precise knowledge of the energy—that is, ΔE would have to be small) the principle requires the process to exist under observation for a correspondingly long time (Δt would have to be large).   It follows that a violation of the energy conservation principle amounting to ΔE can exist unknown to the observer if its duration is no longer than the Δt set by the above uncertainty principle.

A virtual particle is just such a violation of this conservation principle. Its lifetime has to be less than Δt, otherwise it ceases to be virtual and comes into real existence as a detectable particle (as in the emission of X-rays referred to below); the corresponding ΔE can be taken as the energy required for its creation. If a virtual particle having a finite rest mass is created (such as, for example, the Yukawa pion in the nuclear interaction) its total energy after creation is whatever kinetic energy it acquires plus its rest energy $m_o c^2$ so that the particle's lifetime is given roughly by

$$\Delta t \sim \hbar/m_o c^2 \tag{4.12}$$

However, if the virtual particle has a rest mass of zero (such as a photon or a graviton) then ΔE can be taken as $h\nu$, where ν is the frequency of the photon or graviton wave. (We recall that the virtual particle has wave/particle duality. $h\nu$ is, therefore, the energy of the quantum having the wavelength $\lambda = c/\nu$.) The lifetime in this case is evidently given by

$$\Delta t = \hbar/h\nu = \frac{1}{2\pi\nu} \tag{4.13}$$

Whatever the case, a knowledge of ΔE clearly leads directly to an estimate of the virtual particle's lifetime and, indirectly, to its range if the velocity is known.

Taking the case of Yukawa's pion as an example with a nonzero rest mass, the lifetime is found to be about 5×10^{-24} sec and, since the pion's velocity cannot exceed the velocity of light, a rough estimate of about 2×10^{-15} m is obtained for its maximum range. (We do not mean to give the impression that the virtual particles' distribution is sharply cut off. Formal theory shows there is a gradual reduction in the probability of finding a pion as the distance from the nucleon increases. The average pion penetrates about 0.8×10^{-15} m away from the core of the nucleon.) For the virtual particle with zero rest mass, on the other hand, its range follows directly from Eq. (4.13): The particle's speed is c and, consequently, its range is $\lambda/2\pi$ (which is written $\lambdabar$).

Whether or not a virtual particle has a nonzero rest mass produces distinctly different forms of interaction. Specifically, if the intermediary in an interaction has a zero rest mass then, as Eq. (4.13) indicates, the range of the particle increases with wavelength λ of the quantum; so too does the probability of the particle's creation or annihilation, and it is possible to show from all this that an inverse square law of force is the inevitable consequence. However, when the intermediary has a nonzero rest mass, the resulting interaction is found to be an inverse square law at short range, but it rapidly falls off (that is, it becomes say an inverse cube or inverse quadratic—the exact form is not certain) at larger ranges. So the general remark can be made that inverse square laws of infinite range involve virtual quanta having a zero rest mass, but more short-range forces involve quanta having a nonzero rest mass.

Returning to the concept of a virtual particle, we can picture it as one of a cloud of transitory quanta surrounding an otherwise isolated, real quantum particle that is in a constant state of self-interaction through the creation and absorption of these quanta. Of course, this begs the question of what is meant by a real quantum particle. We could say that real in this context implies that, in addition to wave/particle duality, the particulate features of this quantum typify fixed, intrinsic properies like mass, charge, and spin. It is interesting to note, however, that the origin of these seemingly intrinsic properties is intimately connected with the ability to self-interact; a real particle is perhaps no more than a disturbance in an energy field brought about by self-interaction. (The charge of the electron e, for example, is just a measure of the coupling between matter and the electromagnetic field.) Further digression into this philosophical area is outside our scope, but we should note that self-interaction has lead to the distinction between what is termed a bare particle, which is the old-fashioned concept of an inert body with fixed, intrinsic properties, and a dressed particle, which is the real particle taken together with its cloud of transitory virtual quanta. This cloud represents the coupling of the particle to the appropriate field. Taking again the electron as an example, it is its coupling with the electromagnetic field that produces the small but measurable deviations from 2 shown by the gyromagnetic ratio g referred to in Sect. 3.2.1. The distinction between dressed and bare particles will be met again in Sect. 4.3.3 in connection with Fig. 4.6.

We pointed out above that if the lifetime of a virtual particle exceeds the values suggested by Eq. (4.12) or Eq. (4.13) then it comes into real existence in the sense that its effects are detectable by the usual means. Consider specifically an electron surrounded by its cloud of virtual photons, and imagine what will happen if the dynamical balance between their creation and absorption is upset. This is particularly likely to happen when the velocity of the electron changes suddenly; there are various observations that can be attributed to the escape of virtual photons into real existence during such a maneuver. For example, the emission of X–ray photons when electrons are slowed down in matter can be thought of as the additional virtual photons that have been created in their deceleration but not reabsorbed by the electron. The same can be said of the Bremsstrahlung emission observed when an electron is scattered by a proton. Furthermore, because of this possibility of escape by virtual photons, we expect that the interaction between two electrons is more complicated when they are in relative motion than when they are at rest. This is the origin of the difference between the electromagnetic and electrostatic interactions; its manifestation is that an accelerating charge produces a magnetic field that varies with time, and Faraday's law tells that this is indeed more complicated than the field of a uniformly moving charge.

4.3.3 The Feynman Diagram. The two preceding sections have indicated the general conclusion from the quantum theory of fields that all action-at-distance interactions are in effect contact interactions and arise from the creation and annihilation of particles at definite points in space and time. In the late 1940s R. P. Feynman showed that mathematical expressions arising in this theory could be represented in a useful diagramatic way. The power of these Feynman diagrams is not only their exact correspondence to the algebraic expressions, but also the fact that they depict in an easily comprehensible way the physical processes involved. (Only a few of the most elementary forms of the Feynman diagram need be introduced here. The reader seeking more comprehensive coverage is referred to the books by Ford and Blanpied at an elementary level and by Mattuck at a more advanced level, which are listed at the end of the chapter.)

Two important ideas, which are contained in the opening sentence above, point the way to the Feynman diagram: First, only the creation and annihilation of particles is of concern, and second, these events are localized at definite points in space and time and do not span ranges of either of these parameters. Accordingly, a Feynman diagram consists of a space-time plot (Fig. 4.6) in which the three space coordinates of three dimensional space are represented by the one parameter **r**. The diagram is used to illustrate sequences of events in the history of systems of particles. Each event, such as a particle's creation or annihilation, is represented by a space-time point (or vertex); these are joined together by arrows known as the particle's world lines. These world lines do not represent the actual paths of the particles in space and time (although in many cases it is convenient to think of them as such) but through their connectivity they are essential

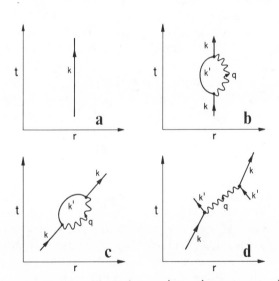

Fig. 4.6 Feynman diagrams illustrating various circumstances of an electron in a state represented by **k**. In the conventional arrangement time is plotted along the ordinate, and the electron's position in space **r** is along the abscissa; normally the axes are omitted. (*a*) The bare particle at rest. (*b*) The self-interacting dressed particle at rest, where a virtual photon has been spontaneously created and then annihilated some time later. (*c*) Similar to *b* except here the electron is in motion. (*d*) an alternate representation of *c* which is discussed in the text.

to illustrate the proper sequence of events. Time t here is plotted along the ordinate, and the particle's position r along the abscissa, so that the diagram is read from the bottom to the top to give the order of events when time is moving in its usual (positive) direction.* The inclination of the world line represents the particle's speed, which can range from zero (a vertical world line) to the velocity of light (which is represented by the gradient of the photon q in Fig. 4.6d). The origin and termination of a world line, which correspond to the creation and annihilation of the particle, are the vertices representing an event; these are shown as dots. Particles that are bosons are represented by wavy world lines to distinguish them from the others, which are fermions. (The significance of this difference is described in Sect. 3.2.4; Table 3 shows the division of the commonly encountered quanta between these two classes.) Note that a particle represented by a world line may be virtual or real; it depends upon its lifetime with respect to the Δt of Eq. (4.12) and Eq. (4.13).

Figure 4.6 shows some elementary examples of a Feynman diagram for a system consisting of a single electron that is initially in a quantum state labeled k. Figure 4.6a represents a motionless bare particle, which we have pointed out is now regarded as an unreal concept; rather, the electron is constantly undergoing self-interaction with virtual photons as depicted in Fig. 4.6b. Here the stationary electron in state k emits a photon q and enters the intermediate state k' only to absorb the photon at a later time and return to its original state. The same process is represented in Fig. 4.6c except that here the electron is in motion throughout the self-interaction. (The world line for k' is usually curved in such a figure for clarity. This is permissible since it really does not represent the particle's actual path.) Fig. 4.6d shows an alternate way of representing the process in Fig. 4.6c. We imagine that at some time during the intermediate particle's lifetime it spontaneously disappears and reappears at some other point in space; but exactly the same interaction is represented here as in Fig. 4.6c.

The form shown in Fig. 4.6d is the main objective. Although it here specifically represents the transition of a single electron from an initial state k to the same final state k via the intermediate state k', it also appears frequently in the context of an interaction between two particles. To see this, it is convenient to take a specific example that has already been encountered in Sect. 3.4.4 and Sect. 3.4.5 and that will appear again in Sect. 4.4. In the case of elastic scattering of an itinerant electron by a split virtual level, the system consists of two electrons. The first, an initially

---

* The convention is not universal and frequently the axes are interchanged. This leaves the possibility of slight confusion since usually it is considered unnecessary to include them in the diagram.

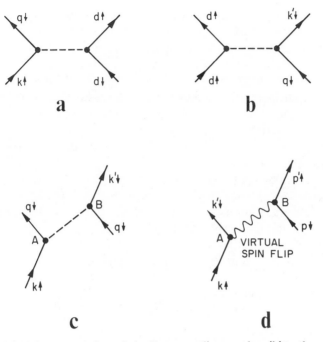

Fig. 4.7 (*a,b*) The construction of the Feynman diagram describing the particular interaction between two electrons that is discussed in the text. (*c*) The interaction, which is a combination of those shown in (*a*) and (*b*); it takes an electron that is incident in a state $k\uparrow$ to a final one $k'\downarrow$. Two such interactions that occur consecutively at the same site.

itinerant electron in a delocalized energy state, say* $\mathbf{k}\uparrow$, is captured into a split virtual level. There it feels the influence of a resident electron in the level—as part of an incomplete d shell, for example—and in a state denoted by $\mathbf{d}\downarrow$ for example. Various consequences of this encounter are possible but here the case is chosen where the itinerant electron ends up in a new state $\mathbf{q}\downarrow$, and the d electron, which must remain localized in this problem, is, therefore, afterwards in the state $\mathbf{d}\uparrow$ required to conserve spin. Such an encounter is said to have scattered an itinerant electron from a state $\mathbf{k}$ to one $\mathbf{q}$ and to have flipped its spin in the process.

Alternatively, we can view this as shown in Fig. 4.7*a*; the interaction, which is represented by the horizontal dotted line, simultaneously annihi-

---

* This is a convenient way to specify a particular spin orbital $(\mathbf{k},\sigma)$; $\mathbf{k}$ specifies the spatial part while $\uparrow$ and $\downarrow$ correspond to the two possible configurations of the spin coordinate σ described in Sect. 3.2.3.

lates electrons in states $k\uparrow$ and $d\downarrow$ and creates others in states $q\downarrow$ and $d\uparrow$. Since it is inherently impossible to distinguish one quantum particle from another of the same kind, it is meaningless to inquire whether the electron appearing in $q\downarrow$ is actually the one that was incident in $k\uparrow$; it is just as appropriate to think of scattering as the creation and destruction of particles in the manner outlined above. In the form of the Feynman diagram shown in Fig. 4.7a, the inward inclination of the particles' world lines with respect to the central axis of the diagram (thus: $\nearrow \vdots \nwarrow$) implies the particle's destruction in the interaction, and their outward inclination ($\nwarrow \vdots \nearrow$) implies a creation.

In the particular scattering process under consideration, the state having electrons in $q\downarrow$ and $d\uparrow$ is only an intermediate one. In q the electron is temporarily captured in the virtual level, but eventually it escapes to a final delocalized state k' by quantum mechanical tunneling (Sect. 3.4.5). Various final steps in this two-step process can be chosen and will be described in Sect. 4.4 in connection with the Kondo effect, but here, to illustrate the application of the Feynman diagram, the final scattering event is taken to be that shown in Fig. 4.7b. In this the captured itinerant electron passes to a final state $k'\downarrow$, and the localized electron remains unaffected and retains the configuration $d\uparrow$. The process represented in Fig. 4.7a and Fig. 4.7b can evidently be combined in one diagram as shown in Fig. 4.7c. An incident electron in a state $k\uparrow$ is scattered into a state $q\downarrow$, where this event at A causes a change in the spin component of the ion that is the scattering center. Subsequently the electron is again scattered at the event B to a final state $k'\downarrow$ the ion's spin component remains unchanged during this second stage of the process. As already described in Sect. 3.4.5, such an event is known as a spin-flip scattering process, and it is one of several ways in which an electron can be scattered in metal when the scattering center has a net magnetic moment.

It is useful to note parenthetically an important result described in Sect. 4.4: With such scattering the probabilities of an itinerant electron going through different intermediate states between the same initial and final ones ($k \rightarrow k'$) are unequal. We will show that this result emerges from the application of the exclusion principle to the electrons in the intermediate state. This gives effectively an internal degree of freedom to the initial state that is eclipsed when the argument is limited to the first-order viewpoint (as already described in Sect. 3.4.5). However, here it is sufficient to realize that these unequal scattering probabilities between given states through different channels indicate that one is dealing with a many-body system. In such, as is pointed out in Sect. 4.2, it is not possible to choose any particle for individual consideration because the mutual interactions between the constituents make all their behaviors interdependent. For example, suppose an itinerant electron is scattered by an ion through a process such as A shown in Fig. 4.7c. Since such scattering depends upon the relative magnitude and orientation of the spins of the itinerant and localized electrons, the next electron to be scattered by this same ion feels the influence of its predecessor because the outcome of the second scattering depends upon the details of the first event. Consequently, there is an effective interaction between the two itinerant electrons in which the ion's spin memory of events is the coupling agent.

This brings the discussion back to the objective of the section, which is the application of the Feynman diagram shown in Fig. 4.7d. It arises in quantum field theory as a convenient representation of exactly the above type of two-particle interaction where the constituents are members of a many-body system. The diagram represents two consecutive processes, each like that shown in Fig. 4.7c. In the first, the electron is scattered from an initial state $\mathbf{k}$ to a final one $\mathbf{k}'$ by an event labeled A, of which the details are left unspecified except that it is required to be a spin-flip process. The ion's spin value consequently changes at A, and its energy moves to a new Zeeman value, but these are all degenerate in energy since the scattering is assumed to be elastic (Sect. 3.4.5). A subsequent scattering of another itinerant electron takes place at the same ion at the event B, and in this the second electron is assumed for illustration to be scattered from a state $\mathbf{p}\!\downarrow$ to $\mathbf{p}'\!\uparrow$ so that the ion's spin is again changed and thus returned to its original configuration. However, this second event depends upon the fact that the previous event A had already taken place, and this can be looked upon as an interaction between the two electrons arising from the transfer of spin information by the ion. It is one of several examples where an intermediary is effectively responsible for producing an interaction between constituents of a many-body system. In the shorthand notation used by many-body theory to describe such a coupling, it can be regarded as a result of the exchange of a virtual spin-flip particle.

Other examples of the use of this concept are shown in Fig. 4.8. First, in the context of general physics, Fig. 4.8a illustrates the Coulomb repulsion between two electrons. This is seen as the exchange of virtual photons, as described in Sect. 4.3.1. Figure 4.8b represents a still controversial view of the gravitational attraction between two bodies of masses m and M, and Fig. 4.8c shows a view of Yukawa's proposed origin of the forces between nucleons. Here a neutron and a proton interact and exchange roles through the exchange of a virtual pion. Finally, Fig. 4.8d represents an interaction in which an electron and a positron are annihilated through the exchange of a virtual electron with the creation of two real photons γ.

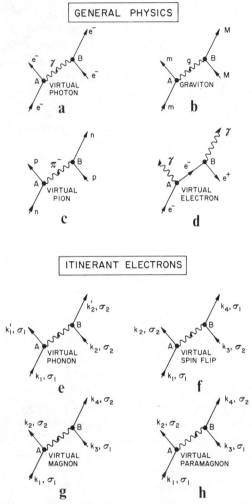

Fig. 4.8 Feynman diagrams illustrating some common two-particle interactions encountered in general physics (upper) and in solid state physics (lower). Each interaction involves an intermediate virtual particle that is a short-lived violation of the energy conservation principle permitted by Heisenberg's uncertainty principle. (a) The electromagnetic interaction between two electrons scattered by the Coulomb force between them. (b) The gravitational interaction between two bodies of masses m and M. (c) Yukawa's view of the origin of forces between nucleons. (d) An interaction in which an electron and a positron are annihilated through the exchange of a virtual electron. The reversed arrowhead in this example indicates that the positron propagates backwards in time; this is part of an important and wider significance of the Feynman diagram that is beyond our present scope. The subject is referred to in Sect. 3.2.4 in connection with Fig. 3.2. (e,f,g,h) Examples that relate to the context of itinerant electrons in solids and that are described in the text.

Turning to the lower half of the figure, this shows four interactions of more immediate interest since they occur between the itinerant electrons in a metal. They reflect the many-body aspect of scattering of itinerant electrons in different circumstances, but their description is deferred until Sect. 4.5; at this point it is more appropriate to continue with the closely related problem introduced first in Sect. 3.4.5 and again in connection with Fig. 4.7d; namely, the scattering of an itinerant electron by the residents in a split virtual level.

4.4 THE KONDO EFFECT

This section continues the discussion begun in Sect. 3.4.5 and interrupted to establish the Feynman diagram in the preceding section. It concerns the elastic scattering of itinerant electrons by a foreign ion having a net magnetic moment due to the uncompensated spins of electrons in an incomplete shell. Phenomenologically, this process is pictured as a split virtual level around the ion into which the itinerant electron is temporarily captured during the scattering. It has been emphasized already in Sect. 3.4.5 and again in the preceding section, that to see all the subtleties this scattering must be considered as a second-order (two-step) process: First, the itinerant electron is captured from an initial state say $\mathbf{k}$ into a temporary and localized one $\mathbf{q}$ and, second, it escapes from $\mathbf{q}$ to reach a final itinerant state $\mathbf{k}'$ by quantum tunneling through the potential barrier forming the virtual state.

From the first-order viewpoint this is just a scattering of an electron from $\mathbf{k} \rightarrow \mathbf{k}'$. Since the electron has two spin alternatives, a total of four different first-order possibilities can be envisaged, and these have been listed in Sect. 3.4.5. However, all of this overlooks the fact that the ion has effectively an internal degree of freedom conferred by the intermediate condition in the scattering process; specifically, at the beginning of the scattering an electron may or may not be resident in $\mathbf{q}$, and during the scattering this resident may or may not have its spin flipped. Each of these alternatives leads to a separate channel for the scattering to follow between any two initial and final states. In fact, each of the four first-order possibilities referred to above has itself four alternative channels.

We choose to illustrate this schematically for two of the four first-order possibilities. Fig. 4.9 shows the four channels existing for the nonspin-flip process $\mathbf{k}\uparrow \rightarrow \mathbf{k}'\uparrow$ (Fig. 4.10 contains the corresponding possibilities for the spin-flip one $\mathbf{k}\uparrow \rightarrow \mathbf{k}'\downarrow$.) The remaining pair, namely $\mathbf{k}\downarrow \rightarrow \mathbf{k}'\downarrow$ and $\mathbf{k}\downarrow \rightarrow \mathbf{k}'\uparrow$, follow straightforwardly and will not be illustrated. These figures

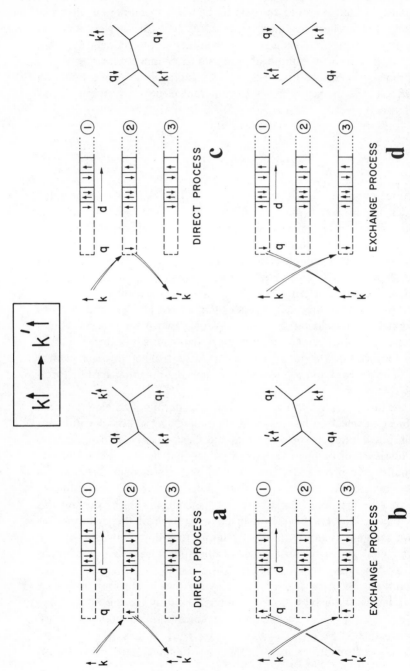

Fig. 4.9 Illustrating the four channels by which an itinerant electron incident in a state $k\uparrow$ can be elastically scattered by a virtually bound state into a final itinerant state $k'\uparrow$. Since the electron's spin orientation is unchanged, the process is known as a non-spin-flip scattering event.

DIRECT PROCESS **a**

EXCHANGE PROCESS **b**

DIRECT PROCESS **c**

EXCHANGE PROCESS **d**

$$k\uparrow \longrightarrow k'\uparrow$$

182

are in the spirit of Fig. 3.12 in which the more tightly bound electrons in the ion's incomplete shell are labeled d, and an electron temporarily localized in the virtual state (and thus relatively weakly bound) is labeled $\mathbf{q}$; $\mathbf{k}$ and $\mathbf{k}'$ are, of course, the initial and final delocalized states of the itinerant electron. Time advances from top to bottom in Figs. 4.9 and 4.10 so that a two-step advance gives rise consecutively to three distinct conditions for the process; these are labeled: Initial ①, intermediate ②, and final ③. The corresponding Feynman diagram is included in each case.

Consider first the nonspin-flip process. In Fig. 4.9a the state $\mathbf{q}$ is initially empty, but subsequently an electron from a delocalized state $\mathbf{k}\uparrow$ is captured into it to give the intermediate condition. Eventually the electron in $\mathbf{q}$ tunnels out quantum-mechanically into another delocalized state $\mathbf{k}'\uparrow$ to give the final condition for the system. The whole is known as a direct process because $\mathbf{q}$ is initially empty. In Fig. 4.9b the order of events is reversed: First, the virtual level $\mathbf{q}\uparrow$, which is initially occupied, is emptied when the electron tunnels into the itinerant state $\mathbf{k}'\uparrow$. Subsequently another itinerant electron enters $\mathbf{q}\uparrow$ from a state $\mathbf{k}\uparrow$. The whole event is known in this case as the exchange of the first process. Note that in neither of these channels was a spin-flip required of any electron localized in the ion. Turning to the remaining channels of Fig. 4.9c and Fig. 4.9d, the former again shows a direct process but in this case the itinerant electron undergoes a spin-flip when captured by the virtual state. A resident electron must correspondingly flip its spin in the intermediate condition, but this is temporary since the electron from $\mathbf{q}\downarrow$ again spin-flips when tunneling out to $\mathbf{k}'\uparrow$ (as it has to because the overall process is a nonspin-flip one); thus the resident electron's spin can flip back to its initial configuration. The corresponding exchange process is shown in Fig. 4.9d, and its interpretation should now be evident. Finally, it is seen how the four channels of a first-order process are distinguished: A channel is either direct or exchange, and its intermediate condition involves either a spin-flip or a nonspin-flip event.

The second example chosen for illustration is the spin-flip case $\mathbf{k}\uparrow \to \mathbf{k}'\downarrow$ shown in Fig. 4.10. The description is quite similar to that given above with the notable difference that here the itinerant electron (and hence a resident in the ion) is left with a permanently reversed spin after the process. In the channel Fig. 4.10a, for example, the state $\mathbf{q}$ is initially empty, but subsequently an electron from a delocalized state $\mathbf{k}\uparrow$ is captured into it and simultaneously suffers a spin-flip. In this case the resident in the virtual state has a spin that initially is not propitiously aligned and that has to flip to give the intermediate condition shown. Subsequently, the resident in $\mathbf{q}$ transfers by tunneling to another delocalized state $\mathbf{k}'$ but without flipping its spin. Hence, the overall process gives a permanent change

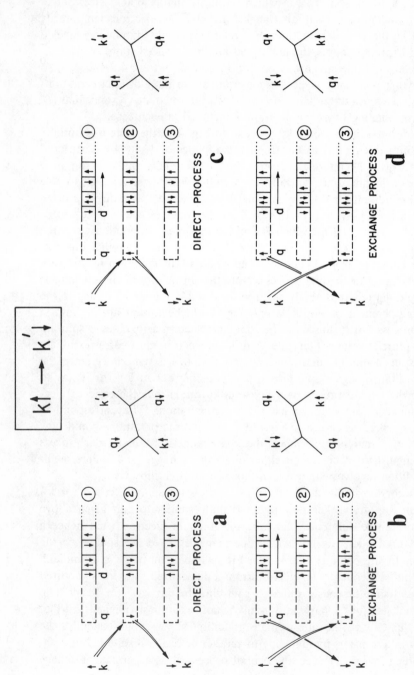

Fig. 4.10 As in figure 4.9 except that here the incident electron goes from a state $k\uparrow$ to a final one $k'\downarrow$. The process, therefore, corresponds to a spin-flip scattering event. The differing probabilities of the event proceeding via one of the four different channels gives rise to the Kondo effect described in the text.

184

n the itinerant electron's spin. The other channels shown in Fig. 4.10b to Fig. 4.10d can be understood in a similar way and seemingly do not warrant separate descriptions, but it is worth noting a detail occurring in the exchange processes of Fig. 4.10b and Fig. 4.10d: When, in contrast to the direct process, the resident in the virtual state does have initially a suitable spin orientation, then a spin-flip is required of some other electron in the ion to balance that made by the itinerant electron during its capture. An example is shown by the second step of the channel in Fig. 4.10b: $k\uparrow$ transfers with a spin-flip to the vacant state q in which the resident happens to have a suitably aligned spin that, therefore, remains fixed during the whole process. A corresponding spin-flip is required of an electron resident in an internal shell.

From these illustrations it seems not difficult to accept that the probability of scattering from any k to k' is different for different channels connecting them. For example, just the comparison of a direct process with its corresponding exchange shows that the former must have q in its intermediate state empty, but in the latter it must be full. These eventualities in general have different probabilities, as do the scatterings through the two channels. The whole effect is to make the order in which events occur important since the scattering of an itinerant electron in k is sensitive to the occupation of all the other electron states such as q.

In a quantitative treatment of this subject* these features are contained in a parameter known as the Fermi factor that appears as a multiplier in the expression for the probability of scattering between given initial and final states. For example, if the probability of a state q being occupied is f_q, then the probability of its being vacant will be $1 - f_q$. A Fermi factor f_q, therefore, appears in the probabilities of scattering via the channels in Fig. 4.9b and Fig. 4.9d; the corresponding factor for the channels in Fig. 4.9a and Fig. 4.9c is $1 - f_q$. We find that when the total probability of scattering between a given $k \rightarrow k'$ is obtained by summing the probabilities of each of the four individual channels concerned, the Fermi factors cancel out for all terms arising from channels for which there is no spin-flipping at any point in the process. The channels in Fig. 4.9a and Fig. 4.9b, for example, when combined give a net scattering probability independent of f_q—which is just what we get if no account is taken of the Pauli principle and the ion is thus denied its internal degree of freedom. [These channels are examples of spin-independent (or potential) scattering (Sect. 3.4.1) where the

* Comprehensive descriptions at a research level have been given by J. Kondo and A. J. Heeger in their articles in *Solid State Physics*, 23, 1969, pp. 183 and 283, respectively. A brief survey of recent progress has been given by P. Nozières, *Physics Bulletin*, October, 1974, p. 457.

exclusion principle does not enter the problem.] But the channels in Fig. 4.9c and Fig. 4.9d when combined give a net probability of scattering for which the Fermi factors do not cancel (and which is, therefore, dependent upon f_q). This is what leads to the Kondo effect, which began as the discovery by J. Kondo* that in the calculation of the electrical resistivity caused by such scattering the appearance of a factor involving f_q gives a term varying logarithmically with temperature. This gave an explanation of experimental results that in some cases had been unexplained in the literature for nearly thirty years. These were observations of what is known as the resistance minimum phenomenon: As the temperature of certain alloys and some seemingly pure metals is reduced, their electrical resistivity decreases in the expected manner, except at the very lowest temperature where it unexpectedly increases markedly with further reduction in temperature. Experimentalists had established at about the time of Kondo's work that this resistance minimum was apparently associated with foreign ions (which are usually present as impurities) possessing a net magnetic moment and was not an intrinsic property of any pure metal; however, Kondo's work was the breakthrough in understanding. In his picture, the resurgence of the resistivity comes from the log T term. Although only one of several components making up the total electrical resistivity of the alloy, it starts to become very significant at lower temperatures. [We recall that the entire discussion here relates only to the resistivity arising from elastic spin-flip scattering. There will, of course, be several other contributions to the total resistivity, such as that arising from potential scattering (with nonmagnetic impurities or imperfections scattering by the host's ions, as well as nonspin-flip processes by the foreign ion, but none of these figure in the Kondo effect.] In several cases Kondo's theory even provided quite acceptable quantitative agreement with various empirical relationships that had previously been established by the experimentalists.

We can say that Kondo's work provided an explanation of the resistance minimum effect but not a complete understanding, for any term containing a factor log T is clearly divergent and leads to a singularity at $T = 0$. This is physically unreasonable because it requires the resistivity to increase indefinitely as absolute zero is approached (Fig. 4.11)—no metals or alloys show such behavior—and theoretically it is neither elegant nor acceptable. It indicates that further sophistication (or, failing that, some ingenious twist) is necessary to remove this singularity without upsetting the body of the theory. Furthermore, experiments show that the logarithmic Kondo behavior is followed by the resistivity as the temperature is

---

* J. Kondo, *Prog. Theoret. Phys.* (*Kyoto*) **32**, 37, 1964.

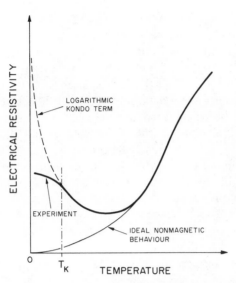

Fig. 4.11 A schematic view of the resistance minimum observed at low temperatures in the temperature dependence of the electrical resistivity of an alloy containing magnetic foreign ions in dilute concentration. A breakthrough in the explanation of this effect occurred when J. Kondo took the trouble to carry to second order a calculation that everyone previously thought was of no interest beyond the first-order approximation. As is explained in the text, the result was a logarithmic contribution to the resistivity that arises from the elastic spin-flip scattering of the itinerant electrons by those resident in the foreign ion and that contributes to its magnetism (Fig. 4.10). Experimentally, however, this logarithmic behavior is followed only above a certain characteristic temperature T_K; below this temperature the alloy's electrical resistivity is relatively independent of temperature. It is now known that this is because well below T_K the ion ceases to be magnetic. T_K varies enormously from alloy to alloy: Typical values for Co, Fe, and Mn ions dissolved in Cu are 700, 30, and 0.01°K, respectively.

The comparable behavior for an alloy containing nonmagnetic foreign ions follows the ideal one illustrated except at the lowest temperatures. The observed resistivity of course never reaches zero but becomes temperature independent in the so-called residual resistance range.

decreased only down to a certain characteristic temperature T_K (now known as the Kondo temperature), below which the resistivity becomes approximately constant (Fig. 4.11). This suggests that something important is happening to the foreign ion's magnetic moment below this temperature: Could this be associated with a breakdown of Kondo's theory and the emergence of some effect that is not covered by it?

In the decade or so since Kondo's work appeared, some theoreticians in solid state physics have spent a great deal of time upon this problem and its associated case* of what happens at $T = 0$. The results are generally beyond our introductory purpose (for the whole affair concerns some rather academic points) and it is sufficient to note here that the salient features of Kondo's suggestion have survived and are regarded as valid for temperatures above T_K. They have even been strengthened through the appearance of better quality data. For example, the effect of the logarithmic term should be pronounced not only in the resistivity but in other physical properties, such as the thermoelectric power, the electronic specific heat, and in the ion's magnetic properties. There has been experimental confirmation of its behavior in all of these properties.

The existence of the characteristic temperature T_K indicates that the conditions are entering a range where Kondo's theory is inadequate and is beginning to break down because of the divergence of the logarithmic term, but is there any physical meaning to be attached to what is happening below T_K? Here the discussion enters a controversial area, but some theoretical work† suggests the answer to the above question is that the logarithmic divergence is associated with the formation of a spin-compensated state (Nagaoka's bound state) about the ion. This is thought to arise in the way described in Sect. 3.3.3 and shown in Fig. 3.9: The s-d (or s-f) mixing gives an antiparallel coupling between the spins of the itinerant electron and any resident in the virtually bound state. As the temperature is lowered towards T_K, the coupling due to this mixing becomes stronger. (This is another consequence of Kondo's logarithmic term; more and more electrons can be imagined to hop per unit time into and out of the virtual state.) As the temperature is lowered beyond T_K, the itinerant electrons in the neighborhood of the virtual state become increasingly spin polarized with their spins aligned predominantly in opposition to the net spin on the ion. In a very crude picture, the foreign ion will become progressively surrounded by an extended cloud of antiparallel spin-polarized electrons as the temperature is reduced. The usual oscillatory RKKY spin density around the ion (Sect. 3.3.3) is thus replaced by a long-range and antiparallel spin polarization that compensates the ion's net magnetic moment and thus reduces its observable magnitude; it is believed that eventually this moment can be completely compensated leaving an effectively non-

---

* It is interesting that the philosophy behind these particular mathematical exercises has met outspoken criticism from within the theorists' ranks; this is rather rare. One author wrote recently that such Kondo studies enrich "neither the practitioners nor the scientific community as a whole," and others have pointed out to the theoretician's "less and less fruitful staggering in the jungle of traditional Kondoism."

† Notably that by Y. Nagaoka, *Phys. Rev.* **138**, A1112, 1965; *Prog. Theoret. Phys.* (*Kyoto*), **37**, 13, 1967, whose name is often attached to the effect.

magnetic ion. This is the Nagaoka bound state; it is a direct consequence of the many-body nature of this problem, for the spin polarization of the itinerant electron cloud is maintained by the ion's memory of spin scattering events, just as described in Sect. 4.3.3. The spin-flip scattering of an electron depends upon the nature of the preceding event at the ion; the Nagaoka spin polarization is thus maintained through the interchange of virtual spin-flip particles between the itinerant electrons around the impurity (Fig. 4.8f).

The idea of such a bound state gained great support when it was demonstrated experimentally that at very low temperatures the moment on certain traditionally magnetic ions (such as Fe dissolved in Au or Cu and Mn in Cu) does seem to disappear. However, Nagaoka's view is not the only possible explanation. In Sect. 3.4.4 and Sect. 3.4.5 we described how the localized spin fluctuations caused in the ion's moment by the s-d mixing modify, and can make effectively nonexistent, the ion's net magnetic moment. In other words, instead of regarding the disappearance of magnetism well below T_K as due to a cloud of compensating electron spins surrounding a well-defined ionic spin, we should think of it as a fluctuating moment on the ion whose average over a time of the order $\hbar/kT$ tends to zero.

Which of these two pictures is more correct—or even if there is any real basic difference between them—is still not yet clear. Furthermore, in alloys of the class under consideration, in which the ions of interest are the dissolved constituents in a simple nonmagnetic host, it is difficult to distinguish the apparent loss of the ions' magnetism due to these temporal fluctuations in the magnitude of their effective moments from those spatial fluctuations produced in their moments' orientations by the mutual interionic exchange interactions inevitably present in the alloy. There are two types of such interaction that the ions may experience. First, is the indirect and relatively long-range RKKY interaction that, as we have seen (Sect. 3.3.3 and Sect. 3.4.3) varies in magnitude and sign as the distance between the ions is changed and, as a result, can couple the moments of pairs of ions into either the ferromagnetic or antiferromagnetic configuration depending upon the ions' separation. Second, is the relatively short-range, direct exchange coupling (Type 2 of Fig. 3.9) experienced between ions that are close enough to have appreciable overlap of their incompletely filled orbitals (the d orbitals in the case of a transition metal ion, for example). Although in a solid this second coupling can also favor either ferromagnetic or antiferromagnetic ordering, the preference depends upon the balance of competing forces operating within the ion; it is, therefore, a property of the ion's species, being fixed for any pair of ions in the alloy and independent of their separation. (For example, Mn ions in different alloys generally show a marked preference for antiferromagnetic coupling produced by the direct exchange, whereas Fe ions rather favor the ferromagnetic one.)

Which of these two interionic interactions is dominant in any given context depends upon the mean separation of the ions and, hence, upon the atomic concentration of the solute in the alloy. In extremely dilute alloys (where the solute's concentration is at most only a few ten or hundred parts per million) the single ion Kondo and Nagaoka effects discussed above are believed to be dominant: Neither the RKKY interaction, which we recall decreases in magnitude as roughly the cube of the distance from the ion, nor the direct exchange from d-d overlap, which can only operate between essentially contiguous ions, is a significant influence. However, as the solute concentration is increased (typically into the range of a few atomic per cent), the RKKY interaction is expected to become the dominant one. A few solute ions in the alloy by chance find themselves with near neighbors of the same species* and, therefore, couple their moments through direct exchange to produce localized clusters of ions having their moments aligned according to the preference of this direct coupling, but throughout the crystal the overall alignment is dominated by the RKKY mechanism. The net effect at any given solute site is the sum of the influences there of all the remaining solute atoms, which in this idealized case are assumed to be distributed at random throughout the lattice. Because of the oscillatory nature of the RKKY interaction, this resultant influence will vary randomly in magnitude and direction from one solute site to another. Consequently, as the alloy is cooled down (assuming no externally applied magnetic fields are present) so that the disruptive thermal motions of the ions' magnetic moments are diminished sufficiently, the moment of each solute ion takes up the position of minimum energy dictated by the resultant RKKY coupling at its site. In other words, the alloy consists at

---

* In an alloy with an f.c.c. structure and a solute concentration of 1 atomic per cent, nearly 12% of the solute ions are grouped in near-neighbor clusters. To see this more generally, suppose the crystal's structure has a coordination number C (this is the number of close neighbors for any given ion in the structure). If the alloy is assumed to be ideal in that the foreign ions it contains are substituted at random upon the lattice sites, then a solute concentration of say x atomic per cent means that any lattice site has an $x\%$ chance of being occupied by a solute ion. Since around any given solute ion there are C close neighbor sites, each with very nearly an $x\%$ chance of being occupied by a solute ion, the total fraction of solute ions which find themselves in near-neighbor conglomerations (that is, pairs, triplets, and larger groups) is very nearly $xC\%$. (C for the f.c.c., b.c.c., and c.p.h. structures is respectively 12, 8, and 12.)

low enough temperatures of a complex magnetic structure having the solute ions' moments frozen in the lattice with almost random orientations. There is no long-range ordering of the moments, and over a macroscopic region of the crystal there is no net magnetization. An alloy in this condition is said* to be in the spin glass regime, because of the analogy with a random collection of spins in a glasslike structure; it is the effect of these fixed but random spatial fluctuations of the solute ions' spins that are so difficult to separate experimentally from the temporal fluctuations in the magnitude of an ion's freely oriented moment described earlier in this section.

Further increase in the solute's concentration eventually increases the importance in the alloy of the direct d-d exchange coupling between the solute ions' moments until ultimately it eclipses the RKKY mechanism. A first effect may be to increase the size, and thus the importance, of the direct-exchange coupled clusters of solute ions, so that the alloy is more properly imagined as a conglomeration of clusters each with a net magnetic moment that is randomly oriented with respect to those of any nearby clusters (in which case the alloy is sometimes called a mictomagnet). However, inevitably as the solute concentration is increased a point is reached (known as the percolation limit, and typically of the order of a few tens of atomic per cent) at which the probability of an infinite chain of direct-exchange-coupled solute ions is finite for all directions in the crystal. For concentrations above this limit, ordinary ferromagnetism or antiferromagnetism can be established throughout the crystal by the now dominant direct-exchange coupling between the solute ions' spins.

The disentanglement of these various possibilities from the observed magnetic and electrical properties of this class of alloys remains an active field of research.

4.5   INTERACTIONS BETWEEN ITINERANT ELECTRONS: MASS ENHANCEMENT EFFECTS

In certain circumstances, which are discussed in Sect. 4.4, a foreign ion's memory of spin orientation can provide an indirect link between itinerant electrons that are scattered by the ion. Through this link the nature of a second electron's scattering depends in detail upon what happened at a first electron's scattering; in this case the scattering of a given itinerant electron cannot be examined independently of the others in the fluid. The ion's

* B. R. Coles in *Amorphous Magnetism*, Ed. H. Hooper and M. de Graaf, Plenum, New York, 1973, p. 169.

memory thus establishes the many-body nature of the problem, and we show in Sect. 4.3.2 and Sect. 4.3.3 how all of this can be visualized in terms of a virtual particle acting to couple the first and second electrons scattered by the ion.

We pointed out that this spin-dependent electron scattering is not the only such manifestation of many-body effects among itinerant electrons in a metal; others are shown in the lower half of Fig. 4.8, and in each of these cases the consequences of the many-body interaction can be expressed in terms of a virtual particle. Furthermore, as is explained in Sect. 4.2, a penalty for trying to retain the concept of an individual constituent in such a many-body system is the sacrifice of the idea that it has a fixed inertial mass. In each of the cases shown in the lower half of Fig. 4.8 there is, therefore, an effective change in the electron's mass as a result of the many-body aspect of the interaction. This change, which is a reminder that the electrons are in fact quasiparticles, is known as the mass enhancement (or mass renormalization) effect* and, together with the corresponding inter-actions of Fig. 4.8, it forms the subject of this section.

4.5.1 Electron-Phonon Mass Enhancement. First to be considered is the electron-phonon interaction illustrated in a Feynman diagram in Fig. 4.8e. This produces an attractive force between itinerant electrons in a metal that meet in opportune circumstances. It is most pronounced between those having opposite spins and roughly equal kinetic energies and arises because an electron moving through the lattice of positive ions produces about itself a local charge polarization of the lattice. The physical situation is represented schematically in Fig. 4.12a. As an excited quasi-electron at X moves through the ionic lattice, it repels slightly the electronic charge localized about the ions that fall within a range of about 10^{-10} m (see Sect. 4.2.4). Some net positive charge on the ion's nucleus is effectively uncovered in this process so that the ions surrounding the electron in question are weakly attracted to it and are slightly displaced as it passes through their environment. Such a disturbance of any part of the ionic lattice does not remain localized indefinitely, however, for it will be propagated throughout the ionic system as the sympathetic oscillatory motion of contiguous ions and is, of course, eventually dissipated into the ions' random thermal oscillations. Motion on such a microscopic scale has,

---

* It is evident for electrons in atoms as well as those in metals. For example, in the hydrogen or ionized helium atoms the effective mass of the electron is seen to be slightly different when it is moved closer to the nucleus where it is subjected to stronger electrostatic fields. (This leads to the Lamb–Retherford shift in the spectrum of the atomic gas.) In this case the mass renormalization can be thought of as due to a change in the cloud of virtual photons that surrounds the electron.

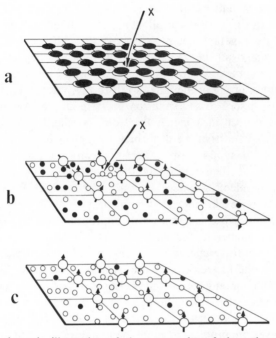

Fig. 4.12 A schematic illustration of three examples of the enhancement of an itinerant electron's effective mass. (*a*) The electron–phonon interaction that displaces the ions in a solid as the itinerant electron X passes between them. This exposes a net positive charge in the electron's vicinity which has effectively to be dragged around with it; so the electron's mass is enhanced. The mechanism leads to a net attraction between pairs of itinerant electrons that meet in opportune circumstances and is the basis of superconductivity (Sect. 4.6). (*b*) The *s* electron–paramagnon interaction thought to arise in incipient ferromagnets. This mechanism envisages a net magnetic moment to be localized upon each ion—a perhaps somewhat unrealistic view in many cases, as the text explains—so that a paramagnon is here the temporary alignment of a number of these ionic spins. The itinerant electrons within the paramagnon's environment are spin-polarized (as the preponderance of open circles indicates) and this polarization has again to be dragged around by some electron X moving in the paramagnon; hence its mass enhancement. (*c*) The electron–magnon interaction that is thought to arise in ferromagnets, such as Gd, for which it is certainly realistic to regard the ions as possessing localized magnetic moments. The itinerant electrons' spins are again polarized through their interactions with the ionic moments (being aligned parallel to them in Gd) and it is these electrons, acting as intermediaries in the indirect exchange interaction (Fig. 3.9) that stabilize the ferromagnetic alignment of the ionic moments. A temporary deviation from ferromagnetic alignment by an ionic spin (as illustrated) is equivalent to the introduction of a localized region of different spin polarization; it constitutes the propagation of a magnon. Mass enhancement arises because a particular itinerant electron will effectively have to drag with it this region of local spin polarization. (Note that in *b* and *c* the itinerant electrons of the same spin alternative are represented by the same symbol, an open or closed circle.)

however, to be quantized and, as is described in Sect. 4.2.5, the basic quantum unit in this case is the phonon. Thus the distortion of the lattice produced by the itinerant electron's passage through it can be described in quantum language as the radiation of phonons throughout the system. Similarly, the interaction between the electron and an ion, which is basically just a slight lifting of the latter's veil to reveal a portion of its nuclear charge, can be described as a collision between an electron and a phonon.

A moving electron in a metal, therefore, constantly creates streams of phonons that radiate from the disturbance it makes in the lattice. Returning to Fig. 4.8e, suppose the itinerant electron is incident in a state k_1, σ_1); Event A then represents the interaction (or electron-phonon collision) as the electron passes an ion. Suppose that a second itinerant electron in the state (k_2, σ_2) arrives in the vicinity in opportune circumstances (which require, in addition to appropriate timing, opposite spins and roughly equal energies for the electrons). In such a case, it is attracted to the effective positive charge about the ion, which is the legacy of event A, and it feels the first electron's presence if it too undergoes a collision (say B) with a phonon generated at A. In other words, the second electron feels the presence of the first through the memory of event A retained in the dynamics of the lattice up to event B. It is this intermediate role of the lattice that gives the problem its many-body nature. We can note in passing that the exchange of this process is also possible (Sect. 4.4). A negative phonon is imagined to be emitted by the electron in a state k_2 before being absorbed by that in the state k_1.

Consequently, there is between suitable pairs of quasielectrons in a metal an effective and attractive force that arises because of the polarization of the ionic lattice produced by their presence. Of course, this is only one component of the total force acting between a pair of such electrons. There is also the normal Coulomb repulsion, which is very small because outside a range of about 10^{-10} m the quasielectron is effectively uncharged (Sect. 4.2.4), and the direct exchange force of Sect. 3.3.1. The latter tends to keep electrons of like spin further apart than would otherwise be the case; thus it favors attraction between pairs of electrons with antiparallel spins— as already mentioned. In appropriate circumstances, when the electron-ion interaction is strong enough, the net force between the itinerant quasielectrons can be attractive. This is why metals with rather mediocre electrical conductivities can make outstandingly good superconductors; for example, a compound of Nb, A1, and Ge remains superconducting at temperatures ranging up to $20.4°K$.

In the spirit of Fig. 4.8e and the simplifications of many-body theory, this attraction between electrons produced by the lattice polarization is described as resulting from the exchange of virtual phonons between events A and B. The attractive force has a short range (which is known as the coherence length) and it is easily suppressed by other influences such as described below, but evidently appropriate pairs of electrons can take advantage of it and, as a result, reduce their energy. Such are known as Cooper pairs, and this whole mechanism is the origin of superconductivity that is described in Sect. 4.6. However, even in metals that do not show superconductivity this electron-phonon interaction plays an important role in determining the response of the electron fluid, particularly at low temperatures where it is manifested as an enhancement of the electron's thermal effective mass. (The distinctions between the different effective masses are outlined in Sect. 6.2.)

Physically, we can see how this comes about from the process represented in Fig. 4.12a. As the itinerant electron moves through the ionic lattice a region of positive polarization moves with it. This corresponds to a drag upon the electron's motion, and it is reflected as an effective increase in the electron's inertial mass. Although this is difficult to demonstrate by direct experiment, it is presumed to account for the fact that the electronic specific heat of several metals is greater than expected from simple theory. In the latter the specific heat of the electron fluid is directly proportional to the electron density of states* at the Fermi energy $n(E_F)$; this can be calculated and compared with experiment. Frequently the result indicates an effective mass for the electrons at the Fermi energy that is several percent greater than expected from simple theory (in Cu and Ag, for example, the difference is about 15%) and this is usually attributed to electron-phonon mass enhancement effect described above.†

4.5.2 Electron-Electron Mass Enhancement: Exchange and Correlation Holes.

The above discrepancy can, however, also arise partly from electron mass enhancement by direct electron-electron interactions, and it is

* Quantitatively this is defined by $n(E) = n'(E)/\Delta E$ where $n'(E)$ is the number of delocalized electrons of both spins per unit volume having energies in the range ΔE about E. Thus $n(E)$ is the total number of delocalized electrons per unit volume of the metal per unit energy interval.

† Customarily such mass enhancement effects are included by writing $m(E_f) = (1 + \lambda_C + \lambda_{ph} + \ldots) m_{st} (E_f)$, where m_{st} is the mass of an electron near E_f obtained from simple theory, and λ_C, λ_{ph}, and so forth are the mass enhancement factors (here assumed additive) arising from the different many-body effects. Thus λ_{ph} is the electron-phonon factor arising from the effect just described, and λ_C is that from the electron–electron Coulomb interaction considered below. For a simple metal λ_{ph} is typically between 1 and 2; for a transition metal it is roughly about half that value. Generally λ_C is thought to be an order of magnitude smaller than λ_{ph} for almost all metals and is usually neglected in comparison with it.

convenient to introduce that point here. The interaction, which stems from the exchange and charge correlation effects upon the electron's motion, is represented schematically in Fig. 4.8a.

It is shown pictorially in Fig. 4.13. First, Fig. 4.13a shows a hypothetical system in which the Coulomb interaction between the electrons is imagined to be turned off. As is described in Sect. 3.3.1, the exclusion principle ensures that electrons of a given spin alternative (here represented by a given symbol) stay out of each other's way. Their separation is about equal to their de Broglie wavelength, which in metals is typically about 10^{-10} m and is represented as the radius of the dotted circle. (In the drawing, no electron approaches within this distance a neighbor of like spin.) Consequently, as far as electrons of like spin are concerned, each is surrounded by a void known as the exchange (or Fermi) hole. From this

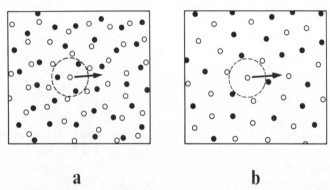

a   b

Fig. 4.13 Schematic illustration of the origin of exchange and correlation holes. (a) The hypothetical case of a noninteracting gas of electrons, like a system in which the Coulomb force between electrons is ignored so that only the exchange force is operative between them. The exclusion principle ensures that electrons of the same spin alternative (here represented by the same symbol) stay apart by a distance of approximately their de Broglie wavelength. In a metal this is typically about 10^{-10} m and is represented here by the dotted circle. In the drawing, no electron comes this close to a neighbor of like spin. Consequently, as far as electrons of like spin are concerned, each is surrounded by a void known as an exchange hole. If an electron moves, as represented by the arrow, the void accompanies it and electrons of like spin must be displaced to let it pass. (b) The more realistic case of an interacting gas of electrons in which both Coulomb and exchange forces are operating. In addition to the exchange hole, there is now an additional void centered upon each electron that is created by the Coulombic correlations tending to keep electrons of whatever spin as far apart as possible. This is known as the (Coulombic) correlation hole; in typical metals its diameter is about half that of the exchange hole, but here we take the two to be equal. The total volume enclosed by the dotted circle in b is, therefore, the exchange and correlation hole. In this case, as an electron moves, all the others in the path of this void must be displaced to let it pass.

hole a net charge equivalent to half the electronic charge e is excluded. If an electron moves as suggested by the arrow, the exchange hole accompanies the electron wherever it goes, and electrons of like spin have to be displaced to let it pass.

Figure 4.13b shows the more realistic case where both the Coulomb and exchange forces are operating. In addition to the exchange hole, there is now centered upon each electron a void created by the Coulomb repulsion. This repulsion correlates the electrons' motion so that whatever their spin they try to keep as far apart as possible. The resulting additional void is called the Coulomb correlation hole (or simply correlation hole). In a typical metal its radius is about half that of the exchange hole, but for convenience in the diagram they are taken to be the same so that in Fig. 4.13b no electron approaches closer to a neighbor than the radius of the dotted circle. The total volume enclosed by this circle is known as the exchange and correlation hole, and it turns out to contain a net deficit equivalent to one electronic charge. A little care is needed not to confuse either the exchange or correlation holes with that part of an elementary excitation also known as a hole and introduced in Sect. 4.2.3. In fact, a moving itinerant electron together with the exchange and correlation hole constitute the Landau quasiparticle of Sect. 4.2.4.

Once again, when the electron moves this hole accompanies it everywhere, and in this case the electrons in its path have to be displaced to let it pass. Thus the hole has to be effectively dragged around with the electron against the resistance of surrounding members of the fluid that may be displaced by it. (An electron X in Fig. 4.14a has to drag the area enclosed by the dotted circle in this two-dimensional example.) Thus the electron's inertial mass is enhanced. Calculation suggests, however, that this electron-electron mass enhancement effect is small compared with the electron-phonon interaction discussed above and is probably negligible in most cases.*

4.5.3 Spin-Flip Mass Enhancement by a Localized Ionic Moment.

The next interaction to be considered is illustrated in Fig. 4.8f. It arises between two itinerant electrons that are successively scattered by a given isolated ion that has a localized magnetic moment due to unpaired electron spins in an incomplete d or f shell. The case was already considered in some detail in Sect. 3.4.4 and Sect. 3.4.5. Known generally as the s-d (or s-f) interaction, it leads to the Kondo effect discussed in Sect. 4.4. The electrons are effectively coupled by the ion's spin memory, which is ex-

---

* N. W. Ashcroft and J. W. Wilkins, *Phys. Lett.* **14**, 285, 1965 and T. M. Rice, *Ann. Physics,* **31**, 100, 1965 reach this conclusion.

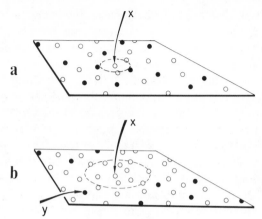

Fig. 4.14 The effective mass of an itinerant electron in a metal deduced from measurements of such as the electronic specific heat or the thermal conductivity is frequently greater than expected from the simple view that treats the electrons as independent particles. This mass enhancement effect is a consequence of the many-body nature of the problem, and its origins can be roughly divided into two categories: those resulting from direct electron–electron correlations and those from indirect correlations via an ionic intermediary. An example of each is shown very schematically here. (*a*) An illustration of how any itinerant electron is surrounded by an exchange and correlation hole (Fig. 4.13) that has effectively to be dragged around with it, thereby increasing the electron's inertial mass through the correlation effect. (*b*) The special case of the itinerant *d* electrons in an incipient ferromagnet such as Pd. A paramagnon, which is here the temporary alignment of the electrons' spins by direct exchange effects, is shown established within the dotted limits. An electron *X* within this paramagnon has effectively to drag this region of spin polarization about with it and so suffers a mass enhancement through the paramagnon's effect. An electron *Y* of opposite spin is also affected by this through the Berk–Schrieffer mechanism discussed in the text.

pressed as a virtual spin-flip particle in the context of Fig. 4.7; in principle, there should be a renormalization of the electrons' masses due to this process. Although it has been referred to by Kondo* in connection with *s-d* scattering, it does not seem to have been demonstrated experimentally that it is an important effect.

4.5.4   Electron-Magnon Mass Enhancement.   This results from an effective coupling which exists between the itinerant electrons moving in a ferromagnetic metal, particularly a rare earth metal, where for most purposes the electrons giving rise to the ion's spin (in this case those in a

---

* J. Kondo, *Prog. Theoret. Phys.* (*Kyoto*) **32**, 37, 1964.

partially filled f shell) can be regarded as localized on the ion and not itinerant in the solid to any significant degree. In such a system the state of minimum energy has the ionic spins in parallel alignment, but if one of these spins is flipped, so that its component along any given direction is modified, this spin modification—when set against a background where all the other ionic spins are aligned—is equivalent to a local and temporary region of spin polarization. Just as in the case of charge polarization encountered above in the electron–phonon interaction, the spin polarization does not stay localized indefinitely. It is propagated throughout the spin system as the sympathetic motion of the ionic spins about their equilibrium positions and is known as a spin wave. Again, such motion is quantized, and the quantum unit in this case is called the magnon and was first encountered in Sect. 4.2.5.

A passing itinerant electron in a state k_1, σ_1, for example, can interact with an ion's spin giving a spin-flip event A (Fig. 4.8g.). We can say that this gives rise to the propagation of magnons throughout the spin system, just as a temporary charge polarization gives rise to a propagation of phonons. The spin of a second itinerant electron in the state k_3, σ_1, for example, can be affected by this polarization if it arrives in the vicinity in opportune circumstances. If it undergoes a spin-flip event at B with a magnon generated at A, it thereby feels the presence of the first electron. Thus the spin system imparts a many-body aspect to the itinerant electron system. Again, rather than consider in detail the intermediate role of the spin system, it is convenient to look upon the spins of the two itinerant electrons as being coupled through the exchange of a virtual magnon so as to polarize the spins of the itinerant electrons localized in the same region of a metal and under the influence of a given ion's spin. It is entirely analogous to the electron–electron interaction through virtual phonons, which is illustrated in Fig. 4.8e, except that here we are dealing with a region of spin polarization produced in the system of ionic spins by an electron's passage as opposed to the charge polarization produced in the lattice in the phonon case.

The corresponding physical situation is depicted in Fig. 4.12c, which could represent a rare earth metal such as Gd. The ions possess local moments due to unfilled f shells, which are located quite deep in the ions' cores; the itinerant electrons are polarized by their direct exchange interaction with these spins. (In Gd the magnetic moment per ion in the solid is greater than that of the free ion; therefore, it is assumed that the electrons' spins are here aligned parallel to the corresponding ionic spin.) This polarization leads to an effective coupling between the ions' spins through the indirect exchange mechanism in which the itinerant electrons are the intermediaries. The overall alignment is, therefore, a ferromagnetic one, as depicted in the figure, although excited states can exist as magnons: The component of an ion's spin along a particular direction may temporarily be modified through a spin-flip collision with an itinerant electron so that a region of local spin polarization is created. The itinerant electron's mass is enhanced in this context exactly as in the analogous case of the electron-phonon interaction, except that here the electron has effectively to drag around with it a region of spin polarization. It has been estimated[*] that in the ferromagnetic transition metals the effect of this enhancement due to the electron-magnon interaction is of the same order as that arising from the electron-phonon type discussed above and is, therefore, quite a significant effect.

4.5.5 Incipient Ferromagnets and Electron-Paramagnon Mass Enhancement.

Finally, Fig. 4.8h illustrates an interaction that closely resembles that described above but that occurs in a class known as incipient ferromagnets or exchange enhanced metals. These are the metals (or alloys based upon them) that occupy positions toward the ends of the $n = 4$ and 5 series of the d block in Fig. 2.1, Pd and Pt being the prototypes. In these metals the exchange interaction between spins arising from unfilled d shells is not quite sufficient to produce their ferromagnetic alignment throughout the whole solid at absolute zero, but it is large enough to produce at low temperatures a temporary ferromagnetic alignment of the spins over localized regions of the metal. Such alignment is, of course, offset by the thermal fluctuations of the spins, but a tiny region can be imagined to exist in the metal containing perhaps a few hundred d spins for which temporarily there is a predominant and ferromagnetic alignment. Such a region is known as a paramagnon or enhanced spin fluctuation, and as the temperature is lowered we can imagine such localized spin fluctuations becoming increasingly persistent in space and time; in other words, the range and lifetime of the paramagnon increases correspondingly.

The physical origin of this enhancement can be seen by considering a single, narrow energy band of itinerant d electrons[†] in the metal. Suppose

[*] H. S. D. Cole and R. E. Turner, *J. Phys. C.* **2**, 124, 1969.

[†] It is important to note that in metals of the type under consideration the d electrons cannot for all purposes be regarded as localized upon their ions; there is sufficient overlap between the d shells of contiguous ions so that their electrons are effectively itinerant. Since the d shells are almost full, this itinerancy is often more appropriately described as the motion of the corresponding holes in the d shells rather than of the electrons themselves. (Just as a rising air bubble in water can be regarded as the downward motion of an equivalent volume of water.) Thus it is that we often see the d holes spoken of as the itinerant magnetic entities.

that in a small region of the metal a predominance of spin-up itinerant electrons is created. This could occur fortuitously during their random spin fluctuations, or it could be the result of a magnetic field acting in the region. As was described in Sect. 3.3, the close approach of any two electrons with parallel spins is inhibited by the exclusion principle; no restriction other than Coulomb repulsion governs their approach when the spins are antiparallel. Consequently, a spin-down electron interacts predominantly with its spin-up counterparts. Since we are considering a region of the metal in which the spin-up electron density temporarily exceeds that of the spin-down, the electrostatic contribution to the total energy of a spin-down electron is greater than for any spin-up counterpart—simply because in a given time interval the spin-down one interacts with (that is, is repelled by) more electrons. As a result, the energy difference between the spin-up and spin-down configurations is reinforced by the exchange correlation of the electrons' motion, and it acts to lengthen the paramagnon's lifetime.

It is worth noting that if the itinerant electrons' spin density is produced by an externally applied magnetic field, the exchange effect clearly contributes to the increase of the spin-up population compared to that of the spin-down. Consequently, the magnetic susceptibility of the electron system, which is known as the spin or Pauli contribution to the metal's susceptibility, is more paramagnetic than expected from the simple free electron theory, which neglects the interactions between the electrons. This difference is known as the exchange (or Stoner*) enhancement of the Pauli spin susceptibility. For strongly paramagnetic metals like Pd it produces an order of magnitude increase of the susceptibility over that expected from the simple theory. So in addition to its effect upon the electrons' inertial properties—which is the main concern in this section—the exchange interaction also modifies the system's overall magnetic properties. Further discussion of this point appears in Sect. 4.7.

Even without an externally applied magnetic field the existence of paramagnons in metals like Pd is manifested in such parameters as the electronic specific heat or the electrical resistivity. This arises in turn from the enhancement in the electron's effective mass that is produced by the interaction with the paramagnon, and we now turn to this.

By the nature of an incipient ferromagnet, the interaction between a given itinerant d electron in state k_1, σ_1, for example, and similar electrons in its vicinity is such that the surrounding electrons have predominantly and temporarily a ferromagnetic alignment. Thus if A in Fig. 4.8h represents an event in which two d electrons in a paramagnon interact, σ_2 is the

---

* After E. C. Stoner (who first discussed it) *Rept. Prog. Phys.* **11**, 43, 1947.

spin orientation appropriate to the paramagnon in which the electron finds itself. A second d electron in the state $\mathbf{k}_3$, σ_1, for example, arriving in the vicinity in opportune circumstances can be influenced by the local polarization represented by the paramagnon (just as in the case of the electron–magnon interaction above) and, through a spin scattering event B, can feel the effects of the polarization initiated by the first passing electron. The origin of the effective interaction between these two d electrons then follows in this description from the exchange of a virtual paramagnon. The analogy with the preceding cases is evident, and there is also an enhancement of the d electron mass that arises here because each itinerant d electron has effectively to drag around with it a region of local polarization in the system of itinerant spins.

This is illustrated in Fig. 4.14b, which relates to the case where the itinerant nature of the magnetic electrons can safely be emphasized, as in Pd, for example. These itinerant d electrons experience an exchange interaction that is strong enough to produce ferromagnetic alignment over a limited region (as with the open circles enclosed by the dotted limits). This is a paramagnon. Suppose the open circles represent electrons having the spin-up configuration. A given electron X has effectively to drag around with it this region of spin polarization with the result that its mass is enhanced by the paramagnon's presence. (Theory suggests* that this enhancement amounts to about a factor of 3 in the case of Pd.) In addition, suppose an electron Y of down-spin is suitably endowed to participate with X in the electron-phonon interaction described in Sect. 4.3.1 and Fig. 4.12a. Since this attractive interaction between X and Y is of short range, in order to take advantage of it X has first to pass through a region of unfavorably oriented spins. (We recall from Sect. 3.3.1 that electrons of similar spin stay out of each other's way to a greater extent than those with unlike ones. Thus an electron experiences the greatest resistance when passing through a crowd having spins antiparallel to its own.) Consequently, the momentary presence of the paramagnon is an effective barrier preventing the operation of the electron–phonon attraction, and, if the associated spin polarization has a large enough extension in space and persistence in time, it can dominate the situation and eclipse the phonon interaction completely.† Here is one reason why incipient ferromagnets such as Pd cannot be made superconducting.

Another form of interaction in incipient ferromagnets that is thought to produce an enhancement of the electron's mass is known* as s electron-paramagnon scattering and is illustrated schematically in Fig. 4.12b. Sup-

* A. I. Schindler and M. J. Rice, *Phys. Rev.* **164**, 759, 1967.
† N. F. Berk and J. R. Schrieffer, *Phys. Rev. Lett.* **17**, 433, 1966.

porters of this concept emphasize the view that the d electrons remain localized upon ions and that only the s electrons from the valence shells of the atoms are itinerant in the solid. (There is evidence to suggest that this view is not very realistic, but it seems to be a necessary assumption required to make the problem theoretically tractible.) A paramagnon in this context is the temporary alignment of the ionic spins over several lattice sites. The itinerant s electrons are spin polarized in the region of the paramagnon, as indicated in the figure by the open circles, through spin-flip scattering with the ionic spins forming the paramagnon. Again, any given itinerant electron X has effectively to drag around with it this region of spin polarization, and this is the origin of the mass enhancement under discussion. It is difficult to demonstrate experimentally that this effect is very significant; theoretically* it has been estimated to produce an enhancement of the s electron mass of about 1–3% in Pd.

4.6 SUPERCONDUCTIVITY

It seems fair to say that superconductivity is perhaps the most dramatic display of a many-body effect encountered in solids; for nearly five decades it was certainly the most basic unsolved problem in the physics of metals. It first emerged in 1911 when H. Kamerlingh Onnes discovered that upon cooling Hg to very low temperatures its electrical resistivity abruptly disappears—apparently to zero—at a temperature close to 4.2°K (Fig. 4.15). Intensive experimental and theoretical efforts on this subject followed (which fill volumes when discussed in detail) but it was not until forty-six years later that the first irrefutable theory of the origin of the effect on a microscopic scale was published.† This theory did not immediately answer all the questions raised in the preceding period, but it was a major step forward in that it established the underlying mechanism in superconductivity. Furthermore, although it was developed for the electron fluid in a metal, it has potentially a much wider application to other systems of fermions, ranging from the internal workings of the nucleus to those of neutron stars.‡

* A. I. Schindler and M. J. Rice, loc. cit.

† J. Bardeen, L. N. Cooper, and J. R. Schrieffer, *Phys. Rev.* **108**, 1175, 1957; work for which they were awarded the 1972 Nobel Prize in Physics.

‡ At the white dwarf stage of its life cycle, a shrinking star may become unstable enough to explode into a gigantic flare (or supernova). Some astronomers believe remnants of this event can continue to shrink to form an object even more dense than a white dwarf. This is the pulsar (or neutron star) and it is thought to consist entirely of neutrons, perhaps in a superfluid state. Should the pulsar stage be itself unable to stabilize, then the continuing contraction of the pulsar leads ultimately to the formation of that controversial entity, a black hole.

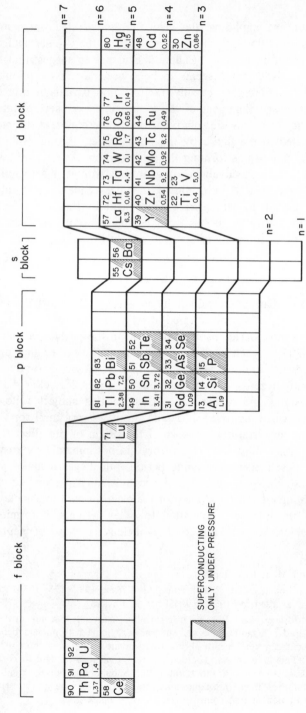

Fig. 4.15 For many years after its discovery superconductivity was regarded as a rare property possessed by only a few metallic elements. Now it is known to be one of the most common phenomena in the condensed state: It is the rule rather than the exception among the metallic elements, and it has also been found in hundreds of alloys and compounds including some very bizarre mixtures. This table, which is based upon one given by B. T. Matthias, shows the occurrence of the effect among the elements. The transition temperature in each case is shown in °K.

In the case of superconductivity, fact turned out to be stranger than the most fanciful fiction. As J. Bardeen has pointed out, if the BCS Theory (as it later became known) had been derived purely by deductive reasoning—before the observation of the effect in nature—no one would have believed that such remarkable properties predicted by it could be possessed by any real metal. But, of course, that is not how it happened. In the years between 1911 and the appearance of the BCS theory there was a constant interplay between experimental results and theoretical attempts to interpret them, and the BCS theory was a culmination of this activity. A most important development in this period came in the spring of 1950 when two groups* working independently in the United States discovered the isotope effect. Briefly, by measuring the superconductive transition temperature for two separated isotopes of Hg they showed that it varies inversely as the square root of the isotopic mass; that is, it depends upon the mass of the positive ions forming the metal's lattice. This was a very telling result because the ionic lattice is known to have characteristic vibrational frequencies that also depend on the square root of the isotopic mass, and the obvious conclusion is that these vibrations must in some way be involved in the superconducting state. The same conclusion had been reached independently at about the same time by the theoretician H. Fröhlich, but it was primarily the clue provided by the experiments that started Bardeen et al. on the course to the BCS theory. It is now well known that the lattice does play a crucial part as it provides a mechanism giving an effective attraction between suitable pairs of electrons, which is essential for the formation of the superconductive state. This is the electron-phonon interaction discussed in Sect. 4.5.1, and that section makes a suitable starting point for the following qualitative description of the superconductive effect.

In the late 1940s it became clear that superconductivity, and the related phenomenon of superfluidity† in liquid He^4, could be explained if somehow in these abnormal states of matter all or part of the itinerant particles (namely, the electrons in the metal and the He atoms in the liquefied gas) could occupy a single energy state. This state would be common to the whole system and would lie in energy somewhat below the ground state of

---

* C. A. Reynolds, B. Serin, W. H. Wright, and L. B. Nesbitt, *Phys. Rev.* **78**, 487, 1950; E. Maxwell, *Phys. Rev.* **78**, 477, 1950. T. H. Geballe has pointed out (*Sci. Amer.*, November, 1971, p. 22) that Kamerlingh Onnes and his colleagues came extremely close to discovering the effects themselves in experiments made in 1927 upon two isotopes of Pb. Their experimental accuracy was just slightly insufficient to make the effect measurable.

* A superfluid can be described as a liquid having no entropy and no viscosity. It can, therefore, slide freely over any surface, or pass freely through any normal liquid with which it may be mixed.

the normal system. Furthermore, in the case of the metallic superconductor at least, the results of certain experiments strongly suggested that a certain minimum energy was necessary before electrons could be excited out of this common state. Put another way, there was evidence of an energy gap in the system's excitation spectrum that the electrons must bridge before escaping from their common energy state.

The formation of a single-energy state became known as a condensation, in the sense that an ordering takes place among the constituents of the system that in energy space results in the compression (or condensation) of the range of their normally permitted energy states into the single state of the abnormal condition. (The latter is often known accordingly as the condensate. It is important to note the distinction between this meaning of condensation, which we emphasize is visualized as occurring among occupied states in energy, or momentum, space, and the meaning used in Chapter 1 where it describes the bringing together in normal space of atoms to form the solid or liquid state.) It was less difficult to conceive such a condensation occurring in liquid He^4 than it was among the itinerant electrons of a metal. The itinerant atoms in the former are bosons and can, therefore, all occupy a single energy level, but the itinerant electrons in a metal are fermions and are prevented by the Pauli exclusion principle from doing this (Sect. 3.2.4). The problem was how to conceptualize charged bosons arising from what is available in a metal.

This difficulty was finally resolved in the following way. H. Fröhlich[*] first pointed out that an attractive force could exist between itinerant electrons that occupy energy states close to the Fermi surface. He showed that this attraction could arise from the electron–phonon interaction, which has already been described in Sect. 4.5.1. Briefly, we recall that an itinerant electron moving through the ionic lattice becomes a quasiparticle because of its constant interaction with electrons bound in the outer shells of the ions it passes. Another itinerant electron can be attracted to the deformation of an ion produced by the passage of the first, and this is equivalent to an attraction between the two quasielectrons. The most favorable circumstances for this attraction to dominate the electrons' mutually repulsive influences occur for electrons separated by not more than about 10^{-10} m and having opposite spins and momenta. L. N. Cooper[†] then showed

---

[*] H. Fröhlich, *Phys. Rev.* **79**, 845, 1950.

[†] L. N. Cooper, *Phys. Rev.* **104**, 1189, 1956. The first suggestion that bound electron pairs could exist as bosons was actually made by a chemist, R. A. Ogg (*Phys. Rev.* **69**, 243, 1946) in quite a different context. But his arguments seemed implausible at the time and his experiments were in any case difficult to reproduce, so the idea made no appreciable impact. Very readable descriptions of this historial question can be found in Blatt's book (p. 86) and in Anderson's review, both of which are cited at the end of this chapter.

shortly afterwards that no matter how weak the net force between the quasiparticles and no matter what its origin (in the application to nuclei or neutron stars it would obviously not be the electron-phonon interaction that is involved) if it is attractive, the system will find it energetically favorable to condense into bound pairs of electrons. Such a bound pair, which in several ways behaves like an itinerant two-electron molecule in the metal, is now called a Cooper pair (Sect. 4.5.1). Furthermore, because such a pair has a total spin of zero, it is a boson and can, therefore, take part in a condensation of the type described above. Since it is also obviously a charged entity, the problem of how charged bosons can arise in a metal is solved. It was from this point that the BCS theory was developed.

The qualitative results of this theory can be summarized pictorially as in Fig. 4.16. First, as described in preceding sections, a nonsuperconducting metal at absolute zero has all its energy states in the conduction band

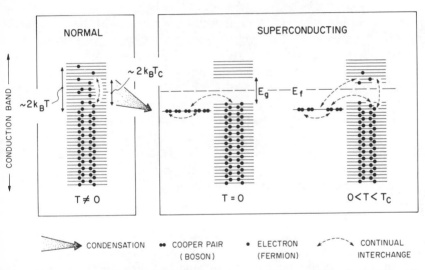

Fig. 4.16 In a superconducting metal below a certain transition temperature T_c and close to absolute zero, a fraction of about 10^{-4} of the itinerant electrons (those within about $k_B T_c$ of the Fermi energy) find it energetically favorable to correlate their motion in pairs (Cooper pairs). These can be treated in several respects as itinerant entities having twice the mass and charge of the electron, and a net spin of zero: They are thus bosons. At $T = 0$ these Cooper pairs are condensed in momentum space into a single collective energy state, and associated with this condensation is the appearance of an energy gap E_g in the quasielectrons' excitation spectrum. At a temperature above absolute zero (as on the far right) both Cooper pairs and uncorrelated electrons coexist in the metal in a state of dynamic equilibrium. The presence of the Cooper pairs provides an essentially superfluid component that supports the superconductive current under the influence of an externally applied electric field.

up to E_f filled as permitted by the Pauli principle. However, as the temperature is increased above absolute zero, various excitations are created: There are different forms of collective excitation (as described in Sect. 4.2.5) which are not of interest here, as well as increasing numbers of electrons that are excited into higher energy states to form Landau (electron/hole) quasiparticles. Consequently, at a nonzero temperature T the situation is as shown in the left-hand side of Fig. 4.16, which could apply equally to the normal (i.e., nonsuperconducting) state of a superconductor. (We recall from Sect. 3.4.4 that in usual circumstances only electrons having energies within a range of about k_BT of E_f can contribute to electron transport effects. Within this energy band-width a dynamic equilibrium exists as electrons are constantly interchanged between states, corresponding to the continual creation and decay of the quasiparticles.)

Suppose the left-hand side of Fig. 4.16 represents a superconductor in its normal state. As the temperature is reduced and approaches that (T_c) at which superconductivity is observed, it is believed that a few of the itinerant electrons that meet in opportune circumstances and that lie in energy within about k_BT_c of E_f form short-lived Cooper pairs having temperature-dependent lifetimes.* As the temperature is reduced further and T_c is passed, the material switches abruptly to the superconducting state. At the theoretical limit of absolute zero it would reach the superconducting ground state. In this condition all electrons within a range of about k_BT_c of E_f—which is roughly a fraction 10^{-4} of the itinerant electrons (that is, about 10^{25} per m³)—form Cooper pairs. In real space this means that their motion is correlated, as shown pictorially in the lower half of Fig. 4.17. The center of mass of each pair is either at rest or, if the material carries an electron current, is moving with a common velocity in the direction of this current. The corresponding situation is shown in the energy diagram for $T = 0$ in Fig. 4.16. The Cooper pairs are condensed into a single energy state that lies in energy about k_BT_c below the Fermi energy. A consequence of this condensation is the appearance of a range of forbidden energies within the conduction band and centered about the Fermi energy. This is the energy gap (E_g) referred to above. (There is a class of superconductors having no apparent energy gap and, therefore, known as the gapless superconductors, but we do not consider it here.) At absolute zero this gap separates the entirely filled electron states from the entirely

---

* In the detailed theory this effect is described in terms of fluctuations of a certain order parameter, so that the experimental manifestations of these vestiges of high temperature superconductivity have consequently become known as superconducting fluctuation effects. A typical example is the abrupt increase in electrical conductivity shown by some materials as the temperature is reduced and approaches T_c.

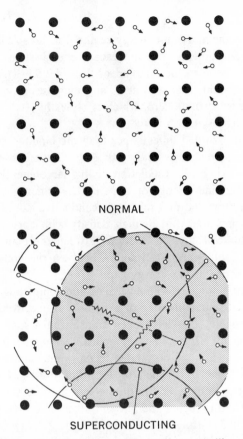

NORMAL

SUPERCONDUCTING

Fig. 4.17 (Top) A two-dimensional model of a normal crystalline metal subject to no external influences. The itinerant electrons have random motions (symbolized by the arrows) and form an electron fluid in which is immersed the lattice of relatively fixed positive ions (large black dots). If an electric field is applied to the sample, it superimposes a uniform drift motion upon the individual random motion of each electron. (*Bottom*) In the superconducting state electrons that meet in appropriate circumstances correlate their motions to form Cooper pairs (Fig. 4.16). The most opportune circumstances occur when electrons of equal and opposite momenta and opposite spins pass within a given distance of each other (the coherence length), which is typically between about 10^{-8} and 10^{-6} m. Electrons meeting in such circumstances temporarily correlate their motions in a manner that maintains an equilibrium number of coupled pairs at any given temperature below T_c. The correlation is such that all the pairs' centers of mass are at rest when the metal is free of external influences, but they move with a common drift velocity if an electric field is applied; it is rather as if the pair were temporarily connected by an elastic spring, as indicated. The figure is an oversimplification of the BCS picture in that the density of pairs is greatly underestimated. On average the volume occupied by one pair (shaded) contains the centers of about 10^6 other pairs. There is, therefore, a very strong interaction between these pairs that is crucial to the BCS theory and leads to the energy gap in their excitation spectrum (Fig. 4.16). (After U. Essmann and H. Träuble.)

empty ones. It is the minimum energy required to break up a Cooper pair and release its constituents into the vacant energy states as two normal quasielectrons (this being the bogolon of Sect. 4.2.5). This energy gap is temperature dependent; it is zero at T_c and increases monotonically to its maximum value at absolute zero, which is typically 10^{-22} to 10^{-23} J (or about 10^{-3} to 10^{-4} eV).

This is about 10^3 or 10^4 times larger than the binding energy of an isolated Cooper pair, which is typically 10^{-27} J. One of the triumphs of the BCS theory was the correct prediction of this large energy gap compared with the small binding energy of a pair. The difference between these quantities comes from the fact that superconductivity does not arise from Cooper pairs in isolation, but in high density with correspondingly large interactions. This collective interaction between the many electron pairs produces an additional strong attraction between the constituents of any pair and so gives rise to the energy gap.

At a temperature above absolute zero, a superconducting metal is, therefore, in an excited state consisting of a number of electrons—mostly from broken Cooper pairs—that have been transferred across the energy gap and into the vacant electron energy states. Simultaneously present are other electrons that are still correlated in pairs much as they were in the ground state (see the case when $0<T<T_c$ of Fig. 4.16). The excited electrons behave essentially as the quasielectrons of a normal metal; they can be easily excited to higher energies and likewise can be easily scattered by the various mechanisms referred to in Sect. 2.2. However, quite the opposite situation holds for the electrons that remain correlated in pairs: Individual pairs can only be excited if sufficient energy is available to bridge the energy gap E_g, and they can only lose energy in circumstances that are so improbable as to be negligible. It is this resistance to change shown by the correlated electrons that is the essence of superconductivity, and it can be understood as follows.

We have said that the Cooper pairs are bosons; thus they jointly occupy a single collective energy state. This requires an energy behavior that is the ultimate in conformism: At all times all Cooper pairs in the sample must have exactly the same energy. Thus, a pair may be given a certain momentum (when its center of mass is set moving in a fixed direction) by an externally applied electric field. But the requirement is that all pairs shall simultaneously have the same total momentum, so the centers of mass of all pairs must experience the same drift velocity under the given influence. The resulting charge transport is the superconducting current, and we see that this can only be established in arbitrarily small steps that simultaneously and coherently increase the momentum of all the Cooper pairs (which is, in fact, how supercurrents are started experimentally). But by the same token, a Cooper pair can only reduce its momentum—for example, by a collision with a phonon or an impurity ion—if the rest of the pairs simultaneously undergo exactly the same loss of energy. Clearly the likelihood of this happening is effectively nonexistent and, therefore, once a supercurrent is established it cannot be stopped by the scattering of the electrons forming the Cooper pairs. This can be phrased in the pictorial language of quantum mechanics: Since the Cooper pairs must move with a common velocity, they can, therefore, be represented collectively by a single de Broglie wave Ψ. And since their density is quite high, only a small drift velocity is necessary to produce a given current density. Consequently, the de Broglie wavelength is large, much larger than the dimensions of any scattering center in the metal. For example, a current density of 10^{10} A/m^2 requires a common drift velocity of about 1 m/sec for the pairs. This corresponds to a de Broglie wavelength of about 6×10^{-4} m, which puts this quantum parameter out of the microscopic range and into the macroscopic one. Since a wave is generally only scattered by obstacles having dimensions greater than or comparable with its wavelength, scattering of this de Broglie wave does not occur. It is this feature—the macroscopic extension of the condensed state over all Cooper pairs—that is responsible for superconductivity. The phenomenon does not arise because a fixed amount of energy is required to break up a pair. Assigning a macroscopic number of electrons to a single wave function Ψ in this way means that $\Psi\Psi^*$ is not the usual probability of finding an electron in a given volume (as in Sect. 3.2.3) but is actually the density of electrons to be found in the volume.

The scattering of individual electrons forming Cooper pairs occurs continually, of course, but it only serves to establish a dynamic equilibrium in the number of electrons correlated in pairs. As Fig. 4.16 shows, there is a continual interchange among members of the correlated class as well as between them and the normal quasielectrons. However, this scattering is not important because, whatever equilibrium number of electrons is established in Cooper pairs, they provide, as we have seen, a lossless current circuit. When an electric field is applied to the metal, these correlated electrons short-circuit any current transported in the normal way by the uncorrelated quasielectrons. It is as if in the superconducting state the fluid of itinerant electrons is separated into two components: One normal fluid (consisting of the quasielectrons) and the other superfluid (consisting of electrons correlated in Cooper pairs). When an electric field is applied, the superfluid component (which has effectively no viscosity) can penetrate the normal component without hindrance and so produce the superconductive current.

In addition to the thermal excitation of a constituent across the energy gap, a Cooper pair can be broken up in other ways. The phonons that effectively hold the pairs together travel with the speed of sound; thus if the correlated electrons are forced through the lattice with a drift velocity that approaches or surpasses that of sound, the attractive mechanism breaks down. Since the number of Cooper pairs in a superconductor at a given temperature is fixed, a demand for increasing current density can only be met by increasing the pairs' drift velocity. Hence, superconductivity can be destroyed if the supercurrent density exceeds a certain threshold value.

An applied magnetic field can have a similarly disruptive effect on a Cooper pair through the influence of the Lorentz force (Sect. 6.3.2) on the pair's constituents. However, besides the vanishing electrical resistance, a superconducting metal is also characterized by a tendency to resist the penetration of any applied magnetic field into its interior. The magnetic penetration depth depends upon the temperature and the material, but it is typically of the same order as the coherence length. The actual relationship between these two quantities distinguishes two types of superconductor: Type I, in which the coherence length is greater than the penetration depth, and Type II, in which the reverse is true. This distinction is more than numerological, however, for the two types show quite different responses in applied magnetic fields. However, to go into these details is to go beyond our present scope.

This subject should not be left without reference to another manifestation of the macroscopic nature of the superconducting state. This is the Josephson effect, named after its discoverer* who was at the time a twenty-two year old graduate student in the Cavendish Laboratory. It relates to the transfer of Cooper pairs that can take place in certain circumstances between separated superconductors.

Imagine two isolated pieces of a superconducting metal. From the preceding remarks we can see that the Cooper pairs in each piece occupy a condensed state that is described by a single wave function of macroscopic dimensions. As long as the two pieces remain sufficiently separated, there is no interaction or correlation between these two wave functions; electrons cannot be made to flow from one piece to the other. However, there are

---

* B. D. Josephson, *Phys. Lett.* **1**, 251, 1962; work for which Josephson later shared the 1973 Nobel Prize for Physics. Josephson's temporary professor of the time, P. W. Anderson, has described this period in a nontechnical article in *Physics Today*, November 1970, p. 23.

various ways of bridging the separation to produce a quantum mechanical coupling between the two pieces so that the possibility of quantum mechanical tunneling by electrons is no longer zero. The most direct approach, and that leading to the strongest coupling, is simply to link the two with a thin superconducting wire. Cooper pairs can then be transferred from one piece of metal to another as a superconducting current in the wire. The whole system becomes essentially a single superconductor and has thus a single macroscopic wave function to describe its condensed state. In other words, the wave functions appropriate to the two pieces of metal have been brought into perfect phase coherence by this coupling.

Josephson proposed—and later experiments verified—that a similar tunneling by Cooper pairs can be made between pieces of superconductor even when only weakly coupled quantum-mechanically. Such weak coupling can be produced either by making the superconductive bridge of very small dimensions—a thin film or a contact of very small area—or by making it a thin film of a material that is not superconducting in the bulk. Josephson's conclusion was that providing the link can transmit phase coherence between the two wave functions on either side of the separation, Cooper pairs can tunnel superconductively through it, that is, without developing any voltage in it. This is expressed in Josephson's now famous relation $J = J_c \sin \phi$, where J is the current able to pass superconductively through the link, ϕ is the phase difference between the two macroscopic wave functions on each side of the link, and J_c is a critical current above which the link's superconductivity breaks down so that a finite voltage appears across it. (What is happening at this breakdown is that sufficient energy is being given to the Cooper pairs to break them up through excitations across the energy gap of the superconductor. The resulting quasi-electrons give an Ohmic component to the tunneling current as they tunnel through the link in the normal quantum mechanical way, which is by means of incoherent quantum jumps across it.)

There are actually two classes of Josephson effect. When the super-current is less than the critical current ($J<J_c$) the d.c. Josephson effect is observed in which no voltage is developed in the link; in the other class ($J>J_c$) a finite voltage is developed in the link and leads to the a.c. Josephson effect in which the supercurrent J oscillates with time. This is an important new field that has many applications, both instrumental and fundamental, but to go into details is again to go beyond our chosen scope.

4.7 THE ENHANCEMENT OF SPIN PARAMAGNETISM

This final section of a chapter concerned with the manifestations of many-body effects in the fluid of itinerant electrons returns to a subject first discussed in Sect. 4.5.5: The exchange enhancement of the fluid's magnetic susceptibility. We described there how the combined influence of the Coulomb and exchange interactions between contiguous electrons has two effects. First, it changes the electron's inertial properties from those expected of a free electron (conventionally accounted for by allowing the quasielectron's effective mass to be variable) and, second, it makes the fluid easier to polarize magnetically than would be expected otherwise. The latter means specifically that when a static magnetic field is applied to the metal the number of electrons that align their spin parallel to the field's direction is greater than predicted by simple theory. In other words, the intensity of magnetization produced by a given applied magnetic field is greater than expected (it being more paramagnetic) and this is equivalent to an enlargement of the spin susceptibility of the electron fluid. This behavior is known as exchange or Stoner enhancement; its physical origin is explained in Sect. 4.5.5; here we are concerned with a more quantitative view leading to a criterion for ferromagnetism that will subsequently be useful.

Parenthetically, we should perhaps recall that the magnetic susceptibility of the electron fluid has two components. The first is a diamagnetic one arising from the electrons' translational motion in the magnetic field (the Landau diamagnetism) and the second is paramagnetic arising from the alignment of the electrons' intrinsic magnetic moments by the field (the Pauli spin paramagnetism). The present discussion is, of course, entirely concerned only with the second component.

Let us consider a sample of unit volume containing N itinerant electrons. Let them occupy energy levels that span a relatively narrow range, like case II of Fig. 2.7 which typifies the situation for the d electrons of a transition metal. When the electron fluid is free from applied fields and is in its lowest energy state, each of these levels is occupied by two electrons of opposite spin as dictated by the exclusion principle. In other words, the electron population is divided equally into two groups, each characterized by one of the internal alternatives available to an electron. This is expressed in $n_\uparrow = n_\downarrow = N/2$, where $n_\uparrow$ and $n_\downarrow$ are the populations of the two groups.

Suppose that a small number Δn is transferred from one of these groups to the other, leaving a net difference between the two groups of $2\Delta n$. As emphasized previously, such a switch can only be achieved at the cost of increasing the kinetic energy of the transferred electrons, and this extra energy has to be supplied by whatever influence causes the transfer. In the present case this is the externally applied static magnetic field H, but in the absence of a field it can be a thermal excitation producing a paramagnon as described in Sect. 4.5.5. This increase in the electrons' kinetic energy can be written down directly. Suppose the field is turned on and, as a result, Δn electrons reverse their magnetic moments to become parallel to the field's direction (the up direction); Δn electrons transferred through a mean energy difference of ΔE gives a total increase in energy of $\Delta n\Delta E$, which written in terms of the density of states at the Fermi level $n(E)$ $(= \Delta n/\Delta E$, Sect. 4.5.1) is just $(\Delta n)^2/n(E)$.

This increase in the electrons' kinetic energy is offset by corresponding reductions in two other components of their total energy. First, turning an electron's moment from an antiparallel to a parallel alignment with a field H yields $2\mu_B H$ in energy (Sect. 3.2.5); clearly a total of $2\Delta n\mu_B H$ of magnetic energy is involved here. Second, the imbalance in the up- and down-spin populations means that more pairs of electrons with parallel spins exist than if the populations are identical. Applying H, therefore, changes the total energy of the electrons due to their mutual interactions, since it is known from the description in Sect. 3.3.1 that for any pair this interaction energy depends upon the orientations of the electrons' spins. The repulsive energy between two electrons of antiparallel spin is U of Eq. (3.22), and the total of such interactions for two antiparallel populations of $n_\uparrow$ and $n_\downarrow$ is evidently* $Un_\uparrow n_\downarrow$. The change in this interaction energy produced by the transfer of Δn electrons is thus

$$U\left[\frac{N}{2} + \Delta n\right]\left[\frac{N}{2} - \Delta n\right] - U\left[\frac{N}{2}\right]\left[\frac{N}{2}\right]. \qquad (4.14)$$

The sum of the preceding three contributions is, therefore, the net increase in the energy of the N electrons produced by the field H and the consequent transfer of Δn electrons. Suppose the field-free value of this total energy is E_o, its value with H applied is given by

$$E = E_o + (\Delta n)^2\left[\frac{1}{n(E)} - U\right] - 2\Delta n\mu_B H \qquad (4.15)$$

The stable solution is obtained by taking the limiting value of E with respect to Δn; and when $\delta E/\delta n = 0$ Eq. (4.15) gives

$$\Delta n = \frac{\mu_B n(E)H}{1 - Un(E)} \qquad (4.16)$$

---

* An electron from the first population can interact with all members of the second, giving a term $Un_1 n_\downarrow$ to the total interaction energy. A second electron from the first population gives likewise a term $Un_2 n_\downarrow$, a third gives $Un_3 n_\downarrow$, and so forth. The sum of all such terms is the total interaction energy and is $\sum Un_n n_\downarrow \equiv Un_\uparrow n_\downarrow$.

The intensity of magnetization of the electron fluid (M) is clearly just the number of unpaired spins multiplied by the electron's magnetic moment ($M = 2\Delta n\mu_B$) and it follows from Eq. (4.16) that the spin susceptibility χ is given by

$$\chi \equiv \frac{M}{H} = \frac{2\mu_B{}^2 n(E)}{1 - Un(E)} \qquad (4.17)$$

It is a standard result of the intermediate theory that, when the mutual interactions between electrons are ignored, the Pauli susceptibility in the corresponding situation is given by $2\mu_B{}^2 n(E)$. The effect of the interactions is, therefore, to modify the spin susceptibility by the factor

$$S \equiv \frac{1}{1 - Un(E)}, \qquad (4.18)$$

which is known as the Stoner enhancement factor. In metals like Cu, Ag, Au, and A1 it is small and generally negligible, but it becomes very important towards the right-hand end of the $4d$ and $5d$ transition series (in Pd, for example, $S \approx 10$) where the elements are known consequently as the strongly enhanced metals. It is in such that the particular collective excitation known as a paramagnon is easily created, as described in Sect. 4.5.5.

There are actually two kinds of enhancement that are distinguished: Host enhancement, which is the case described above where itinerant electrons occupying a narrow energy band in a pure metal are close to a ferromagnetic instability, and local enhancement where a foreign ion incorporated into a strongly enhanced metal is able to spin-polarize a relatively large region of the surrounding electron fluid. If a single foreign ion is incorporated into a host metal, the interaction U_I between the electrons in its neighborhood will, in general, be different from the value U effective in the host. If $U_I > U$, it is expected from the preceding description that a uniform external magnetic field will induce in the region around the ion a spin polarization that is bigger than in the host alone. Theory shows that, if the host is paramagnetic and if $U_I - U$ exceeds some critical value, the paramagnetic state becomes unstable so that a local moment forms on the foreign ion. (Note that if $U_I < U$ the opposite effect can occur: The host's enhancement is quenched in the neighborhood of the ion and the paramagnon excitations of the host are suppressed there.) The total moment associated with a foreign ion in these circumstances is consequently much bigger than expected from the electronic arrangement in its outer shells. This is known as a giant moment. Iron or Co atoms dissolved in Pd are good examples of its occurrence: Their effective moment in the alloy is about $11\mu_B$, but only about $3\mu_B$ of this comes from unpaired spins in their outer shells. The rest is made up* of ferromagnetically aligned moments (of about $0.05\mu_B$) that are induced on about 200 Pd ions within the foreigner's vicinity. These induced moments are in turn the manifestation of the spin polarization in the electron fluid produced by the local enhancement.

The enhancement of spin paramagnetism is one manifestation of how the fluid of itinerant electrons reacts to an external influence—the field H in this case—so as to lower its total energy. There is another reaction that occurs in a very similar context and that it is convenient to introduce here. Specifically, the description again relates to itinerant electrons that occupy a relatively narrow energy band (as the d electrons of the transition metals discussed in Sect. 6.4.2), but there is no externally applied field in this instance; the issue is how can the electron fluid organize itself so as to lower its total energy resulting from the mutual Coulomb and exchange interactions between its constituents. In essence the problem is very similar to that described above. The basic fact is that the exchange interactions between the electrons arising from their Coulomb repulsions always favor the parallel alignment of pairs of electrons' spins (Sect. 3.3.3 and Fig. 3.9) since the partners stay out of each other's way to greater extent than in antiparallel pairs (Sect. 4.5.2). Thus from the point of view of the Coulomb interaction above, a partial or complete ferromagnetic polarization of the electron fluid is energetically the most preferential condition. However, as in the case above, the cost of such a polarization (which arises basically from the fact that ferromagnetically polarized electrons are constrained to move in a smaller volume) is an increase in the kinetic energy of the electrons. (Compare the origin of the energy hierarchy of states in a conduction band discussed in Fig. 2.11.)

It has been said elsewhere that spontaneous ferromagnetic alignment in an electron fluid is thought never to occur at the electron densities encountered in metals; the cost in kinetic energy is just too high. But there is a less drastic reorganization that can be imagined. This has the electrons' spins nearly parallel over short ranges and so decreases the exchange energy—which is in any case a short-range effect—but does not incur the massive cost in kinetic energy that is inevitable in the long range organization of ferromagnetic alignment. This introduces the concept of a nonuniform (periodic) spin polarization in the electron fluid; in other words, the fluid has a net fractional spin polarization, for example, $\mathbf{P}(\mathbf{r})$ at every point $\mathbf{r}$ in the metal, but its direction varies from place to place. In one of

---

* G. G. Low and T. M. Holden, *Proc. Phys. Soc.* **89**, 119, 1966.

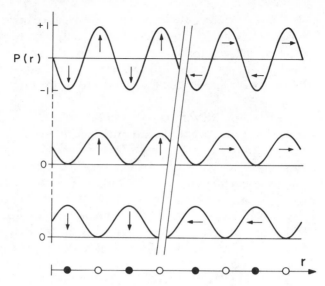

Fig. 4.18 A schematic view of the most elementary form of spin density wave. I•
opportune circumstances the itinerant electrons of a narrow energy band find i•
energetically preferable to split into separate spin-polarized groups; those of on•
spin orientation associate with one set of atoms in the lattice, and those of opposit•
spin associate with another. The lattice is, therefore, effectively split into two sub
lattices (represented by the solid and open circles) in which the spin polarizatio•
$P(r)$ of each sublattice varies sinusoidally with a wave vector Q and in antiphas•
with its counterpart of opposite spin. The net effect (shown by the upper curve) i•
a constant number density in the electron fluid but a periodic spin polarization
This is the spin density wave. In principle, the polarization can take any orientatio•
with respect to Q but in practice only two have been observed in a pure metal
$P(Q)$ normal to Q, giving a transverse polarization to the wave (left-hand side) an•
$P(Q)$ parallel to Q, giving a longitudinal polarization (right-hand side).

the earliest discussions of this idea* it was shown that itinerant electron•
imagined to be hopping from ion to ion divide into two groups to benefi•
from the energy reduction given by a nonuniform spin magnetization. Th•
first is centered about ions of type A in the lattice, and the other about thos•
of type B. Ignoring the cost in kinetic energy, the lowest energy configuratio•
has electrons of one spin orientation in orbitals about A atoms and thos•
of the opposite spin in orbitals about the B atoms, so that the electron flui•
is antiferromagnetically ordered. But the kinetic energy of the fluid is •
minimum when there is no spin polarization at all; that is when equa•
numbers of electrons of both spins are centered about A and B atoms. W•
expect that some balance is struck between these two extremes, so that th•

* J. C. Slater, *Phys. Rev.* **82**, 538, 1951.

actual populations on A and B atoms is not purely of one spin, but clearly such a balance has a periodic spin polarization mixed into it.

This state, in which the number density of electrons of a given spin direction varies periodically through the lattice in antiphase with that of electrons of opposite spin, has since been called a spin density wave (SDW). It is important to realize that such are static oscillations in the itinerant electrons' spin polarization, rather like the RKKY oscillations discussed in Sect. 3.4.3, except that here the amplitude of the spin polarization does not fall off with distance as it does in that case. Also, a more or less uniform electron number density is envisaged in the present case, whereas the RKKY oscillations are associated with comparable oscillations in the charge density. It is important to distinguish these static spin density oscillations from the spin waves discussed in Sect. 4.2.5: We recall that the latter are the elementary excitations supported by a system of fixed spins that are coupled over a long range.

A schematic illustration of the variation in the spin polarization in an SDW is shown in Fig. 4.18. The number densities of the spin-up and spin-down electrons vary sinusoidally with a wave vector $\mathbf{Q}$, so that the SDW can be represented by $\mathbf{P}(\mathbf{r}) = \mathbf{P}(\mathbf{Q}) \exp(i\mathbf{Q}\cdot\mathbf{r})$. As the figure shows, the amplitude of the polarization of one group at a given Q is exactly opposite to that of the other. In general $\mathbf{P}(\mathbf{Q})$ is presumed to be able to point in any direction relative to $\mathbf{Q}$, but in practice only two relative orientations of these vectors have been encountered: In the first $\mathbf{P}(\mathbf{Q})$ is perpendicular to $\mathbf{Q}$, so that the SDW is a transverse spin polarization relative to $\mathbf{Q}$, and in the second $\mathbf{P}(\mathbf{Q})$ is parallel to $\mathbf{Q}$, so that the SDW is a longitudinal polarization relative to $\mathbf{Q}$. Since the wave vector $\mathbf{Q}$ can itself take any arbitrary value, the period of the SDW is not necessarily an integral number of lattice spacings.

When it was first suggested seriously that SDW's may arise in a free electron fluid,* it looked for a time as if much of the established theory of metals might have to be revamped. But it turns out that such waves of polarization can only form under very special conditions, and their existence is by no means as likely as was at first thought. At the present time only the b.c.c. form of Cr (and perhaps that of Mn) is thought to provide the conditions necessary. This point arises again at the end of Sect. 6.4.2.

4.8 CHAPTER SUMMARY

There are roughly 10^{29} itinerant electrons in every cubic meter of a metal. Each interacts to a greater or lesser extent with all the others and with the

* A. W. Overhauser, *Phys. Rev.* **128**, 1437, 1962.

electrons and nucleons that comprise the fixed ions; the result is a many-body system of unimaginable complexity. It seems almost derisory to mention in the same breath that the most complicated many-body problem that can presently be solved—and then only to give an approximate solution—is the three-body problem. Faced with these facts, a newcomer to the field can hardly be blamed for taking a pessimistic view of the likelihood that any progress can be made towards a detailed understanding of the behavior of itinerant electrons in a metal.

However, the success of earlier attempts to understand metals provides a hopeful sign and a clue to the direction that should be followed. The free electron (Sommerfeld) model of the early 1930s, which ignores any interactions between the itinerant electrons, is known to provide a surprisingly good description of the behavior observed for several experimental parameters. In other words, although the electrons are mutually interacting, their response is in many respects not unlike that expected of free particles.

It was L. D. Landau and his collaborators who in the 1940s first described how such features of the behavior of a many-body system on the microscopic scale can be used to advantage. Firstly, it was shown that, when an outside influence is switched on, what is of interest is not the individual behavior of all the particles in the system but the response of the system as a whole. It is the behavior of the excitations of the system above a ground state that describes this response, and it is these that should receive attention. Secondly, he showed that these excitations of the itinerant electron system behave like the free particles in Sommerfeld's metal. Thus an intractable many-body problem can be transformed into a tractable one of noninteracting (or at least very weakly interacting) particles. This transformation is known as the reduction of the many-body problem, and it is described qualitatively in Sect. 4.2; in Sect. 4.2.2 it is applied to the illustrative case of the two-body problem. Descriptions of the principal excitations encountered in the metallic state are given in Sect. 4.2.3 to Sect. 4.2.5.

The quantum theory of many-body systems is, of course, an overwhelmingly mathematical subject, but some feeling for the underlying processes can be obtained from a descriptive treatment that leans heavily upon metaphorical analogies with concepts from the macroscopic scale. A description of the origin of the Coulomb force provides a convenient introduction to some of these, and in Sect. 4.3 the idea is introduced that such action at a distance effects arise from the transfer of virtual particles. These represent temporary deviations from the requirements of energy conservation that are permitted on the microscopic scale by Heisenberg's uncertainty principle. Metaphorically, they can be regarded as undetectable particles that are constantly emitted and absorbed by the real (that is, detectable) quantum particles of everyday experience. As a result, a real particle can be looked upon not as a corpuscular body but as an entity having blurred limits caused by a surrounding cloud of virtual particles. Interactions that can be described in these terms occur in several different contexts, and it is useful to have an elementary understanding of Feynman's pictorial treatment of them (Sect. 4.3.3).

The final sections (Sect. 4.4 to Sect. 4.7) draw on this background for separate descriptions of three important manifestations of the many-body effect among itinerant electrons. First, the Kondo effect is described in Sect. 4.4. In this the many-body interaction is established through the spin memory of an isolated foreign ion having a magnetic moment. Superconductivity is described in Sect. 4.6, and in this the charge memory of the lattice is the mediating influence establishing the many-body interaction. Finally, the exchange enhancement effect is described in Sect. 4.7, and in this the spin memory of a small region of the electron fluid is the link providing the many-body effect.

RELATED READING

Specific research papers are referenced where cited in the text.

MANY-BODY THEORY

M. Ya. Azbel', M. I. Kaganov and I. M. Lifshitz, "Conduction Electrons in Metals," *Sci. Amer.*, January (1973) p. 88. A descriptive review article at an elementary level.

D. ter Haar, "Some Recent Developments in the Many-body Problem," *Contemp. Phys.* (1959) p. 112. A generally elementary account of how many-body theory has come to underline many branches of physics.

Other more specialized books, in order of increasing technical level, are:

D. ter Haar, 1958, *Introduction to the Physics of Many-body Systems* (Interscience: New York).

D. Thouless, 1972, *The Quantum Mechanics of Many-body Systems* (Academic Press: New York).

D. Pines, 1963, *Elementary Excitations in Solids* (Benjamin: New York).

FEYNMAN DIAGRAMS

In order of increasing technical level:

K. W. Ford, 1963, *The World of Elementary Particles* (Blaisdell: New York). A very readable, popular account of this subject, although now somewhat dated.

W. A. Blanpied, 1969, *Physics: Its Structure and Evolution* (Blaisdell: New York). An intermediate level treatise.

R. D. Mattuck, 1967, *A Guide to Feynman diagrams in the Many-body Problem* (McGraw-Hill: New York). A specialists' monograph.

GRAVITATIONAL WAVES

D. W. Sciama, "Gravitational Radiation and General Relativity," *Physics Bulletin* (The Institute of Physics: London), November (1973) p. 657. A nontechnical account of the differences between gravitational and electromagnetic waves.

J. L. Logan, "Gravitational Waves: A Progress Report," *Physics Today,* March (1973) p. 44.

T. J. Sejnowski, "Sources of Gravity Waves," *Physics Today,* January (1974) p. 40. A relatively nontechnical account of some sources of strong gravitational waves, as well as the latest developments in gravitational wave detectors.

ELEMENTARY EXCITATIONS

In order roughly of increasing technical level:

M. Ya. Azbel', M. I. Kaganov and I. M. Lifshitz, "Conduction Electrons in Metals." (See under MANY-BODY THEORY.)

J. R. Schrieffer, "Electronic Excitations in Metals," in *Atomic and Electronic Structure of Metals* (American Society for Metals: Metals Park, Ohio) Ed. J. J. Gilman and W. A. Tiller (1967) p. 153.

L. D. Landau and E. M. Lifshitz, 1958, *Statistical Physics* (Pergamon: London).

P. L. Taylor, 1970, *Quantum Approach to the Solid State* (Prentice-Hall: Englewood Cliff, N. J.) p. 1. Gives an intermediate introduction to the concept of an elementary excitation.

P. W. Anderson, 1963, *Concepts in Solids* (Benjamin: New York) p. 99. Like the following, this book is not designed for beginners.

D. Pines, 1963, *Elementary Excitations in Solids* (Benjamin: New York) p. 1.

SUPERCONDUCTIVITY

I. M. Firth, 1972, *Superconductivity* (Mills and Boon: London). An elementary introduction to the subject, including its microscopic origins.

R. D. Parks (Ed.), 1969, *Superconductivity* (Dekker: New York), Vols. I and II. A handy collection of over twenty reviews by leading specialists. Although the treatment is almost invariably at an intermediate or advanced level, an adventurous student can profit from the less mathematical sections of several contributions, particularly in Vol. I. For an historical and prophetical perspective of this subject, see particularly P. W. Anderson's closing summary (p. 1343 of Vol. II).

J. M. Blatt, 1964, *Theory of Superconductivity* (Academic Press: New York).

JOSEPHSON EFFECTS

I. Giaever, "Electron Tunneling and Superconductivity," *Science* **183**, 1253 (1974). This is Giaever's Nobel Prize address. It is a most enjoyable and nontechnical view of how a "master in billiards and bridge" came to share the 1973 prize for physics.

B. W. Petley, 1971, *An Introduction to the Josephson Effects* (Mills and Boon: London). This is a companion volume to Firth's (see under SUPERCONDUCTIVITY), which also contains a description of Josephson effects.

L. Solymar, 1972, *Superconductive Tunneling and Applications* (Chapman and Hall: London). For the intermediate or advanced student, but still a predominantly descriptive treatment.

J. Clarke, "Josephson Junction Detectors," *Science,* **184,** 1235 (1974).

SPIN GLASS

N. Rivier and D. Taylor, "Spin Glasses: Law and Order in Magnetism," *New Scientist,* March 1975, p. 569.

MASS ENHANCEMENT EFFECTS

N. E. Phillips, "Low Temperature Heat Capacity of Metals," *Critical Reviews of Solid State Sciences,* **2,** 467 (1971). A researcher's article that provides excellent coverage of the literature relating to electron mass-enhancement effects.

5

THE LATTICE OF IONS

5.1 INTRODUCTION

A crystalline metal has two structural units: The delocalized electrons, which form a fluid-like plasma, and the regular periodic structure of the ions known as the lattice. The preceding chapters consider different aspects of the electron fluid—generally without much reference to the organized arrangements of the ions—and here we continue this dichotomous view with a consideration of the lattice types encountered in metals that generally ignores the fact that the lattice is immersed in the electron fluid.

We should always remember, however, that this fluid is responsible for holding together whatever arrangement the ions adopt in a particular metal. The distribution of the itinerant electrons in this fluid provides the attractive bonds that draw the ions together, and it also provides part of the force that resists their unlimited condensation. This force arises primarily from two sources. First, the incompressibility of the electron fluid in the interionic spaces and, second, the mutual repulsion experienced by ions when their outer shells of localized electrons start to overlap. Specific groups of metals exist that provide examples of the extreme cases where one of these stabilizing forces is overwhelmingly dominant. For example, in the lighter alkali metals (particular Li, Na, and K) the ion is relatively small, in so far as an ionic size can be defined (Sect. 2.2). In this case the incompressibility of the electron fluid provides most of the force opposing the closer approach of the ions in the solid. This, in fact, is embodied in the behavior shown in Fig. 2.4 and is further emphasized by the result that the calculated compressibility of the electron plasma in Na, for example, is found to be essentially equal to that observed for the metal as a whole. Consequently, these metals have a relatively open lattice structure in which the ionic radius is small compared with the interionic separation.

The other extreme is exemplified by the group of metals Cu, Ag, and Au. Their ions are relatively large so that when the solid is formed from the condensation of atoms the outer shells of the ions start to overlap (and so repel each other) before the incompressibility of the electron fluid becomes a really significant factor. As a result, the force opposing further condensation of the ions in the normal metal arises overwhelmingly from the repulsion between the ions caused by their overlapping extremities. The lattice structures of these metals, therefore, are the close-packed variety where an individual ion can be in contact with either eight or twelve equidistant neighbors. This is known as the coordination number of the structure (Sect. 4.4).

5.2 MORPHOLOGY AND SYMMETRY

The description given in Sect. 2.3 of how the metallic state arises from a condensation of atoms is concerned exclusively with how itinerant electrons arise from the interatomic interactions. It ignores completely what is, in fact, a far more fundamental manifestation of the net effect of all the interactions that exist between the constituents of a solid: The equilibrium lattice structure adopted by the ions. This structure represents the balance of all the forces acting within both the itinerant electron system and the ionic system, as well as between the two of them. Quite often this balance is very delicate and can be upset simply by changing the sample's ambient temperature or pressure, so that the atoms reorganize themselves to form a new structure. To give a detailed explanation of why one lattice structure is preferred to another in given circumstances is, therefore, a very difficult problem. It involves a minute comparison of quantities that are only grossly understood, and the margin of uncertainty in the final result is usually great enough to make the whole a futile exercise. Consequently, much of the interesting information that is contained in the knowledge of a solid's crystal structure cannot presently be extracted with certainty; therefore, crystallography remains overwhelmingly a descriptive rather than an interpretive branch of science.

So it is for the crystal structures of metals.* The problem is how best to arrange and classify the observed structures so as to show their inter-

---

* In recent years there have actually been important improvements in calculations showing why one metallic structure is preferred over another; therefore, we should not be quite so dogmatic. These have been thoroughly described at the researcher's level in *Solid State Physics,* **24,** 1970 and have been summarized at a more intermediate level in Chap. 13 of P. Wilkes, *Solid State Theory in Metallurgy,* Cambridge University Press, London, 1973. A description is given in Sect. 5.3.

relationships and their relevance to other properties that are studied, such as the electrical or mechanical properties. One distinctive characteristic of the lattice has already been mentioned: The degree of close-packing (or openness) found in the structure. Another is its symmetry.

A system that looks the same from different points of view is said to possess symmetry, and one that is identical under whatever rotations bring a cube into itself is said to have cubic symmetry. The direction about which a rotation brings the system into an identical position is called a rotational axis of symmetry. It is n-fold if there are n identical positions in a full revolution. The larger n, the higher the rotational symmetry of the structure. Of the sixty or so common metallic elements (Fig. 2.1) almost exactly half have an atomic arrangement at room temperature that gives them a crystal structure of cubic symmetry. This class comprises those metals having face-centered, body-centered, or simply cubic structures; the chief noncubic metals have the close-packed hexagonal, rhombohedral, tetragonal, or orthorhombic structures. The distribution of the room temperature structures of the common noncubic metals among the crystal classes is shown in Table 6. The minimum symmetry requirement for each

Table 6   Distribution of the Common Metallic Elements at Room Temperature among the Crystal Classes

Symmetry of Crystal Class	Elements Included
ORTHORHOMBIC	Ga, α-U.
RHOMBOHEDRAL	α-As, Bi, α-Hg, Sb, α-Sm.
TETRAGONAL	In, β-Sn
HEXAGONAL	α-Be, Cd α-Co, α-Dy, α-Er, α-Gd, α-Hf, α-Ho, α-La, Lu, Mg, α-Nd, Os, α-Pr, Re, Ru, α-Sc, α-Tb, α-Ti, α-Tl, α-Tm, α-Y, α-Yb, Zn, α-Zr.
CUBIC	Ac, Ag, Al, Au, Ba, α-Ca, γ-Ce, β-Co, Cr, Cs, Cu, Eu, α-Fe, β-Hf, Ir, K, β-La, Li, α-Mn, β-Mn, Mo, Na, Nb, Ni, Pb, Pd, α-Po, Pr, Pt, Rb, Rh, α-Sn, α-Sr, Ta, β-Tb, α-Th, V, W, β-Yb.

of the crystal classes, and the common convention used to specify a right-handed triad of axes (to be called x_1, x_2, x_3) in each of them is:

1. **Orthorhombic Class.** The minimum symmetry requirement for this class is three orthogonal directions having twofold symmetry, but none having symmetry of a higher order. Each of the axes x_1, x_2, x_3 is taken to be along one of these twofold symmetry directions.

2. **Rhombohedral Class.** The minimum symmetry requirement is a single direction of threefold symmetry. Conventionally, the axis x_3 of the triad is assigned to this and is known as the trigonal axis. In the plane perpendicular to this axis the crystal has also three directions of twofold symmetry (having an angular separation of 120°) and the axis x_1 is assigned to one of these and is often known as the binary axis. The remaining axis x_2 of the triad is sometimes referred to as the bisectrix axis.

3. **Tetragonal Class.** The minimum symmetry requirement is a single direction of fourfold symmetry. Conventionally, the axis x_3 is assigned to this and is known as the tetragonal axis. The crystal also has two directions of twofold symmetry that are perpendicular to each other and lie in the plane normal to the tetragonal axis. The axes x_1 and x_2 are assigned to these directions.

4. **Hexagonal Class.** The minimum symmetry requirement is a single direction of sixfold symmetry. Conventionally, the axis x_3 is assigned to this and is known as the hexagonal axis. As in the case of the rhombohedral class, the axis x_1 is assigned to one of the directions of twofold symmetry that exist in the plane perpendicular to the hexagonal axis.

5. **Cubic Class.** The minimum symmetry requirement is four axes of threefold symmetry. All such systems have three equivalent and mutually perpendicular directions that define a cube and are designated x_1, x_2, and x_3. The axes of threefold symmetry turn out to be the diagonals of this cube.

In terms of the symmetry of their crystal lattice, metals can thus be divided into two classes: Cubic and noncubic, where the latter can itself be subdivided into the first four classes listed above. However, there is another approach to all of this; it refers to the morphology of the atomic arrangement rather than to the symmetry of its representative lattice. In other words, the alternative emphasizes the position of any atom with respect to its neighbors and suggests a way of imagining how the solid is built up.

It starts from the observation that, although there are approximately seven hundred different known crystal structures among the metals, alloys, and semiconducting compounds, they can be looked upon as arising from the stacking one on the other of two-dimensional layers of atoms (also known as nets or rafts). (The atoms in these are not necessarily exactly coplanar, but we shall temporarily ignore that complication.) We find* that about half of the known structures can be obtained by stacking together layers having one of three distinct arrangements of atoms (or morphologies):

1. **Triangular Layers** where an atom is located at every interstice of a space-filling array of identical triangles.

2. **Hexagonal Layers** where an atom is located at every interstice of a space-filling array of identical hexagons.

3. **Kagomé Layers** (so-called) where an atom is located at every interstice of a space-filling array of hexagons and equilateral triangles.

These different morphologies are illustrated in the periphery of Fig. 5.1; at the center of the diagram is shown one of the many ways such layers can be combined by stacking. A typical crystal structure is, therefore, visualized as layers of atoms, each layer having one of the above morphologies and stacked one on the other with an appropriate separation. Figure 5.2 shows two examples of compounds that are directly comparable with the arrangement shown in the center of Fig. 5.1: AlB_2 has alternate layers of Al atoms in a triangular layer and B atoms in an hexagonal one, and $CaCu_5$ has alternate layers with triangular and Kagomé morphology.

The situation is greatly simplified when only the metallic (and semi-metallic) elements are considered, for in their crystal structures all the stacked layers have the same morphology. Furthermore, in the majority of cases these layers have the atoms located in a plane with triangular arrangement. Fig. 5.3a, with the regular hexagonal arrangement, shows the close-packed case that leads to the close-packed structures, notably the face-centered cubic (f.c.c.) and close-packed hexagonal (c.p.h.) structures. Layers of atoms with the geometry of Fig. 5.3b, which can be regarded as a distortion of the close-packed arrangement, lead, on the other hand, to the more open body-centered cubic (b.c.c.) and body-centered tetragonal (b.c.t.) structures.

Even the more uncommon structures can be considered to arise from stacked layers of atoms, but these layers have still further distortions from

---

* W. B. Pearson, *The Crystal Chemistry and Physics of Metals and Alloys,* Wiley-Interscience, New York, 1972.

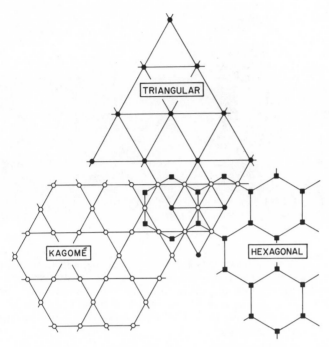

Fig. 5.1 About half of the approximately seven-hundred different crystal structures found among metals, alloys, and semiconducting compounds can be obtained
by stacking (with an appropriate separation) layers of atoms having one of the
three morphologies shown in the periphery of the diagram. The center of the
diagram shows one of the many ways such layers can be combined by stacking.
(After W. B. Pearson.)

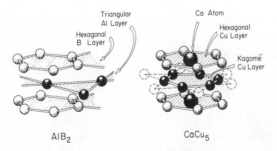

Fig. 5.2 Two examples that are directly comparable with the arrangement shown
in the center of Fig. 5.1: AlB_2 has alternate layers of Al atoms in a triangular
arrangement and B atoms in a hexagonal one; $CaCu_5$ has alternate layers of the
triangular and Kagomé arrangements.

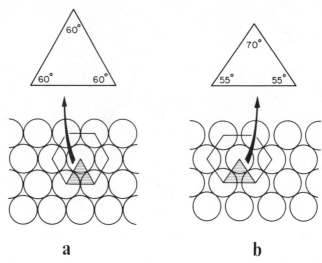

a   **b**

Fig. 5.3 Two atomic arrangements in the hard-sphere model that, when stacked one upon the other in the appropriate fashion, gives rise to the structures possessed by the overwhelming majority of metals. (*a*) The close-packed case, which leads notably to the f.c.c. and c.p.h. structures. (*b*) A deviation from it which leads to the b.c.c. and b.c.t. structures. The aim is to emphasize how metals can be regarded generally as layered structures.

the close-packed hexagonal arrangement of Fig. 5.3*a*. For example, these may consist of a distortion of the hexagon within the plane of the layer (rather like that which produces Fig. 5.3*b* from Fig. 5.3*a*) or a layer may even be corrugated so that the atoms in it are no longer coplanar. Examples of both are seen in Sect. 5.2.3.

5.2.1 The Close-Packed Structures. These are formed from the stacking of identical layers each with the morphology of Fig. 5.3*a*. As we have said, the principal close-packed structures are the f.c.c. and c.p.h. arrangements, but there are also less familiar ones that will be described below. This multiplicity arises from the fact that, once two close-packed layers have been stacked, any subsequent layer in the structure can be placed in two different ways. To see this, suppose the open circles in Fig. 5.4*a* are the ions of the first layer, and the shaded circles are ions of the second layer. Looking at this plan view, we see two possible positions for the third layer: Its ions can either occupy positions like P_1, which are centered over the interstices of the first layer, or they can occupy positions such as P_2, which are centered over the ions in the first layer. When they occupy positions P_2, the third layer is indistinguishable from the first. Since the position of

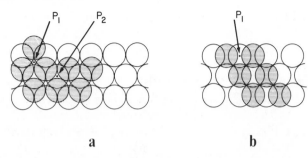

a   b

Fig. 5.4 A plan view to illustrate the different ways in which three identical layers of atoms having the arrangement of either Fig. 5.3*a* or Fig. 5.3*b* can be stacked one upon the other. These lead to the different stacking sequences described in the text.

the first layer with respect to some axes in its plane is conventionally denoted by A, the third layer will also have the position A, and in this case the stacking sequence will be $ABA \ldots$, where B will denote the position of the second (shaded) layer in Fig. 5.4*a*.

A, B, and C are the only distinguishable positions which the triangular nets of ions can take as the structure is built by the addition of successive layers. However, innumerable different sequences of these positions are possible when there are a substantial number of layers in the structure. This is illustrated in Fig. 5.5 for the case of the first seven layers of a

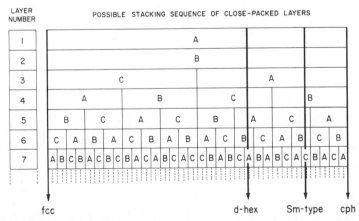

Fig. 5.5 Illustrating the possible stacking sequences for the first seven layers of a lattice built up in the manner shown in plan view in Fig. 5.4*a*. Any vertical line in the figure reads off a possible stacking sequence, but of all these possiblities only four are found for the pure metals. These are shown by the heavy vertical lines in the figure, together with the name of their corresponding crystal structure.

close-packed lattice. Any vertical line drawn through the figure (like the heavy lines shown) reads off a possible stacking sequence. Of all these possibilities, however, only four are found in nature for the pure metals. Two of these are by far the most common and are given by the extremities of the figure: $ABCA$ . . . and ABA . . . give, respectively, the f.c.c. and c.p.h. structures. Perspective views of their arrangements are shown in Fig. 5.6, which ignores the rhombohedral distortion there, and in the upper part of Fig. 5.7. The other two sequences: $ABACA$ . . . and $ABABCBCACA$ . . . are the less familiar double close-packed hexagonal (d-hex) and samarium-type (Sm-type) structures, respectively. Perspective views of these are shown in Fig. 5.7 and Fig. 5.8; the d-hex is found in the lighter rare earth metals such as α-La, α-Nd, and α-Pr; the structure of α-Sm is unique among the pure elements at normal pressure. It is a member of the rhombohedral class.

The d-hex structure has the notable feature (compare Fig. 5.6 and the lower part of Fig. 5.8) that the atoms on the A layers have neighboring planes in the typical cubic sequence . . . CAB . . . , but those on the B and C layers have neighboring planes in the typical hexagonal sequences . . . ABA . . . or ACA . . . Consequently, the atoms in this structure are dis-

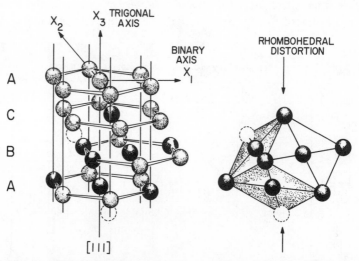

Fig. 5.6 Illustrating the atomic arrangement of α–Hg. This can be regarded as a rhombohedral distortion of an f.c.c. structure in which a compression along the [111] direction has forced the close–packed hexagonal planes (left) closer together but not upset their stacking sequence $ABCA$. . . . The arrangement of the rhombohedral cell within the structure is emphasized by the darker atoms and is also displaced for clarity to the right-hand side of the figure.

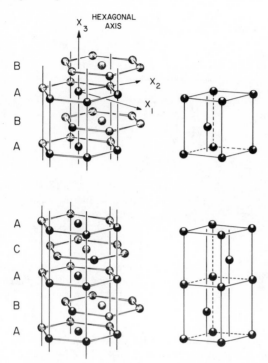

Fig. 5.7 Illustrating two atomic arrangements typical of the hexagonal class of metals. The well-known close-packed structure is illustrated in the upper part of the figure; the lower half shows the much rarer double close-packed structure (possessed notably by certain lighter rare earth metals). In each case the unit cell is emphasized by the darker atoms and is also displaced for clarity to the right-hand side of the figure.

tributed between two types of environment: One cubic and the other hexagonal. This has important consequences for the orbital magnetism of the ion, as in the case of Pr referred to in Sect. 3.2.5.

5.2.2 The Loosely-Packed Structures.   These are the relatively open or spacious structures. With the exception of the simple cubic (s.c.) structure,* where each ion occupies the corner of a cube, they are formed by stacking layers with the morphology of Fig. 5.3b. As we have said, they comprise the b.c.c., b.c.t., and f.c.t. (face-centered tetragonal) structures. The stacking arrangement is illustrated in Fig. 5.4b. It has alternate layers such that the ions of a second (shaded) layer occupy positions over the

---

* This structure is possessed by only one element in nature, Po.

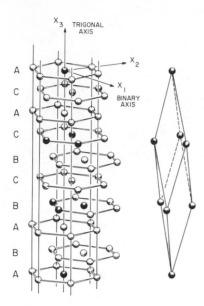

Fig. 5.8 Illustrating the atomic arrangement of α–Sm at normal pressures. Again, this structure can be regarded as being formed from stacked close–packed layers in the sequence shown (left). The rhombohedral cell is emphasized by the darker atoms and is also displaced for clarity to the right-hand side of the figure.

interstices of a first; thus the ions in a third layer occupy positions like P_1. However, when the stacking is such that the ions in one of these layers are in contact with those in the neighboring ones, then a moment's thought shows that the resulting structure does not have cubic symmetry: The separation between, for example, the first and third layers is greater than that between any two ions within a layer.

Such packing in fact gives rise to a tetragonal structure as illustrated in Fig. 5.9. The layers shown in Fig. 5.3b form the $\{110\}$ layers in the structure, and the interatomic spacing along the tetragonal axis x_3 (denoted by c) is then less than that a along either of the two other orthogonal axes. Note that this structure can be regarded as either f.c.t. or b.c.t., since they are equivalent. This is shown in Fig. 5.9b; the arrangement of ions giving an evident b.c.t. structure with an interatomic spacing ratio of c/a is illustrated in the lower part of the figure; however, when a few more atoms are added, as in the upper part of the figure, the equivalent f.c.t. structure emerges. This has a spacing ratio of $c/a\sqrt{2}$. Conventionally the structure of In is regarded as a very slight distortion of the f.c.c. and is, therefore, looked upon as f.c.t., but β-Sn is more appropriately viewed as b.c.t.

The b.c.c. structure is closely related to the tetragonal ones. To obtain it, instead of stacking the layers in a close-packed manner so that the ions of adjacent layers are in contact, an appropriate spacing has to be maintained between them so that the tetragonal distortion occurring along x_3 in

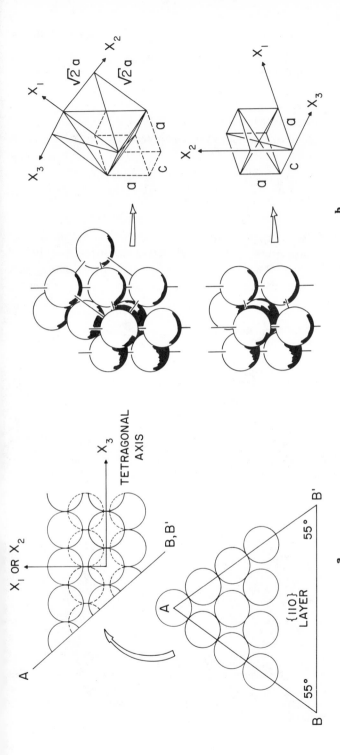

Fig. 5.9 Illustrating the formation of a tetragonal atomic arrangement. (*a*) Layers of atoms having the arrangement of Fig. 5.3*b* that are stacked with close packing one on top of the other. These form the {110} planes of a tetragonal structure in which the interatomic forces acting along the tetragonal axis (x_3) are less than those acting along either of the two other orthogonal axes (x_1 and x_2). (*b*) The result, which can be regarded either as a body–centered or face–centered tetragonal structure, since they are equivalent. The arrangement giving an evident b.c.t. structure with an interatomic spacing ratio of c/a is shown in the lower part of the figure, but when a few more atoms are included, as in the upper half, an equivalent f.c.t. structure emerges having the spacing ratio $c/a \sqrt{}$ [2].

235

Fig. 5.9 is eliminated. By allowing an appropriate separation of the layers ABB', the construction shown in Fig. 5.9 can be made to yield the b.c.c. structure.

5.2.3 Structures with Low Symmetry.

With the single exception of Sm, the structures encountered in the two preceding sections fall into three of the crystal classes listed in Sect. 5.2: Cubic, hexagonal, and tetragonal. The remaining classes—rhombohedral and orthorhombic—contain the elements listed in Table 6, and their structures form the subject of this section.

These structures can also be regarded as being formed from the stacking of layers of atoms having a morphology derived from that shown in Fig. 5.3a. Figure 5.10 illustrates the arrangement characteristic of As, Bi, and Sb, which are examples of the rhombohedral class. (These substances are actually semimetals; that is, they would be insulators having completely

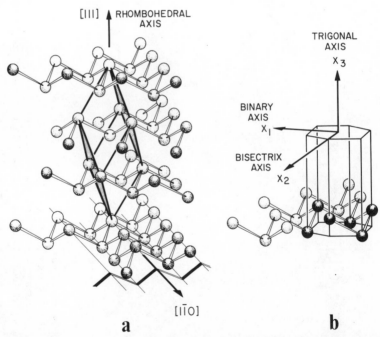

Fig. 5.10 Illustrating the atomic arrangement characteristic of As, as an example of the rhombohedral class of semimetals As, Sb, and Bi. (a) Equivalent corrugated layers of atoms that are stacked along the [111] direction in a sequence giving a rhombohedral primitive cell. (b) A view down the [111] direction in which the arrangement has hexagonal symmetry.

illed energy bands separated by a forbidden energy gap (Sect. 2.3.3) if it were not for an overlapping in energy of these bands along particular crystallographic directions.) They have equivalent corrugated layers of atoms stacked along the [111] direction in a sequence that produces a rhombohedral primitive cell, one of which is illustrated in Fig. 5.10a. But when viewed along the rhombohedral axis, the atoms have a regular hexagonal arrangement in the (111) plane; in terms of this symmetry the triad of axes x_1, x_2, x_3 is defined as shown in Fig. 5.10b.

Another example of the rhombohedral class is shown in Fig. 5.6. It is the structure of α-Hg, which can be regarded as a rhombohedral distortion of the f.c.c. structure in which a compression along [111] has taken place forcing the close-packed layers closer together but not affecting their sequence ABCA. . . . The arrangement of the face-centered rhombohedral cell within this structure is illustrated by the darker atoms; for clarity this cell is displaced to the right of the figure.

Finally, and least symmetric of all, are arrangements that fall into the orthorhombic class. Only two examples of this occur among the pure elements: Ga and α-U. Figure 5.11 shows the arrangement of ions in the former. Each layer consists of atoms located at the vertices of distorted hexagons; the layers are stacked as shown in perspective in Fig. 5.11a and in plan view in Fig. 5.11b. The orthorhombic primitive cell abc is illustrated, together with the arrangement of the axes x_1, x_2, x_3. The arrangement of ions in α-U differs in detail from that shown in Fig. 5.11; in some respects it is simpler because, instead of the distorted hexagons in the x_1x_3

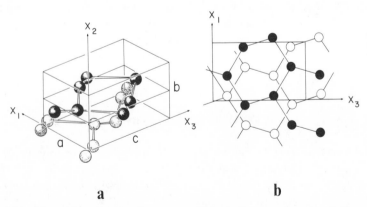

a   b

Fig. 5.11 Illustrating the atomic arrangement of Ga, an example of the orthorhombic class; that of α–U is similar except that instead of distorted hexagons of atoms stacked along the x_2 direction it is more appropriately regarded as being formed from crooked chains of atoms aligned along the x_3 direction.

plane, α-U has simply chains of atoms aligned in a zigzag fashion along the x_3 directions and packed so as to give orthorhombic symmetry.*

An important point to stress in this context is that, apart from the orthorhombic class (two metals out of a total of about 32 non-cubic ones) all the noncubic metals have a single axis of highest rotational symmetry. This is known as the principal axis; it is either the rhombohedral, the tetragonal, or the hexagonal axis, and it has been designated throughout by x_3. Such metals are known as the axial metals. Because of the layered structures of these metals, an itinerant electron moving in the lattice can find different circumstances depending upon its direction of motion. Electrons moving within a layer find equivalent conditions whatever their direction of motion (as in a cubic metal, for example) but different circumstances prevail if their motion is directed along the principal axis. The macroscopic manifestation of this is that the electrical properties (at least) depend upon the direction in which they are measured in the crystal. Such a crystal is said to be anisotropic in the particular parameter under discussion.

The physical explanation for this anisotropy lies ultimately in the details of the interatomic bonding, since the itinerant electrons are also importantly involved in the bonding in the metal. For example, if this bonding within a layer is significantly stronger than that between atoms in adjacent layers—a condition not unknown in practice—then the simple expectation is that the electrical conductivity measured in the layer will be higher than that measured perpendicular to it. Such experimental parameters and the atomic arrangement that is adopted are, therefore, complimentary manifestations of the interatomic bonding. We must emphasize that this is a very naive view, however, for in any real metal at an arbitrary temperature not only is the effect of the crystallographic structure upon the electronic motion significant, but various electron scattering mechanisms must also be accounted for, and they can completely dominate the motion of the itinerant electrons.

5.3 COHESION AND STRUCTURE

5.3.1 Consequences of Periodicity. The formation of energy bands in a solid is described in Sect. 2.3 as the result of the mixing (or hybridization) of the electrons in the outer orbitals of the atoms that are condensed to

---

* It seems unnecessary to provide a perspective drawing here. Such can in any case be found in W. B. Pearson's book cited in Sect. 5.2, in E. S. Fisher and H. J. McSkimin, *Phys. Rev.* **124**, 67, 1961, and in J. Donohue, *The Structure of the Elements*, Wiley-Interscience, New York, 1974, Fig. 92.3.

form the solid. Although this is a plausible picture, it does not emphasize the fact that the energy gaps between the bands are forbidden ranges of energy that may not contain any possible energy states for the itinerant electrons. It is as if, by a judicious choice of their distribution in the inter-ionic spaces, the itinerant electrons are able to make a cut in their otherwise continuous range of kinetic energies. States just below the cut are lowered in energy by the electrons' redistribution, and those just above it are correspondingly raised. Consequently, the cut widens into a gap, which we shall call a band gap so as not to confuse it with the term "energy gap" used in a different context in Sect. 4.6. Since the electron states are occupied only up to some finite energy E_f, the net effect of introducing a band gap is always to lower the total energy of the itinerant electron system; the bigger the gap, the greater this reduction in energy. This is illustrated in Fig. 5.12 for the case of the s/p band considered in Sect. 2.3.3.

There is another convenient and well-known picture to describe the origin of these bands of forbidden energies. This is in terms of the diffraction of the electrons by the regular planes of ions in the lattice. Being quantum particles, the electrons have wavelike properties, and the wavelength of an itinerant electron with the kinetic energy E_f turns out to be typically only a few angstroms. This is just the order of magnitude to be diffracted by the regular planes of ions that exist in the crystal lattice. As a consequence, if an electron is imagined to be accelerated along a given direction in the crystallographic lattice, its kinetic energy will progressively increase (and so its wavelength will shorten) until the Bragg reflection condition is satisfied at some critical wavelength λ. This depends upon the spacing of the lattice planes and the angle at which the electron strikes them; in general it depends upon the particular direction of the electron's motion in the lattice. Since this reflection occurs at a critical value of the electron's momentum p (or wave vector k, since $\mathbf{p} = \hbar\mathbf{k}$) it is conveniently described in k space by constructing all the planes upon which the wave vector has its critical value. (Since the wavelength λ of a conduction electron is typically about 10^{-9} to 10^{-10} m, k is of the order of 10^9 to 10^{10}m^{-1}.) These planes enclose a volume in momentum space known as the Brillouin zone.* When the component of the electron's velocity normal to these reflecting planes reaches its critical value, it is reversed. So an electron's motion normal to these planes is reversed elastically by the reflection. In fact, the electron's energy in that direction cannot be increased beyond the critical value at the reflection; the electron effectively bounces

* Drawings of the Brillouin zones of fourteen different crystal structures are shown in A. P. Cracknel and K. C. Wong, *The Fermi Surface,* Clarendon Press, Oxford, 1973, p. 16.

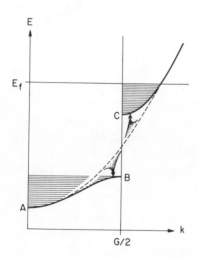

Fig. 5.12 Showing the variation of the energy E of an itinerant electron with its wave vector **k** for a simple, nearly free–electron metal. The dotted line is the parabolic variation expected for electrons moving freely without hindrance by the ionic lattice. However, when the periodicity of the potential produced by this lattice is accounted for, the band gap BC appears in the electron's range of permitted energies. Different views of the origin of this gap can be taken. One is that it arises from the Bragg diffraction of the electron's wave by the lattice planes; lattice planes perpendicular to **k** and having a spacing a give rise to a Bragg reflection whenever $\mathbf{k} = \mathbf{G}/2$, where **G** is the (physicist's) reciprocal lattice vector ($= 2\pi/a$). Alternately, the band gap can be viewed as a way of lowering itinerant electron's energy. For example, in the case of the s/p energy band considered in Sect. 2.3.3 the states below the gap, which are entirely s-like at A, are able to lower their energy from the free–electron parabola by hybridization with states of p character. At B in this example the states are entirely p-like bonding ones (Sect 2.3.3) and those at C, which have been pushed upwards in energy by a corresponding amount, are entirely s-like antibonding ones. These displacements produce a net reduction of the energy of the electrons in the band only because the band is filled up to a finite level E_f (indicated by the shading) and this typically lies close to the energy gap as shown.

off a perfectly elastic wall in the crystal and takes up a new direction of motion.

Whatever view is taken, the energy gap must not be thought to arise from an abrupt effect occurring only at a particular value of **k**. As an electron nears a Bragg diffraction condition, its motion is affected and the electron starts to diffract before actually reaching the Bragg plane. Correspondingly, the lowering of energy levels away from the free electron parabola produced by the electrons' redistribution in the interionic space is appreciable at values of k shifted substantially from the critical one of $G/2$ (Fig. 5.12). The point is that the energies of all the itinerant electrons are affected to some extent by the introduction of a band gap. It is just that electrons with motion very close to the critical condition are most strongly affected and show the most marked deviation from the behavior expected of a free particle. In fact, the diffraction picture only works at all because the interaction between itinerant electrons and the regular lattice of ions is weak and produces a relatively small pertubation of the electron's motion. For most of an itinerant electron's lifetime between successive collisons in an NFE metal (Sect. 2.3.4) its behavior is very close to that expected of a free particle. This is a reason why the very elementary Sommerfeld model was so successful in this context, and it is the reason why the simplifying approach described in the following section can be taken.

5.3.2. Pseudization of Nearly Free-Electron Metals.

We have said that the effect of the regular lattice of ions upon the itinerant electron's motion can be described as a diffraction phenomenon only because the interaction is basically a weak one. Indeed, it is an established experimental fact that the itinerant electrons in the s/p bonded metals (Sect. 2.3.4) move for the majority of their lifetimes as free particles; only a small fraction of them experience an interaction with the lattice that seriously disrupts their free motion. This is a surprising result, considering the density of the ionic lattice and the number of possible encounters between ions and an electron during the latter's lifetime (Sect. 2.2), and it is one which was originally misinterpreted. It is not, as was originally thought,[*] caused by a very weak intrinsic scattering power of an ion in a metal, but rather it is caused by the fact that the attractive and repulsive forces between the electrons and ions almost cancel each other leaving the ion with only a weak net scattering influence. It is this near cancellation that is the physical basis of the nearly free-electron behavior that is observed: It is thus responsible for the success of the Sommerfeld model and the validity of the diffraction picture (Sect. 5.3.1). Perhaps more important, however, is the fact that it presents an opportunity in a quantitative approach to ignore the details of an ion's intrinsic (and strong) scattering power, and all other counteracting influences, and to represent the whole by an imaginary (but weak) scattering potential that nevertheless has the same overall effect from the itinerant electron's point of view. This imaginary construction is called a pseudopotential (or sometimes a model potential when it is derived not from first principles but by using bits of experimental data). The whole approach is known—to the annoyance of some purists—as pseudization.

Pseudization involves recognizing that an electron scattered by an ion does not penetrate deep inside the ion's core. It is thus largely unaffected by the details of the ion's electrostatic potential in this region. However, this is just the region in which the potential is most atomic, being shielded by the outer shells of electrons from outside disturbances, and shows characteristic atomic-like oscillations in its magnitude. Since these are not only irrelevant to the scattering of the electron but also a great technical inconvenience in any calculation, it is a relief to be able to ignore them by constructing a pseudopotential. The reason for introducing this subject here is that it arises in the calculation of the energy differences between different lattice structures, and we turn to this topic in the following section.

The procedure is to construct an imaginary but working model of the electrostatic potential that exists around an ion in the metal being examined. This pseudopotential $v(\mathbf{r})$ is assumed to be spherically symmetric. As is often the case in solid state physics, it is convenient to work both in normal space (spanned by the vectors $\mathbf{r}$) and in reciprocal* space (spanned by the vectors $\mathbf{q} = 2\pi/\mathbf{r}$). In this case the physical properties of interest are directly related to the reciprocal space version. For example, when an electron is diffracted by an ion, the strength of the diffraction depends upon the Fourier transform of the ion's electrostatic potential, $v(\mathbf{r})$. From standard theory this Fourier transform of $v(\mathbf{r})$ is just $v(\mathbf{q})$ where:

$$v(\mathbf{q}) = \frac{1}{\Omega} \int v(\mathbf{r}) \exp\left(-i\mathbf{q}\cdot\mathbf{r}\right) d^3r \tag{5.1}$$

and Ω is the atomic volume. (This is just the volume in normal space effectively occupied by an atom of the metal. It is given by the molar volume divided by Avogadro's number.) The transform $v(\mathbf{q})$ can be thought of as the amplitude or strength of a process that scatters an itinerant electron from a state $(\mathbf{k},\sigma)$ to another $(\mathbf{k}',\sigma)$ as a result of encountering the ion's potential $v(\mathbf{r})$. Thus, in the usual symbolism:

$$v(\mathbf{q}) = \langle k' | v | k \rangle \tag{5.2}$$

when $\mathbf{k}' = \mathbf{k}+\mathbf{q}$. The second important point is that the band gap (BC of Fig. 5.12) which, of course, arises only at a particular value of the recipro-

---

* The reader is assumed to be familiar with this space in which any cell has dimensions inversely proportional to those of the corresponding cell in normal space. It is also known as momentum space, since $p \sim 1/\lambda$, or Fourier space since a function defined in normal ($\mathbf{r}$) space is expressed in reciprocal ($\mathbf{q}$) space as its Fourier transform.

cal lattice vector $\mathbf{q} = \mathbf{G}$, turns out to be just $2|v(G)|$ to a first approximation.

The arguments relating to the stability of different crystal structures pivot about the shape of $v(\mathbf{q})$. Figure 5.13 shows schematically the typical shape of $v(\mathbf{r})$ and its corresponding Fourier transform $v(\mathbf{q})$ for an isolated ion (Al^{3+}, for example). R_c in normal space (Fig. 5.13a) corresponds roughly to the limit of the ion's central core. Outside this range the potential is dominated by the Coulombic part (which has a variation of the form Ze^2/r, for a charge Ze on the ion) and so varies monotonically in the manner shown. In the Fourier space representation (Fig. 5.13b) the Coulomb part becomes correspondingly dominant as $q \to 0$; one consequence of this is that $v(0)$ has the fixed value shown. As q increases from zero, corresponding to the approach towards the ion's core, the pseudopotential $v(\mathbf{q})$ passes through a zero at q_0 followed by a local maximum at q_m. The position of q_0 has no immediate physical significance, although it is connected with the size of the ion's core of closed shells and with the positions of the atomic energy levels in the free atom; when the ion under consideration has a small core this tends to increase q_0. However, it is not necessary at this point to worry about such details. It is sufficient to note that the pseudopotentials for many elements have been derived and put to use, and their qualitative form is almost invariably that sketched in Fig. 5.13b.

The importance of this shape of $v(\mathbf{q})$ in the present context arises from the fact that in practice, and in the crystal structures under consideration,

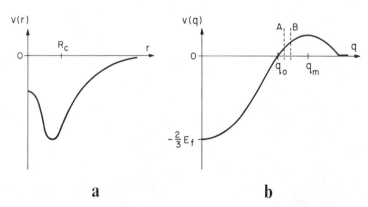

a   b

Fig. 5.13 (a) The shape of a typical pseudopotential in normal space. (b) The same in Fourier transform space. R_c corresponds approximately to the radius of the ion's core. Beyond this length the electrostatic potential is dominated by its Coulombic part and so varies monotonically in normal space.

the first few reciprocal lattice vectors lie close to q_0, as when G falls at A or B in Fig. 5.13b. Since the corresponding band gap BC (of Fig. 5.12) is proportional to $v(G)$ and since the total energy of the itinerant electrons depends upon the size of the band gap (Fig. 5.12 and Sect 5.3.1) it is obvious that the total energy of the electrons is sensitive to the positions of the reciprocal lattice vectors with respect to q_0. For example, suppose in a given structure the reciprocal lattice vector G was the value of q corresponding to A in Fig. 5.13b. If the lattice could undergo a change of structure that would allow G to move away from q_0 (say to B) then, assuming all other things remain equal, the corresponding increase in $v(G)$ would give a bigger band gap in the electrons' energy states and a lower total energy to the electron system. Thus the second crystal structure would be energetically preferable to the first. This can be expressed in Fourier space as the inclination of reciprocal lattice vectors to arrange themselves as far away from q_0 as possible.

This contribution to the itinerant electrons' total energy that is dependent upon the size of the band gap is called the band structure energy U_{bs}. In addition to the influence that sets the reciprocal lattice vectors as far away from q_0 as possible, U_{bs} is also strongly affected by the position of these lattice planes with respect to the Fermi surface* of the metal. Figure 5.14 should make this clear. Suppose some imaginary metal has lattice planes arranged in normal space as in Fig. 5.14a. The corresponding lattice vector G in reciprocal space (that is, along k_x or q_x) is $2\pi/a$. If $G/2 > k_f$, the radius of the Fermi sphere (Fig. 5.14b), none of the itinerant electrons is energetic enough to satisfy the Bragg diffraction condition for motion along the x direction. There is, therefore, no band gap in an electron's normal range of energies along this direction. However, if the lattice in normal space can be dilated so that a is increased, the reciprocal lattice vector shortens correspondingly and eventually the situation in Fig. 5.14d is reached where $G/2 < k_f$. In this case a band gap appears for motion along the x direction; the magnitude of U_{bs} is reduced by its appearance, and so the dilated structure becomes energetically preferable (again, assuming all other considerations remain unchanged).

This gives immediately a qualitative explanation for the variation of U_{bs} with q (Fig. 5.14c) which is the principal objective of this section. Firstly, following the description given in connection with Fig. 5.13b, we see that U_{bs} passes through a local minimum of zero when $q=q_0$. This is an implicit

---

* This is the surface that in reciprocal space separates the electrons' momentum states that are occupied at absolute zero from those that are vacant. The usefulness of the construction is illustrated in more detail in Sect. 6.2; it was first encountered in Fig. 4.4c.

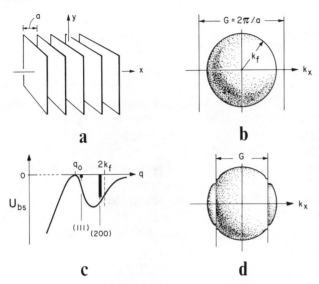

Fig. 5.14 Showing two extreme positions for the Brillouin zone planes with respect to the Fermi surface of a nearly free-electron metal. (*a*) A set of lattice planes in normal space. Depending upon their separation *a* and upon the radius of the Fermi sphere k_f in reciprocal space, the corresponding planes of the Brillouin zones may or may not intersect the metal's Fermi surface. (*d*) An intersection that produces a band gap introduced into the itinerant electron's permitted energy states for motion along the *x* axis that reduces the crystal's total energy. (*c*) The variation of this band structure component U_{bs} of the crystal's cohesive energy (explained more fully in the text). The vectors corresponding to the (111) and (200) planes that are marked arise in the discussion of the energy band structure of Al given in Sect. 6.2.

feature of the shape of the pseudopotential and has otherwise no straightforward physical interpretation. Secondly, once *q* is larger than $2k_f$, so that the Brillouin zone planes no longer cut the Fermi surface, then the magnitude of U_{bs} must start to fall off appreciably since the band gap is eliminated. Actually, the derivative dU_{bs}/dq turns out to be a maximum at $q=2k_f$, and, as a result, the curve has the general form shown in Fig. 5.14c.

5.3.3 Structural Factors: Some Examples. In a few of the pure s/p metals the crystal structure adopted from the range described in Sect. 5.2 is thought to reflect the dominance of the band structure contribution U_{bs} (Sect. 5.3.2). The aim of this section is to give a qualitative description of this class. We will see that, although a generally close-packed structure with high electron density in the interionic spaces is expected from purely electrostatic considerations, the particular refinement chosen can sometimes be attributed to U_{bs} and the effects that are described in Sect. 5.3.2. We must first, however, broaden the base of the discussion from simply the band structure energy to all the terms that are thought to contribute significantly to the energy of a metal's crystal structure.

Suppose a metal is formed in the following imaginary way. First, the separated atoms that will make up the sample are ionized to the extent that the z valence electrons are detached. This produces the precursors of the metal's two structural units, but it requires an expenditure of energy I per atom, where this is the total ionization energy for the removal of the valence electrons. Now suppose that these ionized constituents are brought together to form the metal. Since the metallic state is stable, energy is recovered in this process. Let its value be U_0 per atom, where this is the cohesive energy per atom (Sect. 2.2). The total binding energy of the metal B (which is the energy required to separate it into isolated ions and electrons) is thus $I - U_0$ per atom.

The cohesive energy U_0 cannot be calculated exactly. The best that can be done is to introduce a simplified model that is both plausible physically and tractable mathematically. A great problem is the dependence in principle of U_0 upon the crystal's structure, but in the extremely simplified models that are generally discussed this is ignored on the experimental grounds that modifications of crystal structure in a pure metal frequently occur with very little change in atomic volume. We can infer from this that, at least to a first approximation, the cohesive energy can be calculated with a model that is independent of the crystal's structure. It does not serve our present purposes to go into the details of such a model, or of the quantitative results, but it is useful to cite the terms arising in one of the simplest approaches because they serve to underline an important remark made in Sect. 5.2: Any calculation of a crystal's structure involves a minute comparison of many terms, most of which are known only very approximately.

Perhaps the simplest calculation of U_0 imagines an ion to be put into a fluid of free electrons. A pseudopotential is chosen for the ion which in normal space is that of a square potential well of radius R_m. The calculation is restricted to a single atomic cell (Sect. 2.3.2) which, although it must be a polyhedron in a real metal, is taken to be a sphere of radius R_a. The total energy of this system, which becomes the cohesive energy per atom U_0 when summed over all atomic cells, is the sum of: The kinetic energy of the itinerant electrons in the cell; the electrostatic energy of the ion immersed in the cell; the exchange energy of the itinerant electrons in the cell (which is the correction for the effect of the Pauli principle of Sect.

3.3.3); and the correlated energy of the itinerant electrons in the cell (which is the correction for the electrostatic correlation in their motion of Sect. 4.5.2). The calculation of these quantities gives* for these terms, respectively

$$U_0 = \frac{3}{5} z E_f + \left(\frac{3}{2} z^2 R_m^2 / R_a^3 - A z R_m^3 / R_a^3 - 0.9 z^2 / R_a \right)$$
$$- \left(\frac{0.458\ z}{r_s}\right) + (0.0311 \log r_s) \qquad (5.3)$$

Here A is the depth of the square potential well and $r_s \equiv z^{-1/3} R_a$.

Eq. (5.3) is only that part of the cohesive energy that is assumed to be independent of the crystal structure. There obviously must be a structurally dependent component that determines the crystal structure in any given circumstance, and this has to be added to Eq. (5.3) to give the total cohesive energy. Although the contribution of Eq. (5.3) is larger than the structure-dependent one, it is the latter that is our principal interest in this section and the subject to which we now turn.

The structure-dependent component has itself two parts: U_{bs}, the origin of which is described in Sect. 5.3.2, and the Ewald energy U_E. The latter is the electrostatic energy arising from a set of ions each with a net positive charge of † z and fixed in a regular array that is immersed in the electron fluid. For example, suppose that in a sample U_{ii} is the total energy of the Coulomb repulsion between the ions, U_{ie} is the total energy of the interaction between the ions and the itinerant electrons, and U_{ee} is that from the mutual repulsion between the itinerant electrons. Then the total Ewald energy (the electrostatic energy per atom) is given by

$$U_E = \frac{1}{N} \left(\frac{1}{2} U_{ii} + U_{ie} + \frac{1}{2} U_{ee}\right) \qquad (5.4)$$

where N is the number of atomic cells in the sample. (The factors of $1/2$ occur because each charge is counted twice in the calculation of U_{ii} and U_{ee}.) The Ewald energy demands a close-packed structure with a high itinerant electron density in the interionic spaces. It, therefore, favors

---

* V. Heine and D. Weaire, *Solid State Physics* **24**, 249, 1970.

† Actually the net charge is slightly greater than the ion's simple valence because some of the charge within the ion's core (R_c of Fig. 5.13) is expelled outside and piled into the region between R_c and the limits of the atomic cell. This expulsion, which arises from the requirements of the wavefunction's orthoganality within the core (Sect. 3.2.4) leaves a vacancy in the core's charge which is called an orthoganality hole. Outside the core the ion's effective charge is thus $z(1 + \alpha)$, where α turns out to be about 0.1.

structures of high symmetry such as f.c.c., b.c.c., and c.p.h. As far as is known, it is actually quite small in these structures, but it is thought to increase very rapidly if they are distorted. It is thus responsible for those features shown by the crystal structures described in Sect. 5.2, but it cannot account for the finer details that explain why one structure is preferred over another.

These finer points have in some cases been attributed to the role of the band structure energy U_{bs}. As described in Sect. 5.3.2, the contribution of this term to the crystal's stability arises because a band gap can reduce the total energy of the itinerant electrons. This is manifested in two ways: First, the chosen structure has the greatest weight of reciprocal lattice vectors lying below $2k_f$ so as to produce the most profound intersection of the Fermi surface by the planes of the Brillouin zone (Fig. 5.14d) and second, the structure avoids as far as possible having any reciprocal lattice vectors close to q_0 (Fig. 5.13b). (Again, this is to maximize the band gap.) Figure 5.14d illustrates why the reciprocal lattice vectors lie preferably within $2k_f$ in a one-dimensional case, but we must remember that a metal has electronic motion in three dimensions that are usually not equivalent. Therefore, the real situation is quite a bit more complicated than Fig. 5.14d suggests, and some way is needed to show the different contributions to the energy reduction that can arise from electron motion along different directions in the lattice. This is the purpose of Fig. 5.15, which shows for the f.c.c., c.p.h., and b.c.c. structures the magnitudes and structural weights of their reciprocal lattice vectors of interest (assuming equal atomic volumes in each structure). These are compared with the corresponding value of $2k_f$ appropriate to a given valence z, and so the diagram is effectively a concise description of the band gap reductions that are possible for the three structures considered.

To understand the diagram it is perhaps helpful to describe briefly how it is constructed. Consider the case of the f.c.c. lattice, for example. Let the unit cell in normal space have a side of length a. The planes with the closest spacing in normal space form the $\{111\}$ set and have the spacing $a(3)^{-1/2}$. (The lattice planes in normal space involved here and in subsequent discussions are sketched in Fig. 5.16.) The corresponding reciprocal lattice vector $\mathbf{G}$ has, therefore, a length $2\pi(3)^{1/2}/a$. There are eight such vectors in this set, one arising from each direction along the cube's four diagonals. Therefore, a structural weight of 8 is assigned to this reciprocal lattice vector whose position on the scale of $2\pi/a$ units of Fig. 5.15 is at approximately 1.73. The planes having the next closest separation in normal space form the $\{200\}$ set. Their spacing in normal space is just $a/2$, and thus the corresponding reciprocal lattice vector $\mathbf{G}$ has a length

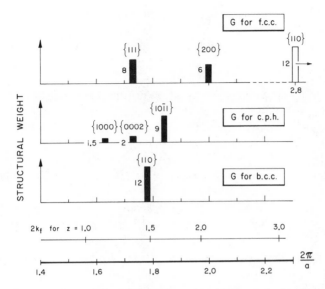

Fig. 5.15 Showing the magnitudes and structural weights of the shortest reciprocal lattice vectors **G** for the three common crystal structures. (Equal atomic volumes are assumed throughout.) The sets of lattice planes in normal space that generate these **G**'s are indicated in each case. The corresponding diameter of the free-electron sphere ($2k_f$) is included for four different values of the valency z. Further details of this table's construction are given in the text.

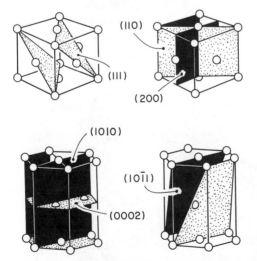

Fig. 5.16 A sketch showing the lattice planes in normal space that gives rise to important reciprocal lattice vectors in the cases in Fig. 5.15 and Fig. 5.17.

249

$4\pi/a$ (or 2 in the units of Fig. 5.15). There are six such vectors in this set, one arising from each direction along the three coordinate axes in normal space. The corresponding reciprocal lattice vectors for other sets of planes can be obtained by similar arguments (the {110} set is shown) but it is only those having values close to the appropriate $2k_f$ value of the metal that are of interest here. Turning to the other structures, the same arguments are followed except that the atomic volume is held constant (at $a^3/4$). For example, in the b.c.c. structure there are only two atoms per unit cell (as opposed to four in the f.c.c. structure) and so the length of the unit cell in this case has to be $a/(2)^{1/3}$ to maintain the fixed atomic volume between the structures. In this case the {110} planes are separated in normal space by $(2)^{5/6}a$ so that the corresponding reciprocal lattice vector has a length $(2)^{5/6} 2\pi/a$, which is approximately 1.78 in the units of Fig. 5.15. There are twelve of these in the set; one for each direction along the six <110> directions. Although the geometry in the c.p.h. case is a little more complicated, similar arguments lead to the situation shown; again, the atomic volume has been held constant. The values of $2k_f$ shown towards the bottom of the figure follow from elementary theory: It is a standard result for a free electron metal that $k_f=(3\pi^2 C)^{1/3}$, where C is the number of electrons per unit volume in normal space. Since the valence is z, $C=4z/a^3$ in this case; rewriting the expression for k_f in the units $2\pi/a$ of Fig. 5.15 gives $k_f=(3z/2\pi)^{1/3}$.

Figure 5.15 shows at a glance which of these three common structures is preferred for a given valence in terms of this contribution to the band structure energy. Thus for $z=1$, $2k_f$ is just less than the smallest $\mathbf{G}$ for the c.p.h. structure. Of the three, this structure, therefore, comes closest to having a band gap contribution to the total energy and is thus the predicted structure. For $z=1.5$ we see that the b.c.c. structure has the greatest structural weight below $2k_f$ (12 as opposed to 3.5 for the c.p.h. and 8 for the f.c.c.) and is the predicted structure for this valence. For $z=2$ it is again the c.p.h. structure that is preferable (although by seemingly a narrow margin over the b.c.c.) and, finally, for $z=3$ the f.c.c. structure is preferred. The success of this approach can be judged from the following list (the asterisk indicates that the element has the predicted structure; in the cases of Li and Na it is a low temperature phase):

$z=3$	(f.c.c. predicted)	Al*, Ga, In, Tl
$z=2$	(c.p.h. predicted)	Be*, Mg*, Zn*, Cd*, Hg
$z=1.5$	(b.c.c. predicted)	β–brass type alloys*
$z=1$	(c.p.h. predicted)	Li*, Na*, Cu, Ag, Au.

We recall that as far as we have gone with the above description it applies only to the nearly free-electron (s/p bonded) metals. The elements Cu, Ag, and Au are not truly in this class because of the participation of their d states in the cohesion (Fig. 2.9b and Fig. 6.2). This almost certainly accounts for the fact that they do not have the predicted c.p.h. structure.

The foregoing is otherwise believed to describe the essential physics behind the choice of the crystal structures from the range given in Sect. 5.2. In the cases of Ga, In, Tl, and Hg, the failure to show the predicted structure is thought to arise from the interference of the second influence referred to above: The avoidance as far as possible of reciprocal lattice vectors close to q_0 (Fig. 5.13b). It turns out that—unless perhaps the reader has previously been converted to pseudization—the qualitative justifications of this aspect for these wayward metals are not as convincing as in the preceding cases, as we shall see below. To attempt a more convincing argument involves long and quantitative descriptions for each metal that are not appropriate here. But some idea of the underlying argument can be obtained by showing the stability of the f.c.c. structure against two kinds of distortion: Rhombohedral and tetragonal.

This is illustrated in Fig. 5.17, which is constructed in the manner of Fig. 5.15. It shows the effect upon the reciprocal lattice vectors close to q_0 of a tetragonal and rhombohedral distortion of the f.c.c. lattice in normal

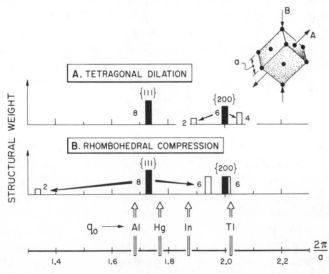

Fig. 5.17 Constructed in the manner of Fig. 5.15, this figure shows the effect upon the reciprocal lattice vectors of the f.c.c. structure of two distortions at constant atomic volume: a tetragonal dilation A equivalent to that of In ($c/a = 1.083$) and an 11% rhombohedral compression B. The corresponding values of q_0 from the ion's pseudopotential are included; the text explains how this diagram illustrates the possible origins of the distortions observed in In and α–Hg.

space. A tetragonal distortion has no effect upon the **G** vectors arising from the {111} planes, but it splits as shown those arising from the {200} set. (In fact the diagram is drawn specifically for the case of In where the small f.c.t. dilation gives a lattice parameter ratio $c/a = 1.083$.) The rhombohedral compression, on the other hand, splits in the manner shown the **G** vectors arising from the {111} planes, and those arising from the {200} set are very slightly shifted (without splitting) to a larger value of **G**. (This is exaggerated for clarity in the figure; it is constructed for the arbitrary case of an 11% compression of the cube's diagonal B and so corresponds to no particular metal.) The corresponding values of q_0 for Al, Hg, In, and Tl which, of course, are obtained from the corresponding ion's pseudopotential (Fig. 5.13b) are aligned for comparison in the bottom of the figure. In the case of In, the argument is* that by adopting the f.c.t. structure it is able to split up the set of lattice vectors of weight 6, which lie fairly close to q_0, into an arrangement that has a smaller weight close to q_0. Thus, although a set of weight 2 is shifted closer to q_0, the set of weight 4 is moved farther away. The net effect is said to be a reduction in the total weight close to q_0 so that the f.c.t. structure is energetically favored.

The reader was warned that these qualitative arguments are not altogether convincing, for who is to say that the detrimental effect of the approach of a weight 2 towards q_0 does not outweigh the benefits obtained from the expulsion of a weight 4? The answer lies in the more quantitative details. It is not just the relative positions of **G** with respect to q_0 that are important but also the slope of the $v(\mathbf{q})$ curve (Fig. 5.13b) at the different positions. To go further leads into details beyond our present scope, but, frankly, it often requires close inspection of the form of the pseudopotential to reveal the slight wiggles or singularities required to support the quantitative arguments.

Turning to the rhombohedral distortion, which is observed in α–Hg (Fig. 5.6), we see that in the f.c.c. structure the set of lattice vectors arising from the {111} planes lies very close to the corresponding q_0 value for this metal. As already described, a tetragonal dilation has no effect upon this set, but the rhombohedral one splits them up as shown in Fig. 5.17. Clearly, in terms of these very crude arguments, the rhombohedral distortion leads to a bigger band gap in the itinerant electron's energy distribution and, therefore, to a preferred crystal structure. Perhaps this is the physical cause for the occurrence of this structure in α–Hg, but why does Al not show the same distortion? Again, the **G** vectors arising from the

---

* V. Heine and D. Weaire, *Phys. Rev.* **152,** 603, 1966.

{111} lattice planes are close to q_0 (Fig. 5.17) and by the same token we would expect to find the f.c.c. structure unstable with respect to the rhombohedral compression. Here the reply is that competition against any distortion from a close-packed structure is provided by the Ewald term, as indicated above, and this is larger in Al than in Hg, because the Ewald energy depends upon the density of the itinerant electrons in the interionic spaces, which is largest in Al because that metal has the smaller atomic volume and larger valence. For this reason Al maintains its f.c.c. structure, but Hg (at room temperature) is able to benefit from the rhombohedral distortion. The **G** values arising from the {111} and {200} planes in Al actually lie in the $U_{bs}(q)$ diagram as shown in Fig. 5.14. Correspondingly, the band gap along the [200] direction is expected to be greater than that along [111]. This is confirmed by calculation of the band structure, which leads to the results summarized in Fig. 6.2.

5.4   CHAPTER SUMMARY

The classification of the crystal structures that are observed in solids is a vast subject, but it is very much simplified when, as in the present case, the emphasis is upon just those structures found in the pure metallic elements. There is then a good case for considering these solids to be built up of layers of atoms that are stacked one upon the other in particular ways. Sect. 5.1 and Sect. 5.2 attempt to outline this approach.

Although it is certainly useful to have such an orderly view of crystal structures, the whole subject is, of course, no more than a neat bookkeeping of the mass of empirical data. In general it says nothing about the detailed reasons why one structure is preferred over another in given circumstances. Modern attempts to do this revolve around the concept of pseudization. This approach treats the interaction between the itinerant electrons and the fixed lattice of ions without taking much interest in the true circumstances, which are acknowledged to be too difficult to handle. It involves instead the construction of a model that is tractable and yet gives the correct behavior of the interaction in the circumstances of interest. The principal parameter in this model is the form of the electrostatic potential that is associated with a single model ion; it is called the ion's pseudopotential. It can be constructed in two ways: Either entirely from first principles or using appropriate pieces of experimental data; however, it almost invariably has the form shown in Fig. 5.13.

We can show without recourse to pseudization that the crystal structure of a metal is preferably a relatively close-packed one having a high density of itinerant electrons in the interionic spaces. This expresses simply the requirement of a minimum Coulomb energy for the combined electronic and ionic systems (the Ewald energy). However, this requirement embraces numerous minor modifications or distortions in the observed structures, and these—at least in the cases of the NFE (or s/p bonded) metals—can be understood for the most part qualitatively in terms of the extra stability produced by the introduction of a band gap into the energy range available to the itinerant electrons. Physically, the extra stability comes from the reduction in the electrons' energy achieved by their non-uniform distribution in the interionic space as they search out the regions of lowest electrostatic potential. (Exactly as described for the origin of the band gap in the imaginary metal of Sect. 2.3.3 and Fig. 2.11.) Depending upon the relative energies of the atom's valence states, the more energetic electrons in the resulting metal may prefer to concentrate within the interionic spaces, avoiding as far as possible the regions close to the ions, or alternately they may prefer to pile up close to the ions.

Whatever the case, the result of this rearrangement is the appearance of the energy gap, and its manifestation in the pseudopotential approach forms the subject of Sect. 5.3. Briefly there are two effects (which are sometimes competitive). First, the reciprocal lattice vectors of the structure are by preference shorter than the diameter of the Fermi sphere. Second, these vectors avoid to the maximum extent possible values lying close to the pseudopotential's crossover point (q_0 of Fig. 5.13b). In terms of these influences it is possible to give a plausible description of the origins of the crystal structures in the NFE metals, and this forms the subject of Sect. 5.3.3.

RELATED READING

Specific research papers are referenced where cited in the text.

CRYSTALLOGRAPHIC ASPECTS

J. F. Nye, 1957, *Physical Properties of Crystals* (Oxford University Press: Oxford). A most readable description of the subject of Sect 5.2 at an intermediate level.

W. B. Pearson, 1972, *The Crystal Chemistry and Physics of Metals and Alloys* (Wiley-Interscience: New York). A specialist's treatise describing the variety and classification of crystal structures encountered in metals and alloys. All of the ideas and many of the diagrams appearing in connection with Sect. 5.2.1 to Sect. 5.2.3 are taken from this source.

PSEUDOPOTENTIAL THEORY

Not much of this relatively modern subject has yet found its way into the elementary or nontechnical literature. Perhaps the least demanding descriptions currently available are:

P. Wilkes, 1973, *Solid State Theory in Metallurgy* (Cambridge University Press: London); especially Chap. 13.

W. Harrison, 1967, "Simple Metals and Pseudopotential Theory," in *Atomic and Electronic Structure of Metals* (American Society for Metals: Metals Park, Ohio) Ed. J. J. Gilman and W. A. Tiller; p. 1.

Advanced descriptions in the specialists' review journals are much more common. For example:

J. M. Ziman, *Advances in Physics* **13**, 89 (1964).

V. Heine, M. L. Cohen, and V. Heine and V. Heine and D. Weaire, consecutive articles in *Solid State Physics* **24**, 1, 38, and 250 (1970). Most of the ideas described in Sect. 5.3 are taken from these articles.

D. Weaire and W. B. Pearson, 1973, consecutive articles in *Treatise on Solid State Chemistry* (Plenum: New York) Ed. N. B. Hannay; pp. 43 and 115.

ELEMENTARY REVIEW ARTICLES

J. M. Ziman, "Thermal Properties of Matter," *Sci. Amer.*, September (1967) p. 181.

J. M. Ziman, "Understanding Solid Matter" *International Science and Technology*, July (1967) p. 44.

6

THE WHOLE

A LATTICE OF IONS IMMERSED IN AN ELECTRON FLUID

6.1 INTRODUCTION

Preceding chapters have attempted to maintain as far as possible the view that the two structural elements of a crystalline metal—the system of itinerant electrons and that of fixed ions—can be treated separately. Although this is pedagogically convenient, it is a primitive approximation that can lead only to the most rudimentary model of a metal. Now we turn to a more realistic description of aspects of electronic motion in the whole system consisting of the ionic lattice immersed in the fluid of itinerant electrons. This chapter brings together the preceding descriptions that bear upon the dynamic properties of the particular quasiparticles in this composite system that have been called (Sect. 4.2) the quasielectron and quasihole pairs. In the following we will use their more colloquial names, conduction electron and hole, and we shall emphasize their particulate properties, but we should always remember that they are part of a metaphorical description of the low-energy excited states of the composite system.

The aim of the chapter is to illustrate aspects of the dynamic response of these quasiparticles to outside influences, such as electric or magnetic fields, so that some understanding of the various macroscopic consequences will emerge. (These are often known by the rather cumbersome title of the galvanomagnetic effects.) We will consider two distinguishable contexts. First, in Sect. 6.3.1 to Sect. 6.3.3, aspects of electron motion in the simpler case of s/p bonded (NFE) metals are introduced. Their relative simplicity results from the fact that all the itinerant electrons belong to one energy band (of s/p symmetry) and are scattered into and out of states within the same band. In the more complicated context considered in Sect. 6.4 and following sections, d or f bands exist in which the electrons have itinerancy to varying degrees so that conduction electrons not only occupy separate energy bands but are also inextricably bound with the metal's magnetism. The chapter is thus a return to the subject of Sect. 2.3.4 but, of course, with the benefit of the subsequent description.

An indispensable tool in all of this is the Fermi surface construction. This subject has already appeared in previous sections (notably Sect. 4.2.3) but continuity and completeness require that at least its salient features be introduced more formally here (bearing in mind that it is a very well-aired subject that is described in almost every textbook of solid state physics).

6.2   THE FERMI SURFACE

Energy level diagrams like Fig. 3.2b tell little about an itinerant electron's motion in three dimensions. This is the purpose of the Fermi surface construction, which arises in the following way. The dynamic properties of a conduction electron are a function of its energy E and its quasimomentum $\mathbf{p}$. The form of this function is known technically as the dispersion relationship. To express this relationship visually and concisely for the system's ground state, a three-dimensional surface can be constructed for each metal. It is one of a class of surfaces each of which is made up by the representative points in reciprocal space that correspond to some selected value of the conduction electron's kinetic energy. As this chosen value is increased, the electron passes—in principle by discrete steps—through one surface after another in momentum space. The result is a set of nested surfaces each corresponding to a fixed energy (Fig. 6.1). At absolute zero, when the system is in its ground state and no quasiparticles are excited in it, there is a sharp boundary between the representative momentum states that are occupied and those that are vacant. This boundary is the outermost member of the set of nested surfaces; it is the Fermi surface. Clearly, for a system of free electrons where all momentum states up to E_f are equally probable, the surface will be a sphere (Figure 4.4c).

There is something very satisfying about a construction that can render a great deal of intangible information—like the motions in three dimensions of innumerable electrons—into a single surface. Although this has often a complicated or quaint shape, it can be modeled, held in the hand, oriented at will, and so provides a tremendous perceptual aid.

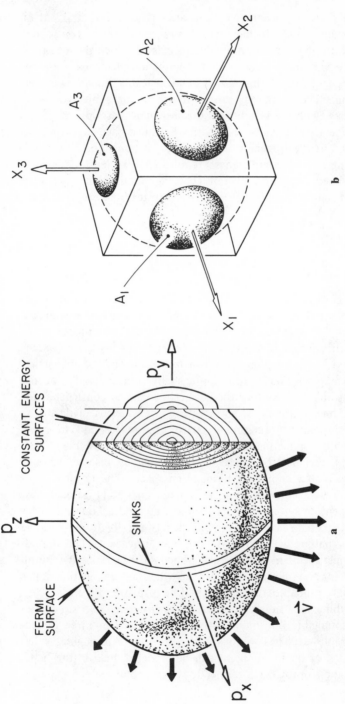

Fig. 6.1 (*a*) The Fermi surface is a three-dimensional construction in momentum space related to the dynamic properties of the itinerant electrons in the manner described in the text. There exists a set of discrete, nested surfaces each made up of the representative points corresponding to a fixed kinetic energy of an electron. The outermost member of this set—rather like the skin on a Spanish onion—is the Fermi surface. In the metal's ground state it encloses all the occupied energy states. When these surfaces are cut by a plane of a Brillouin zone (as in the cutaway portion along the y axis) indicating that the conditions are right for a Bragg reflection to occur there, their shape is modified near the plane as shown. (*b*) An example of the Fermi surface of a nearly free-electron metal. The surface extends slightly beyond the boundaries of the Brillouin zone that in this case is the cuboid of a tetragonally distorted metal like β–Sn.

258

The energies and inertial properties of the metal's conduction electrons are represented by this surface in a well-known manner. The velocity of an electron in any representative state on the surface $f = E(\mathbf{p})$ (or $f = E(\mathbf{k})$, since $\mathbf{p} = \hbar\mathbf{k}$) has a magnitude and direction given by the gradient $\nabla_p(f)$ [or $\nabla_k(f)$ when dealing in wave vector space] which is, of course, normal to the surface at the representative point. (∇_p is defined in connection with Eq. 4.2.) The inertial response of the electron to an external electric field is characterized by its dynamic effective mass, which has a value at any point given by the curvature $\nabla_p^2(f)$ of the surface at that point.

Frequently this is referred to simply as the effective mass—a convention that is also followed here—but it is rather a loose term for the following reason. Since mass is made the variable parameter in the formation of a macroscopic analogy with the quasiparticle's microscopic motion (Sect. 4.2) it can occur in different contexts and can mean different things. Thus the dynamic effective mass expresses the quasiparticle's response to electric fields; it is a property of each point on the Fermi surface. However, when the context is changed to the case of the particle's response in a strong magnetic field (Sect. 6.3.2) we find it appropriate to define the cyclotron effective mass as the variable parameter establishing the analogy; this is the property of a path on the Fermi surface and not of just some particular point on it. When the response of the quasiparticle to a thermal excitation is considered, the appropriate dynamic mass in the analogy is the thermal effective mass, which incorporates the quasiparticle's interactions giving the electron-phonon mass enhancement effect described in Sect. 4.5.1. These different parameters are from separate metaphorical descriptions of the many-body problem, and there is no reason why they should necessarily be the same. Consequently, the precise effective mass intended should always be specified in any description, but usually the author's intention is considered to be self-evident from the context.

When the Fermi surface is a sphere, so that the momentum $\mathbf{p}$ and velocity $\mathbf{v}$ vectors are always colinear, the effective mass is a scalar quantity. For other shapes, where $\mathbf{p}$ and $\mathbf{v}$ are generally not parallel, it becomes a second-rank tensor. When the Fermi surface is cut by a plane of a Brillouin zone, as in Fig. 5.14 or Fig. 6.1a, indicating that conditions are right for a Bragg reflection to occur, the shape of the Fermi surface is modified near the plane as shown; the surface has to intersect the plane at normal incidence. Physically, this can be interpreted as a consequence of the Bragg reflection: On the plane, corresponding to the point of reflection, the component of the electron's velocity normal to the lattice plane is zero.

The shape (or topology, as it is generally known in the specialists' literature) of the Fermi surface clearly tells a lot about the behavior of the conduction electrons. In the case of Fig. 6.1b, for example, which shows a nearly free-electron metal having a spherical Fermi surface extending slightly beyond the boundaries of a Brillouin zone (which in this case is a cuboid since the figure represents a tetragonally-distorted metal like β-Sn) the electrical conductivity parallel to the x, y, and z directions arises principally from the electrons that are represented on the surfaces A_1, A_2, and A_3, respectively. However, the shape of surface does not alone tell the whole story. It says nothing about the symmetry of the wave functions of the electrons that are represented on it. This is an important omission because this symmetry specifies, in the manner described in Sect. 5.3, the way in which the conduction electrons distribute themselves in the interionic spaces. It, therefore, determines the response of these electrons to various scattering centers. For example, a foreign ion may be of a species that scatters preferentially those conduction electrons that pile up in the spaces away from the ions (p-type symmetry of Sect. 2.3.3) in which case the s-type conduction electrons are less bothered by it.

The shape of the Fermi surface can sometimes be obtained directly by experiment.* Although this shape is often a direct consequence of the existence of regions on the surface that represent a predominant electron symmetry, these can only be assigned when the metal's energy-band structure has been calculated for different directions in the crystal lattice (as described below). When this calculation can be carried out and the symmetry information extracted, it combines with the shape of the Fermi surface to give a quite complete picture of the response of the conduction electrons to externally applied fields or temperature gradients. However, of the thirty-odd pure elemental metals this has been achieved to any degree in only three: Cu, Ag, and Au. Note that even a complete knowledge of the shape of the Fermi surface and the symmetry of the electron states represented on it are only a small fraction of the total information needed to calculate most properties of a metal (its optical, magnetic, and lattice vibrational properties, for example, need more than this) but in principle it is sufficient to determine such properties as the heat capacity and Pauli paramagnetism of the electron fluid, as well as most of the metal's properties arising from the itinerant electrons' motion.

---

* The first measurement of the shape of the Fermi surface of a metal was made upon Cu by A. B. Pippard, *Phil. Trans. Roy. Soc. A.*, **250**, 325 1957. Since then a new branch of solid state physics has grown up. It is known as Fermiology and is concerned with establishing in quantitative detail the shape of the Fermi surfaces of metals, alloys, and compounds.

To see how regions of the Fermi surface can represent conduction electrons of a predominant symmetry, it is convenient to return to the comparison of Al and Cu that first appears in Fig. 2.9. These metals have several dissimilarities, but the important one here is that Al is a nearly free-electron metal with s/p bonding (Sect. 2.3.4) and Cu is not because of the presence of its d bands. Both metals adopt the f.c.c. structure, however, and this allows the comparison of their energy band structures shown in Fig. 6.2, which illustrates the effect of mixing the d states into the s/p conduction band.

The drawing is similar to Fig. 5.12; it shows the variation of the conduction electron's energy as it moves along two specific directions in the lattice: The [100] and [111] directions. For motion along the former direction, the Bragg condition is satisfied at a value of $\mathbf{G}/2$ determined by

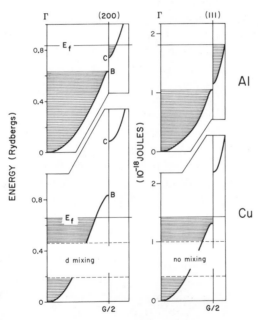

Fig. 6.2 A comparison of the conduction electron's energy for motion along two directions [100] and [111], in the f.c.c. lattices of Al and Cu. (Compare Fig. 5.12.) In Cu the hybridization of the d energy levels is a potential complication: It is relatively large along the [100] direction and nonexistent along [111], and the outcome is the distortion of the metal's Fermi surface from the free-electron sphere that is shown in Fig. 6.3a. In Al there can, of course, be no such hybridization with d levels. The higher valence of Al causes all the energy states in the first Brillouin zone to be filled so that electrons are forced to occupy states in zones of higher energy. 1 Ryd. = 13.60 eV = 21.8 × 10⁻¹⁹ J.

the separation of the (200) planes in normal space and gives the band gaps BC shown in the left-hand side of the figure. The corresponding situation along the [111] direction is shown in the right-hand side. A first point to notice has already been encountered in Sect. 5.3.3: The reciprocal lattice vectors corresponding to the (200) and (111) diffractions in Al lie on the band-structure energy curve as shown in Fig. 5.14c. The band gap along the [111] direction is, therefore, expected to be less than that along [100], as is observed in the figure. Turning to the case of Cu and the mixing (hybridization) of its d bands into the s/p conduction band states, we should realize that where such mixing occurs—giving hybrid wave functions of $s/p/d$ origin—the quantum symmetry requirements produce an effective repulsion between such states and the surrounding d-like functions of the same symmetry. In other words, the cost of mixing in some d-symmetry is a selective repulsion to higher energies of the resultant hybrid. However, the strength of such mixing varies from one crystallographic direction to another, and so, correspondingly, does that of the hybrid state's repulsion. For example, the symmetry of the d wave functions (Appendix A) leaves none of the d electron's distribution along the $<111>$ directions. Consequently, there is no significant mixing of the d and s/p states along this direction; the s/p states are not displaced in energy and the conduction band retains its normal symmetry variation within the zone, which is from a purely s-like symmetry at the zone's center (conventionally designated Γ) to a purely p-like one at the Brillouin zone's face (B of Fig. 6.2). Along the [100] direction, on the other hand, there is a considerable admixture of the d functions into the s/p band. Because of the hybrid's increasing p content as the face of the Brillouin zone is approached, this d admixture decreases progressively towards the zone's face (the symmetry properties of the p and d atomic functions preclude their admixture—a point that emerges from Fig. 2.10a.) But the admixture is nevertheless strong enough to push up appreciably the energies of the s/p band, as the comparison of the left-hand and right-hand sides of Fig. 6.2 shows.

We see from all of this that the hybridization between the d states and those of the s/p band is anisotropic—it depends upon the chosen direction in the crystal's lattice. The strength of the repulsion of the hybrid wave functions is correspondingly anisotropic. At absolute zero the electron states are occupied up to the fixed energy E_f, which, of course, is common to all crystallographic directions; as a result, this repulsion leads directly to a distortion of the Fermi surface in momentum space from that of a free-electron sphere. We see from Fig. 6.2 that along the [111] direction the Fermi surface is intersected by the (111) plane of the Brillouin zone. However, along the [100] direction the repulsion has lifted the point B so far above the Fermi level that the Fermi surface clearly lies some distance from the (200) zone plane; it is as if electrons forced into higher energy states by this repulsion spill over into the lower energy states and produce the corresponding distortion of the Fermi surface from a sphere. The predominant symmetry of states on the different regions of the Fermi surface emerges directly from all this: Along the <111> directions the conduction electron states are expected to be predominantly p-like with no appreciable d component, but along the <100> directions they have a significant d component mixed into their s/p character. (Actually, the latter turns out to be an intermediate case because similar arguments show that the greatest d admixture, and hence the greatest repulsion of the hybrid states to higher energies, occurs along the <110> directions.) These qualitative arguments thus account for the well-known shape of the Fermi surface of Cu that is shown in Fig. 6.3a.

Figure 6.3b shows the major part of the Fermi surface of Al. There are of course, no complications from hybridizing d bands—there is simply a conduction band of s/p hybrids—but the valency of 3 (compared with 1 for Cu) puts the Fermi level at a correspondingly higher energy. The result is that the first Brillouin zone is completely filled, as Fig. 6.2 indicates, so that electrons are forced to occupy states in higher zones. Some states in the second, third, and fourth zones are occupied as a result, but we ignore here the full details and show just the case of the second zone's occupancy. The caption explains how this figure is obtained by folding inward the appropriate pieces of the Fermi surface in what is known as the reduced zone scheme.

6.2.1 Bands and Zones. The preceding description given in Sect. 6.2 relates entirely to the case of an incompletely filled energy band, like the $3s/p$ band in Al of Fig. 2.9. We have pointed out that the Fermi surface is the boundary in reciprocal space between occupied and unoccupied representative points when the metal is in its ground state. However, the capacity of a Brillouin zone to accommodate representative points is not unlimited. In fact it is a standard result of the theory that this capacity is numerically equal to the number N of unit cells in the crystal's lattice. There are, therefore, N representative points in a Brillouin zone each of which corresponds to an energy state that can be occupied by at most two electrons, one of each sign of spin. (This is not true in the magnetic case, which is considered below.) Thus the number of allowed electron states per energy band within Brillouin zone is $2N$. Suppose that the metal has the equivalent of x atoms per unit cell and that each has a valency z. ($x = 1$ for the f.c.c. and b.c.c. structures, and $x = 2$ for the c.p.h. one.

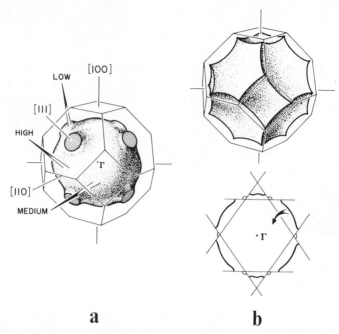

a   b

Fig. 6.3 (*a*) Showing the Fermi surface of Cu inscribed within the first Brillouin zone of the f.c.c. lattice. The distortion from a sphere arises from the anisotropic hybridization of *d* band energy states with those of the *s/p* conduction band. This effect is nonexistent along the [111] directions (Fig. 6.2) giving the regions of representative states on the surface with a low *d*–symmetry content (marked LOW); but it is very great along the [110] directions and so gives regions there of high *d*–symmetry content (HIGH). The hybridization is of intermediate strength along the [100] directions (Fig. 6.2). (*b*) A plan view (lower) of the central (110) plane showing the intersection of the Fermi surface of Al with adjacent planes of the Brillouin zones. Because of the metal's high valency none of the surface lies within the first Brillouin zone, but the sections of it that are emphasized lie within the second zone. These are conventionally folded inwards, as indicated, to reduce the figure into a polyhedron that is identical to the first Brillouin zone. The result in three dimensions is the second zone surface shown in the upper part of the figure. Note that because of the curvature of this surface, which is everywhere negative, it represents states containing quasiparticles having a negative effective mass (holes).

Here the valence is the number of electrons per atom that lie outside the ion's closed shells.) The number of valence electrons contributed per unit cell is xz, and the total number of them in a given sample made up of N unit cells is xzN. These have to be accommodated in the various representative states in reciprocal space, and it is clear that they occupy a volume of the Brillouin zone equivalent to $xzN/2N = 1/2$ (xz). The occupied fraction of a Brillouin zone is thus independent of the size of the sample, since it is independent of N. This is a very important property of the Fermi surface construction, for it would detract considerably from the concept's usefulness if the surface depended upon the size of the sample chosen.

We have remarked that the Fermi surface lies in one or more Brillouin zones. It is probably only in the particular cases of the monovalent metals—specifically the alkali metals and Cu, Ag, and Au—that the surface lies entirely in a single zone. Because of the restricted capacity of the first Brillouin zone, metals of higher valency are forced to accommodate electrons in higher zones. When the metal has a complicated $E(\mathbf{k})$ structure, there may be several partially filled energy bands giving rise to different sheets of the Fermi surface in different zones. It is obviously meaningless to talk of a Fermi surface in a filled band like the $2p$ or $3d$ bands in Al and Cu, respectively (Fig. 2.9) since there is no boundary between occupied and vacant states in the corresponding zones; the Brillouin zone's faces are themselves the Fermi surface in these cases. On the other hand, it is meaningful to talk about a Brillouin zone in such a case—indeed, we have just done so. Whenever there is overlap and hybridization of the atomic wave functions of contiguous ions, there is the possibility of itinerancy of the electrons contained in them, and itinerancy in a periodic potential leads directly to the Brillouin zone concept. Thus in the case of Al in Fig. 2.9, it is meaningless to talk of either a $2s$ Fermi surface or a $2s$ Brillouin zone: There is inadequate overlap of $2s$ wave functions to produce any energy band width and thus the possibility of appreciable itinerancy for the resident electrons. In the case of the $2p$ levels there is an energy band, and hence a Brillouin zone, but they are filled; so it is meaningless to talk of a Fermi surface for the $2p$ zone. Finally, in the $3s/p$ band there are vacant states and so the full range of constructions in reciprocal space is applicable: A Fermi surface exists for this s/p band that, because of the metal's valence, extends beyond the first Brillouin zone (Fig. 6.3). In fact the surface in this metal has sheets in four Brillouin zones.

The description so far applies to the case of a nonmagnetic metal when the electrons of spin-up and spin-down configurations in a given state have the same energy. In a magnetic metal (that is, a ferromagnetic or antiferromagnetic one) the exchange effects between the electrons in a given band produce an energy difference between the two spin configurations (Sect. 6.4.2). Both the spin-up and spin-down configurations have their energies shifted with respect to the nonmagnetic values: Here the spin-up electrons are assumed to have theirs lowered, and the spin-down ones to have theirs raised by the same amount.* Each representative point in

---

* This is not a universal convention; some authors let it be the spin-up electrons that have the higher energy.

reciprocal space now no longer corresponds to an energy state that can contain two electrons. It is as if all these points had been split and displaced in energy to new positions in reciprocal space. To a first approximation the effect of the metal's spontaneous magnetization is to retain the shape of the Fermi surface but to expand and contract its separated spin-down and spin-up components, respectively. The net result is that there are separated Fermi surfaces in separate Brillouin zones for the itinerant spin-up and spin-down electrons. (In the band picture this, of course, corresponds to an energy splitting into separate spin-up and spin-down bands.) In these separate Brillouin zones—which are sometimes called spin zones—a representative point corresponds to a state that can contain only one electron that must have the appropriate spin configuration. A spin zone's capacity for representative points is thus still equal to N, but this corresponds now to a maximum of N electron states per zone.

6.3 QUASIPARTICLE MOTION IN S/P BANDS

6.3.1 Quasiparticle Motion in an Electric Field. The qualitative effects produced by an externally applied electric field are described in Sect. 2.2. We recall that in the absence of electric and magnetic fields the random motion of any one itinerant electron is always cancelled by that of another that is moving in exactly the opposite direction with the same momentum. However, the application of an electric field superimposes a steady drift motion upon each itinerant electron. (Since the electron carries a negative charge, this preferential drift is exactly opposed to the direction of the applied electric field.) The drift motion produced by the field gives a net transport of charge along the field's direction in any time interval. This, of course, is the electric current.

In the absence of a magnetic field, the electrons follow near-linear paths between their inevitable collisions with such objects as foreign ions that are present as impurities in a pure metal, lattice ions displaced temporarily from their equilibrium positions (as in a lattice vibration), isotopic ions, the inside walls of the sample, other itinerant electrons, spin waves in a magnetic metal (Sect. 6.4.2) or any other deviation from a perfectly periodic lattice. It is such collisions that limit the current density flowing in given circumstances under the impetus of an applied electric field and so produce the electrical resistivity. And yet ironically it is such collisions that permit the metal to carry any net current at all. For with no collisions whatsoever a conduction electron is simply accelerated up to the Bragg reflection condition for its direction of motion and then is elastically reflected to the same ate in the opposite direction. In such a situation the electron is trapped into making the so-called Zener oscillations in normal space, and thus it can make no net contribution to any current density. It is the resistive collisions that break up this repetitive motion to make a net current possible.

Scattering events can be regarded as the creation and annihilation of representative points in momentum space. In general, both the energy and the momentum of a conduction electron are changed in a collision, but there are many cases in which the change in energy is so small compared with the energy of an average conduction electron that it can be ignored. Such elastic scattering, which is expected from point defects like impurity ions or lattice vacancies, obviously corresponds to the annihilation and subsequent creation of a representative point on the same energy surface— the Fermi surface, assuming the temperature is close to absolute zero. The main interest here is the change in the electron's momentum (which is, of course, equivalent to a change in its **k** value; $\mathbf{p} = \hbar\mathbf{k}$). If the angular separation of the points annihilated and created on the Fermi surface is small, then the event is known as a small-angle process; otherwise it is a large-angle one. This angular separation is clearly a measure of the effectiveness of the scattering event in diverting the electron's momentum and so removing the particle from the current flow. Furthermore, the angular separation—like the scattering cross section—is inversely proportional to the size of the scattering center. Thus small scattering centers like impurity ions, isotopic ions, lattice vacancies, or other point defects, give rise to large-angle scattering. This is also known as catastrophic scattering when the electron's representative point has an equal probability of reappearing anywhere on the Fermi surface, indicating that the electron retains no memory of its previous collisions. This contrasts with the scattering produced by extended scattering centers such as dislocations or lattice vibrations of very low energy; this is small-angle (or diffusive) scattering in which the electron's momentum is only slightly deviated by the collision so that the motion of the electron after the collision strongly resembles its motion before the collision.

In fact, the scattering by lattice vibrations (phonon scattering) is a rather different matter from scattering by point defects. Energy is always exchanged between the electron and the phonon (if the latter has a local vibrational frequency v, then an energy quantum $\hbar v$ is exchanged) and the scattering process has to express the energy balance between the electron and the phonon. The scattering is thus never really elastic, and it can only be approximately elastic if $\hbar v$ is very small compared with the initial energy of the electron. Furthermore, phonon scattering can give either small- or large-angle processes depending upon whether or not the electron simultaneously undergoes a Bragg reflection. Those electrons that have energies close to the Bragg condition for a given crystallographic condition can sometimes acquire sufficient energy from the phonon collision to satisfy the requirements of the reflection; the electron's momentum in such a process is then turned through a very large angle—typically up to 180°—and this is known as an Umklapp process (or U process) (Fig. 6.4). These processes are clearly most probable at higher temperatures where the higher energy phonons needed for a U process exist in the system. At lower temperatures (typically $T<<\theta_D$), where U processes are most unlikely, the density of phonons is known from simple theory to vary as T^3. Since the collisions can only produce small-angle deviations from the electron's forward motion (the angle is actually of the order of T/θ_D) the resistivity arising from them is found to vary as the square of this angle (that is, as T^2/θ_D^2). This simple theory, therefore, predicts a temperature dependence for the metal's resistivity of T^5 due to this scattering by low energy phonons.*

At intermediate and higher temperatures ($T>\theta_D$) the only significant electron scattering processes leading to electrical resistivity in a pure metal are of the Umklapp variety; small-angle processes do not add any significant contribution directly to the resistivity although they may be important indirectly in bringing the electron's motion into a situation in which a U process is possible. The picture of scattering in terms of the Fermi surface construction (Fig. 6.4) is thus of a given representative point being moved in a random manner over the surface with a Brownian motion as the corresponding electron undergoes constant small-angle scattering processes. None of these shift its momentum vector significantly from the direction of the main current's drift, and as a result, none contribute appreciably to the electrical resistivity. Eventually this Brownian motion brings the representative point close enough to a Brillouin zone boundary to permit a U process (in other words, the electron's energy happens to approach the critical value required for a Bragg reflection in the direction in question). From this point a large-angle process is possible, and then the electron is removed from the current's mainstream; it is said to be scattered resistively. When U processes dominate the scattering, the electrical resistivity of the pure metal at any but the lowest temperatures is a measure of how efficiently these small-angle processes can bring a representative point up to a Bril-

---

* This T^5 law has been described as the worst-obeyed law in solid state physics. No careful measurements have ever shown that it is obeyed by any metal; the exponent is always found to be somewhat above or below the value 5, no matter how much fudging the experimentalist makes with his ruler through the data points.

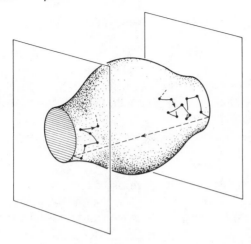

Fig. 6.4 Scattering of an itinerant electron in a metal occurs when its trajectory is changed as a result of a collision with some perturbing influence. In terms of the Fermi surface construction, elastic scattering events can be regarded as the creation and annihilation of representative points on the Fermi surface. Repeated small angle scattering events can be pictured as a kind of Brownian motion of a representative point as it moves in a random manner over the Fermi surface in response to the represented electron's interrupted motion. Eventually such Brownian motion brings the representative point close enough to a Brillouin zone's boundary so that an Umklapp process is possible. This is one in which during the scattering the electron is also able to simultaneously undergo a Bragg reflection. In such a scattering event the electron's momentum is generally turned through a very large angle.

louin zone boundary. As the temperature is reduced towards absolute zero, the small-angle processes become less efficient as the lattice vibrations diminish, and so the electrical resistivity falls. But no crystal is ever ideal; since it always contains impurities and imperfections, ultimately the scattering from these sources becomes dominant. It is typically large-angle scattering, as we have seen, and is also independent of temperature. This is the origin of the temperature-independent component of the metal's resistivity that is observed in the so-called residual resistance range (Fig. 4.11).

Since it turns the electron's momentum through what is usually a wide angle, and so diverts the electron from the direction of any general drift, a *U* process is a resistive one. However, it is not the only kind of resistive process. In fact, any scattering which randomizes the electron's motion— so that after the collision it has an equal probability of going with or against the general drift—is a resistive one. There are regions on the Fermi surface where this randomization occurs, and if a scattering event terminates with a representative point on one of these sinks, as they are called, then the electron has been resistively scattered. A sink may arise from two sources: It may be a region for which the scattering probability is higher than for surrounding regions, which is discussed below, or it may be a region such as the equatorial region of the Fermi surface shown in Fig. 6.1a. If an electric field $\mathbf{E}$ is applied along the y direction in this figure, the electrons in states represented on this equatorial belt make no contribution to the net current in the sample because their velocity is normal to $\mathbf{E}$. When an electron is scattered out of one of these states, there is an equal probability that it will acquire a velocity component that either opposes or supports the general current drift. (The Fermi surface is symmetrical with respect to this equator; it has equal areas on either side of it.) Consequently, any electron that is scattered into such an equatorial state—whether by a large- or small-angle process—is resistively scattered. In normal circumstances such sinks are thought to play only a minor role in determining the electron transport properties, for they are rather sparsely distributed (in the example, the equatorial band is strictly only a line of representative points) but when a magnetic field is applied their importance is believed to increase considerably. This aspect is considered further in Sect. 6.3.2.

The mean time spent by an electron in a given state—that is, the mean free time between its momentum-changing collisions—is known as the relaxation time and is denoted by τ. It is obviously also the mean lifetime of the corresponding representative state on the Fermi surface. In general this is not a constant but varies from region to region over the surface. Conventionally this is expressed as $\tau = \tau(\mathbf{k})$ and is known as an anisotropic relaxation time. Such anisotropy can arise in two ways. First, it can arise from the anisotropic shape of the Fermi surface itself. This brings some regions closer than others to the Brillouin zone planes, so that the electrons represented on them are more liable to experience U processes than those on other regions. (This expresses an intrinsic anisotropy in the phonon scattering.) Second, even if the Fermi surface is spherical, the relaxation time can be anisotropic if the dominant scattering process has an inherent anisotropy. For example, to recall a case mentioned in Sect. 6.2, if a foreign ion having localized electrons with predominantly p symmetry is introduced into Cu, it scatters preferentially those electrons in predominantly p-type states, which are around the $<111>$ directions. What an anisotropic relaxation time means in terms of the Fermi surface construction is that there are regions on the surface that act like the sinks described above: These are the regions where τ is short (high scattering probability) so that, if a representative point wanders—or is swept, as will be described later—onto one of them, the corresponding electron has a higher probability of being removed from the conduction process by having its velocity randomized by a scattering. Because of their higher scattering probability, these regions are inevitably less populated than the rest: The higher scattering rate effectively drains away electrons from them and redistributes their representative points around the Fermi surface—so they are aptly described as sinks.

Finally, we should mention the well-known description of the net current flow in a metal in terms of the Fermi surface construction. Since an electric field adds a fixed drift component to the velocity of each conduction electron, it is not difficult to see that the effect in momentum space when τ is isotropic is a bodily displacement of the Fermi surface along the applied field's direction. (Again, since the electron is negatively charged, the displacement is actually exactly opposed to the applied field's direction.) The origin of this displacement can be looked upon in two essentially equivalent ways: Either as a transfer of representative points right around the Fermi surface to build up the occupancy of states on one side at the expense of those on the other or as the creation of quasielectrons on one side of the surface and quasiholes on the other (Sect. 4.2.3). Whatever the view, the outcome is the displacement described above. This displacement increases with time as long as the conduction electrons continue to gain energy from the applied field, but ultimately a dynamic equilibrium is reached with the scattering processes, which offset the displacement of the Fermi surface, and so the displacement is stabilized.

It follows from this that if the relaxation time is anisotropic, different regions of the Fermi surface are displaced different amounts by a given electric field. In other words, there is a distortion of the Fermi surface as well as its displacement in momentum space. Qualitatively, this is easy to understand from the preceding paragraph: The scattering offsets the displacement because it reduces the drift velocity acquired by the electron. Hence, regions on the Fermi surface having a relatively short τ are displaced less by a given field than those with a long τ.

6.3.2 Quasiparticle Motion in a Magnetic Field.
When a magnetizing force **H** is applied to a metal and establishes a magnetic flux density **B** in it, each itinerant electron moving with a velocity **v** experiences the Lorentz force that is proportional to:

$$e(\mathbf{v} \times \mathbf{B}) \tag{6.1}$$

Because this force is always perpendicular to the electron's motion, it can do no work upon the electron. It is thus an imaginary force in the Newtonian sense. It is unable to change the electron's energy, but it adds a fixed curvature to the electron's path in the plane perpendicular to **B**. The well-known consequence of this force in terms of the Fermi surface construction is to cause each representative point on the Fermi surface to circumnavigate the path formed by the intersection of the surface with the plane that contains the point and is perpendicular to **B** (Fig. 6.5). This path is known as the cyclotron orbit. It is usually a closed curve that, for the monovalent metals at least, is generally enclosed within a single Brillouin zone, but in special cases described below it can be an open curve that extends indefinitely throughout the repeated zone scheme.*

The resulting motion of an itinerant electron in normal space is rather complicated because only the component of the motion normal to **B** is affected. In general the electron follows a tilted helical or hypocycloidal† path about **B**. The picture viewed along **B** is much simpler: The electron's path in normal space is an exact replica (although rotated through $\pi/2$ and scaled in proportion to B^{-1}) of its corresponding cyclotron orbit. There is a seemingly harmless tendency to refer either to the electron's path in normal space or to its replica traced out by the representative point in reciprocal space as the cyclotron orbit.

Returning to reciprocal space, a representative point circumnavigates its cyclotron orbit as long as the lifetime of the electron in the corresponding state permits. If the electron is subject to strong scattering, the representative point is able to complete only a small segment of the orbit during the electron's lifetime between collisions, whereas in the absence of collisions it makes continuous revolutions of the circuit for as long as **B** is applied. It is a standard result that the frequency of these complete revolutions, which is known as the cyclotron frequency ω_c is given by

$$\omega_c = \frac{eB}{m} \tag{6.2}$$

where m is here the cyclotron effective mass (Sect. 6.2). Whether the conditions permit complete revolutions of the orbit or merely infinitesimal segments of it depends upon the electron's relaxation time τ for the particular scattering process compared with the magnetic field strength. This

---

* Like the Wigner–Seitz polyhedra (Sect. 2.3), Brillouin zones are space-filling constructions when stacked together. It is often convenient to do this (as in Fig. 6.8, for example) so that the trajectory of a representative point is not constantly jumping from one side of a zone to the other at each Bragg reflection.

† A hypocycloid is the path traced out in a plane by a point on a circle as the latter rolls inside a second circle. If this second circle is cut and opened to make a straight line, then the path traced out is a trochoid, a curve that will be encountered subsequently.

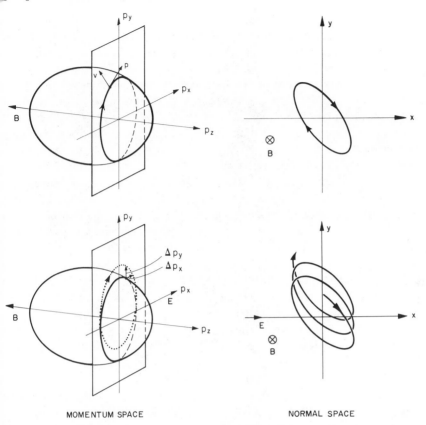

MOMENTUM SPACE   NORMAL SPACE

Fig. 6.5 (*Top*) Showing a cyclotron orbit in both reciprocal (momentum) and normal spaces when a magnetic flux **B** only exists. (*Bottom*) The same situation when an electric field **E** is added in the direction shown. Then the cyclotron orbit in momentum space is perturbed to a new position (dotted) and the electron in normal space follows a trochoidal path giving a net displacement along the direction perpendicular to both **E** and **B**.

is usually expressed by the convenient criterion for an orbit: $\omega_c\tau \gg 1$ permits full revolutions of the orbit; $\omega_c\tau \ll 1$ permits only small segments of it. These specify* what are known as the high-field and low-field conditions, respectively; somewhere between these two conditions lies the ill-

---

* With the field strengths available in the average laboratory it requires very pure samples held at temperatures in the liquid He range to achieve $\omega_c\tau$ values greater than 1. Even in ideal conditions to achieve say $\omega_c\tau \approx 10$ for an orbit would be noteworthy. Consequently, experimentalists generally interpret $\omega_c\tau \gg 1$ to mean $\omega_c\tau > 1.0$.

defined intermediate-field region. The latter is not defined theoretically, for it is a mixture of the two limiting conditions above and so fits neither of them; yet experimentally it is an important region in which many studies of alloys and the less pure metals have been made. Roughly speaking, it is the region in which the behavior observed is typical of neither the high- or low-field conditions but is evidently a mixture of both of them. We shall return to this point subsequently.

At these two extreme limits the effect in momentum space of the application of **B** can be viewed first, in the high-field limit, as the continuous circumnavigation of their cyclotron orbits by all points on the Fermi surface. In this theoretical limit, where the effect of **B** is taken to be so completely overwhelming compared to any scattering effects that the latter are effectively nonexistent, there is no longer any Brownian motion of a point on the surface nor any annihilation and creation of representative points. The shape of the Fermi surface dominates all since it determines the shape of the cyclotron orbits and, hence, the form in real space of their corresponding electron trajectories; the behavior of the system is perturbed from its field-free condition beyond all recognition. At the low-field limit, on the other hand, where the electron completes only an insignificant segment of the cyclotron orbit during its lifetime in a state, the Brownian motion of the representative point on the Fermi surface is still prominent, as is the point's continuous creation and annihilation caused by the electron's scattering. So the system is very little perturbed from its field-free state, but the existence of **B** has imposed about its direction a rotary drift motion of all the representative points; this is superimposed upon their Brownian motion. There is now a general sweeping of points around the surface that continues as long as **B** is applied. If $\tau(\mathbf{k})$ is anisotropic, this means a sweeping of points into and out of regions on the surface having different scattering probabilities. This has interesting consequences, some of which are discussed below, but one is seen immediately. The effect of the sinks is greatly emphasized by this sweeping since they now ultimately capture all points on cyclotron orbits that pass through them, whereas in the field-free situation they had to rely on the chance of a Brownian motion terminating upon them.

Even if there is no scattering present, the itinerant electrons still experience the regular, periodic electrostatic potential of the ionic lattice. When this is weak (as in the cases of the NFE metals, for example), we pointed out in Sect. 5.3.1 that the resulting interaction can be treated as a diffraction phenomenon: When the Bragg reflection condition is satisfied by the electron's wave vector (as it is on the Brillouin zone planes) the electron is elastically reflected by the lattice planes. In a polyvalent metal the Fermi

urface may be intersected by a dozen or more Brillouin zone planes (com-
are the case of Al in Fig. 6.3) and electron states lying close to these
ntersections are modified by the diffraction effect. In an NFE metal, how-
ver, these amount to only a small fraction of the total number of states
epresented on the Fermi surface; thus they behave predominantly like free
lectrons. However, in the presence of a magnetic flux, which we have said
urves the electrons' linear trajectories, this diffraction takes on a new
spect. Fig. 6.6 illustrates schematically how in normal space a reflection
f an otherwise free electron by a set of lattice planes can give a different
ffect depending upon the curvature of the electron's incident trajectory
ith respect to the lattice planes. In the absence of lattice planes, the
lectron's trajectory with the field directed out of the figure is the dashed
ircle, but when reflecting planes are introduced two fundamentally differ-
nt modifications of the trajectory shown at AA and BB are possible. Their
lifference may at first sight seem trivial—one can be described as a convex
eflection and the other as a concave—but when the complete trajectories
nade up of either type of scattering are compared (Fig. 6.7) the impor-
ance becomes evident. Reflections of type BB in Fig. 6.6 give a trajec-
ory that is circumscribed by the electron in the sense expected for a nega-
ively charged particle (to give the electron orbit of Fig. 6.7) but reflections
f type AA give the direction of circumscription expected from a positively
harged particle (hole orbit of Fig. 6.7). In both cases the electron behaves
nicroscopically as expected for a negatively charged particle (both orbits
re made up of segments of the free electron's trajectory); it is the inter-
ction with the lattice that produces on a more macroscopic scale the

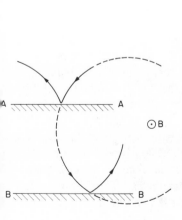

Fig. 6.6 Illustrating in normal space how
the reflection by a set of lattice planes of
an otherwise free electron moving in a
magnetic flux **B** can give a different effect
depending upon the curvature of the electron's incident trajectory with respect to the
lattice planes. The dashed circle represents a
free electron's trajectory in the absence of
lattice planes. To get an idea of its radius,
suppose $B = 1T(10 \text{ kG})$. Then $\omega_c = 17.5 \times$
10^{10} cycles/sec and for a typical Fermi
velocity of 1×10^6 m/sec for a free electron
the radius of its circular path is 5.7×10^{-6}
m. This increases as B decreases, of course,
so that if B is of the order of 10^{-3} T (that
is, just a few gauss) then the radius is a
few millimeters.

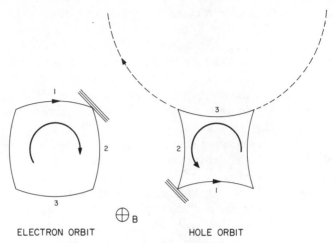

ELECTRON ORBIT   HOLE ORBIT

Fig. 6.7 A view in normal space down the direction of an applied magnetic flux **B** in a nearly free-electron metal. A free electron in the high-field condition follows the circular trajectory shown dashed, but if the electron actually suffers repeated Bragg reflections by the planes of ions (like those represented by the parallel lines) two types of closed trajectory are possible, corresponding to the two types of reflection illustrated in Fig. 6.6. An electron circumnavigates the orbit on the left in the sense expected for a particle carrying a negative charge (that is, in the same sense as the free-electron trajectory); it is said to be on an electron orbit. However, the sense of rotation of the electron on the right-hand orbit is opposite and thus typical of a positively charged particle; it is said to be on a hole orbit. Although both of these electron paths are made up of segments of the free-electron trajectory (like those labelled 1, 2, and 3) upon which the electron is behaving as a normal free electron, the nature of the Bragg reflections around the orbit makes all the difference to the electron's overall response. The hole orbit is typical of those resulting from a Fermi surface like that of Fig. 6.3(*b*). The text explains how as the low-field condition is approached, the response of electrons on such hole orbits changes to become more like a free electron.

effectively different behaviors that are typical of the orbit and not of the microscopic body that traces them. In other words, the quasiparticle—the itinerant electron interacting with the lattice—has variable dynamic properties. As pointed out in connection with Fig. 6.1, it is a matter of convention whether the quasiparticle's effective mass or its charge is regarded as the variable parameter.

In polyvalent metals, and in the monovalent ones having highly distorted Fermi surfaces, both electron and hole orbits coexist in the high-field condition, and the response of the electron fluid to external fields is, therefore, a balance between these characterisitcs. The Fermi surface of Cu illustrates this nicely (Fig. 6.8). When **B** is along a $<110>$ direction

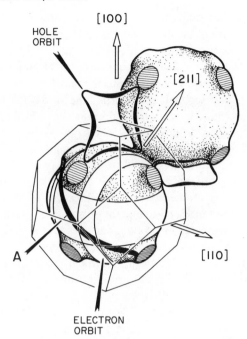

Fig. 6.8 The Fermi surface of Cu shown in the repeated zone scheme illustrates the different responses to the externally applied fields that can be expected of the itinerant electrons in a metal in the high-field condition. With a magnetic flux **B** along a <110> direction, for example, two bands of electron orbits exist around the belly of the surface (left blank) together with the four-cornered bands of hole orbits (shown black). With **B** along a [211] direction there exists a band of open orbits *A* that lies in a <211> plane and extends indefinitely in the [111] direction in reciprocal space.

for example, two bands of electron orbits exist around the belly of the surface (left blank) together with the four-cornered bands of hole orbits (shown black and in the repeated zone scheme) that are qualitatively like those of Fig. 6.7. Similarly, with **B** along a <100> axis there exists a wide band of electron orbits around the middle of the surface together with associated bands of hole orbits (again shown in black; these are known as the dog's bone orbits because of their shape).

In the presence of a magnetic flux, the conduction electrons (that is, the quasiparticles) can, therefore, respond either as electrons or holes depending upon the curvature of their trajectories with respect to the lattice planes. But all of the preceding relates to the behavior characteristic of a complete orbital and, hence, is appropriate only to the high-field condition: It has tacitly assumed that the representative points on the Fermi surface are able to complete at least one cyclotron revolution in the electron's life time so that the nature of the electron's response can be assessed. However as the conditions move towards the low-field limit (experimentally this can be achieved by lowering the field strength or by increasing the temperature for example) and the representative point is unable to complete more than a short section of any segment of its orbit, the free-electron, microscopic behavior of the particle becomes more pronounced and the behavior typi cal of the whole orbit is completely washed out. This can become clear from a consideration of Fig. 6.7 (for it is immaterial here whether the figure represents trajectories projected in normal space or cyclotron orbits in reciprocal space, since one is the replica of the other). The orbits there are made up of free-electron segments (1, 2, 3, etc.). Whenever the sec tion traced out by the representative point during the electron's lifetime lies entirely on such a segment, the conduction electron responds like a free electron. Only when the electron actually undergoes a Bragg reflection in its lifetime—so that the trajectory in reciprocal space includes a point of intersection of two segments—does it respond to the external field like an electron that is nonfree. In the low-field limit, where perhaps tens of thousands of electron lifetimes are required for a point to traverse one seg ment of the orbit, the overwhelming majority of electrons represented on the orbit obviously respond like a free electron. This is illustrated simply by the comparison of the total length of free-electron segments of the orbit with the nonfree-electron ones.

Therefore, as the conditions are changed from the high- to the low-field limit, conduction electrons represented on an electron orbit are expected to change their response from one typical of a negatively charged particle (al though not one particularly like a free electron) to one much more typical of a free electron. The conduction electron on a hole orbit meanwhile changes correspondingly its response from one typical of a positively charged particle to one typical of a negatively charged one that is further more like a free electron. In this way, it is possible in NFE metals, by changing the value of an applied magnetic field compared with the conduc tion electron's relaxation time, to change completely the response of some itinerant electrons in the metal. This effect can be seen quite clearly in the field dependence of the Hall effect (Sect. 6.3.3) of Al, for example.*

Finally, we should describe the circumstance mentioned in the opening paragraph that leads to an open cyclotron orbit. This arises in the special case when a Fermi surface is intersected by Brillouin zone planes in such a manner that it is possible to orient **B** to give a cyclotron orbit that ex

---

* K. Forsvoll and I. Holweck, *Phil. Mag.* **10**, 921, 1964.

ends indefinitely throughout the repeated zone scheme in reciprocal space. Figure 6.8 shows how this is possible in the case of Cu. With **B** along [211], for example, there exists a band of cyclotron orbits (labeled *A* in the figure) that lies in a <211> plane and extends indefinitely in the [111] direction in reciprocal space. In the high-field limit, which is the only circumstance when an open orbit is effective, an electron represented on the orbit is in normal space effectively chanelled between parallel lattice planes at which it suffers repeated Bragg reflections. [Except for the curved trajectories imposed by the Lorentz force, an open cyclotron orbit is a form of Zener oscillation in a component of the electron's motion (Sect. 6.3.1).] The corresponding situation in normal space is shown in Fig. 6.9 for the cyclotron orbits *A*. The flux **B** is applied along a [211] direction, and an electric field is also applied along the direction [110] normal to **B**. The electron makes a generally trochoidal motion (which is broken by its repeated reflections at {111} planes) and has a net displacement along [110]. This illustrates a characteristic difference between open and closed cyclotron orbits in the high-field limit: The average value of the velocity vector taken around a closed orbit is zero in any direction normal to **B** (Fig. 6.1) but for an open orbit it is generally nonzero in directions normal

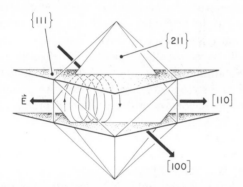

Fig. 6.9 An open cyclotron orbit is one that in the high-field limit extends indefinitely along some direction in reciprocal space. In normal space an electron in an open orbit is effectively channeled between parallel lattice planes at which it suffers repeated Bragg reflections. This sketch shows the behavior in normal space of an electron moving in the band of open orbits *A* of Fig. 6.8. The magnetic flux **B** is applied into the figure along a [211] direction (which for clarity is tilted away from normal to the page by about 28°) and an electric field **E** is applied along the [110] direction normal to **B**. The electron follows the broken trochoidal path in a {211} plane as it repeatedly reaches the Bragg reflection condition for motion in the [111] direction. Electrons that are trapped in such open orbits are clearly constrained to move in a direction specified by the reflecting planes and so are unable to make their normal response to externally applied fields.

to **B**. Electrons that are trapped in such open orbits are constrained in the high-field limit to move in a direction specified by the reflecting planes and so are unable to make their normal response to externally applied fields.

6.3.3 Quasiparticle Motion in Electric and Magnetic Fields.

When both a magnetic flux **B** and an electric field **E** exist simultaneously, the problem soon becomes extremely complicated. Of the innumerable combinations of orientations of **B** and **E**, two are of particular experimental importance: That with **B** and **J** parallel (to be called the longitudinal arrangement) and that with **B** and **J** perpendicular (the transverse arrangement). **J** is the current density produced by the applied field **E**.) In general **E** and **J** are not parallel, and it is more convenient to describe the galvanomagnetic effects with respect to a fixed direction of **J** (along the axis of a cylindrical or parallelepiped sample, for example) rather than **E**. We turn first to effects observed in the longitudinal arrangement.

Since the Lorentz force acts only on the component of an electron's velocity normal to **B**, an electron traveling exactly along **B** experiences no disruption in its trajectory. Most electrons making up **J**, however, are moving at some inclination to **B**. Their general motion in normal space is, therefore, along either a helical or hypocycloidal trajectory with **B** as axis, depending upon the particular conditions. (Hypocycloid is defined in a footnote appearing at the beginning of Sect. 6.3.2.) In terms of the Fermi surface construction this corresponds in the high-field condition to a continuous rotation of the representative points on their cyclotron orbits about the direction of **B**. In the low- or intermediate-field condition the effect of **B** can be viewed as a sweeping of the representative points (Sect. 6.3.2). Some are swept into regions of lower scattering probability, so that the metal's electrical resistivity is decreased, but at the same time others are swept into regions of higher scattering probability (the sinks) so that the resistivity is increased. However, the result is always an overall increase of resistivity when **B** is applied because the regions of the excited distribution having a higher scattering probability depart least from the equilibrium position and so are inevitably less populated than the rest, and as a result, more electrons are transferred to regions of high scattering than would have been the case if no magnetic field had been applied. An increase in resistance produced by a magnetic field is known as a magnetoresistive effect.* In this longitudinal arrangement, where **B** is parallel to **J**, it is known as the longitudinal magnetoresistive effect. Note that there has

* The fractional change in the sample's resistance caused by the application of **B** $(\rho-\rho_0)/\rho_0 \equiv \Delta\rho_0/\rho_0$, is called the magnetoresistance. (ρ_0 is the resistance when **B** = 0.) This is the quantity that is measured experimentally.

to be a variation of the scattering probability over the Fermi surface if a magnetoresistance is to exist in this model of the low-field condition. If the Fermi surface is spherical and also has a completely isotropic electron relaxation time (Sect. 6.3.1) the rotation of representative points clearly has no effect and corresponds to a case where the longitudinal magneto-resistance is zero.

Even if the electron's relaxation time is isotropic, a longitudinal mag-netoresistance can exist in the high-field condition if the Fermi surface is distorted enough to support hole orbits. The origin of the magnetoresist-ance in this case lies in the fact emphasized above: That in the high-field condition it is the average velocity vector for a given cyclotron orbit that expresses the corresponding electron's response to outside fields. It has been said that the component of average velocity perpendicular to **B** tends to zero for a closed orbit as the high-field limit is approached, but what is important here is that for a hole orbit the component parallel to **B** also tends to zero. Physically, the electron is constantly being reflected by the lattice planes (four times per cycle in the hole orbits of Fig. 6.8) so that in each cycle its motion along **B** is divided equally between the forward and reverse directions. (In the case of the dog's bone orbits of Fig. 6.8, for example, it is easy to see that the orbit's velocity component along [100]—that is, along **B**—is alternatively positive and negative on each of the four segments between Bragg reflections.) Electrons that are represented on such hole orbits are obviously prevented from making any net forward motion along the current's direction as **B** is increased up to the high-field limit. The application of **B** thus effectively introduces extra resistive scatter-ing as the high-field limit is approached; hence, the magnetoresistance.

Even with an anisotropy in the shape of the Fermi surface or in the electron's relaxation time, the longitudinal magnetoresistance will not in-crease indefinitely with **B** when the surface supports only closed cyclotron orbits. As the high-field limit is approached, so that the scattering becomes insignificant, electrons are forced to make their helical or hypocycloidal trajectories without interruption. Increasing **B** decreases the radius of the trajectory in normal space, but does not increase the metal's resistivity. Furthermore, when all the hole orbits contributing to a given orientation of **B** have reached the high-field condition, their electrons are removed from the conduction process, and any further increase of **B** is also ineffective there. Consequently, the longitudinal magnetoresistance saturates when **B** is increased sufficiently.

In the transverse arrangement, when **B** and **J** are perpendicular, the situation is considerably more complicated. We are interested here in the motion in normal space of conduction electrons in a metal subjected to the

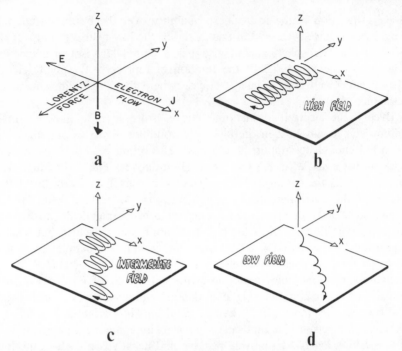

Fig. 6.10 (a) A schematic two-dimensional view of an electron's trajectory in normal space when subjected to a crossed electric field **E** and magnetic flux **B**. (b) Under high–field conditions. (c) Under intermediate–field conditions. (d) Under low–field conditions. These conditions are discussed in the text.

arrangement of fields shown in Fig. 6.10a. First, an electric field is applied that has a component **E** along the $-x$ direction and produces the electron flow **J**. Suppose, in this idealized case, that all the electrons move with a common drift velocity having the x component v_x. If a magnetic flux density **B** is now established along the $-z$ direction, it gives rise to a Lorentz force on each electron that acts along the $-y$ direction; the magnitude of this force is proportional to ev_xB. Similarly, if there is a component v_y of the electron's drift velocity along the $+y$ direction, a corresponding Lorentz force of ev_yB acts along the $+x$ direction. It is simple enough to write down the equation of motion for the x, y, and z directions. These are, respectively:

$$m\ddot{x} = eE + ev_yB$$
$$m\ddot{y} = -ev_xB \qquad\qquad (6.3)$$
$$m\ddot{z} = 0$$

where m is the electron's cyclotron effective mass and where the double dot denotes the second derivative with respect to time.

These equations can be combined into a single vector form. If the xy plane of Fig. 6.10a defines an Argand diagram, then the velocity of an electron in this plane is given by $v = v_x + iv_y$, and the sum of the first two parts of Eq. (6.3) is:

$$m\ddot{x} + im\ddot{y} = eE + ev_yB - iev_xB$$

or

$$m\dot{v} = eE - ievB \qquad (6.4)$$

where $\mathbf{v}$, $\mathbf{E}$, and $\mathbf{B}$ are vectors. Equation (6.4) can be solved in the following way. First, it is rewritten as

$$\frac{dv}{(v - E/iB)} = -\frac{ieB}{m} dt \qquad (6.5)$$

which upon integration gives

$$\log (v - E/iB) = -\frac{ieB}{m} t + \log C \qquad (6.6)$$

The constant of integration is $-E/iB$, since $v=0$ at $t=0$; this is written as $-iv_\perp$, where the meaning of $v_\perp (=-E/B)$ will emerge subsequently. It is then straightforward to reduce Eq. (6.6) to a final form in terms of $v_\perp$ and the cyclotron frequency ω_c [Eq. (6.2)]. Eq. (6.6) can be written as

$$\log \left[\frac{(v - E/iB)}{-iv_\perp} \right] = -i\omega_c t \qquad (6.7)$$

Removing the logarithm gives

$$v = iv_\perp - iv_\perp \exp (-i\omega_c t)$$

and finally, since $\exp(-i\omega_c t) = \cos \omega_c t - i \sin \omega_c t$, the equation becomes

$$v = -v_\perp \sin \omega_c t + iv_\perp (1 - \cos \omega_c t) \qquad (6.8)$$

The components of the electron's velocity along the x and y directions are thus given by $v_x = -v_\perp \sin \omega_c t$ and $v_y = v_\perp (1- \cos \omega_c t)$, where it is recalled that $E/B = -v_\perp$.

Physically, Eq. (6.8) corresponds to a trochoidal trajectory for the electron. It is composed of a circular orbit of angular frequency ω_c together with a fixed displacement in the $-y$ direction (of Fig. 6.10a) at a rate $v_\perp$. (To see this, consider the electron's motion with the time-dependent component removed; put $t=0$, for example.) Eq. (6.8) expresses a rather surprising result: The effect of crossed electric and magnetic fields upon a free electron is to produce an overall drift in a direction perpendicular to both applied fields; the electron's forward motion along the electric field's direction is, therefore, completely quenched (Fig. 6.10b). But this is only true in the high-field limit, which cannot be realized in practice! When electron scattering is admitted, these trochoidal electron paths are continually and catastrophically terminated by a scattering event, after which the electron starts afresh with no knowledge of its previous history. The result, which is typical of the intermediate-field region (Sect. 6.3.2), is to stop some of the electron's transverse drift and bring it into an effective forward motion along the applied electric field's direction (Fig. 6.10c). Thus, whatever forward motion is achieved in this condition arises because such randomizing collisions are able to disrupt the trochoidal motion. Here is another example of how scattering of electrons can be responsible for a net current flow (Sect. 6.3.1). Finally, when the mean free lifetime of a conduction electron between collisions is shortened with respect to the applied magnetic field's strength to the point that the low-field condition is approached, then there is never sufficient time between scattering events for the trochoidal motion to establish itself, so that Eq. (6.8) is no longer a valid description. Microscopically, the electron's trajectory in this case can be viewed as a succession of curved segments (actually parts of a trochoid but never long enough to be very distinguishable from a circle) joined as the result of repeated, randomizing collisions made by the electron (Fig. 6.10d).

All of the above has tacitly assumed that the sample in question has infinite extension, at least in the xy plane. Of course, a real sample has boundaries. Consequently, whenever a transverse drift of electrons is set up (that is, along the y direction in Fig. 6.10) it ultimately builds up a transverse concentration gradient of electrons that in turn gives rise to an electric field opposing any further transverse drift. In simpler language, the electrons pile up against an inside face of the sample and repel further arrivals. The situations depicted in Fig. 6.10 therefore, apply in a real metal only to a transient phase that lasts from the time **B** is applied until the electron's concentration gradient is established so that further deflections are prevented. Thereafter, assuming that the metal is ideal so that all its conduction electrons have identical inertial properties, a dynamic equilibrium is established with the Lorentz force on any electron exactly balanced by the repulsion produced by the piled-up electrons. This leads to a drift of the electrons in straight lines along **J**—as in the case with no magnetic flux—but here with a transverse concentration gradient.

The electric field that this concentration gradient sets up and that offsets an electron's diversion from the mainstream, is called the Hall field. In

Fig. 6.10 it acts along the $-y$ direction and has a magnitude ev_xB. If in the steady state, after the transient phase has passed, the Lorentz force and the force from the Hall field are not exactly balanced for every conduction electron, then the electrons experiencing the unbalanced forces are deflected along a y direction. This is a resistive effect caused by the application of $\mathbf{B}$ (since it diverts electrons from their otherwise forward drift) and is the cause of the transverse magnetoresistive effect. When these two forces are exactly balanced for every conduction electron, then there is no transverse magnetoresistance. For this to occur requires all electrons to have identical inertial properties: There can be no anisotropy in either the shape of the Fermi surface or in the electron's relaxation time for the dominant electron scattering effect. Since this is not the case in any real metal, some transverse magnetoresistance is always observed, but it is instructive to consider this ideal case because it provides an entrance to the slightly more realistic case known as two-band conduction. Figure 6.11 shows a geometrical construction to describe the effect of crossed electric and magnetic fields for the ideal, free-electron case. With no magnetic field applied, suppose the total electric field in the sample $\mathbf{E}$ lies along the $-x$ direction and produces the electron flow $\mathbf{J}$ in the opposite sense ($\mathbf{E} = \mathbf{J}\rho$, where ρ is the resistivity measured along the x direction). Since $\mathbf{J}$ is assumed to be fixed in direction, along the axis of the sample, it is convenient to regard the application of $\mathbf{B}$ as producing the movement of $\mathbf{E}$ away from $\mathbf{J}$. When resolved into components $E_\perp$ and $E_\parallel$ as shown, where $E_\perp$ is the Hall field set up by the piling up of electrons along the $-y$ direction, and $E_\parallel$ is the component remaining along the x direction and producing the corresponding electron flow $\mathbf{J}$, we see that $\mathbf{E}$ shifts through an increasing Hall angle $\theta(= \tan^{-1} \omega_c\tau)$ as the Hall field increases. $E_\parallel$ decreases concomitantly, and so does the total current density $\mathbf{J}$, but no transverse magnetoresistance can be measured because the ratio $E_\parallel/J$ remains fixed: The application of $\mathbf{B}$ reduces $\mathbf{J}$ by an amount that is just that expected from the corresponding reduction in the component of $\mathbf{E}$ along $\mathbf{J}$ when the sample's resistance remains fixed.

The application of this construction to a more realistic case is described below, but first it is important to note a difference between the longitudinal and transverse magnetoresistances that has emerged in the preceding. Longitudinal magnetoresistance has been seen to arise basically because the effect of the flux $\mathbf{B}$ is to put the electrons into a more vulnerable position from the point of view of being resistively scattered than they would experience with $\mathbf{B}=0$. The transverse magnetoresistance, on the other hand, arises from the inability of a Hall field to exactly balance the Lorentz force acting on each electron in the conduction process. This imbalance can arise from

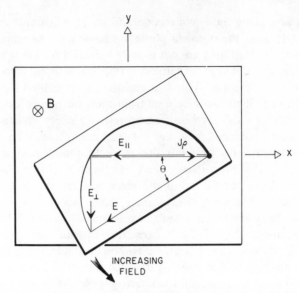

Fig. 6.11 A geometrical construction showing the effect of applying a transverse magnetic flux **B** to a free-electron metal carrying an electron current density **J**. The latter is taken to be along a fixed direction, such as the axis of the sample. The arrangement of fields and fluxes is thus exactly that shown in Fig. 6.10*a*. **B** is applied normal to the figure with the result that a Hall field ($E_\perp$) is set up along the $-y$ direction. $E_\parallel$ is the component of the total field E that remains along the direction of **J**. As **B** is increased, corresponding to the rotation of the template as indicated, the Hall angle θ ($= \tan^{-1}\omega_c\tau$) increases so that $E_\parallel$ and J are reduced correspondingly.

an anisotropy over the Fermi surface—as is essential for the longitudinal effect—but for the transverse effect it is not mandatory; for example, if the itinerant electrons can be divided into two groups, each having entirely free-electron behavior and isotropic properties (but different inertial values) then the transverse magnetoresistance is nonzero. This is because whatever Hall field is set up to oppose exactly the Lorentz force for one group cannot satisfy the corresponding requirement for the second. Hence some electrons will inevitably be deflected from the conduction flow when the magnetic flux is applied, thus giving rise to a nonzero magnetoresistance in this two-band model.*

A prominent example of a case where two groups of conduction electrons exist is provided by a metal, such as one from the NFE class (Sect.

---

* Although this name is universally used in the literature, it is a poor choice in the case of the s/p metals since all the itinerant electrons in them belong to one energy band; two-group model would be more appropriate in this case.

6.2) in which quasiparticles showing both electron and holelike character-istics exist. In a very crude approximation, which is nevertheless frequently made, these groups can be regarded as independent of one another so that conduction in the metal results from the sum of their separate contributions to the process. In this so-called two-band model, the numbers per unit volume of electron-like and hole–like particles (N_e and N_h, respectively) may be equal (in which case the metal is said to be compensated) or un-equal (uncompensated).* The two groups are taken to behave as free particles, and so have representative points on spherical Fermi surfaces, but their differing inertial properties are expressed in the separate mobili-ties of each group (μ_e and μ_h).

The effect of applying a transverse magnetic field to a metal showing such two-band behavior is illustrated in a vector diagram in Fig. 6.12. The orientation of fluxes and forces is exactly that shown in Fig. 6.10a except that in Fig. 6.12, of course, there is also a hole flow established. This lies along the $-x$ direction and gives rise to a pile-up of holes along the $-y$ direction with a corresponding Hall field directed along the $+y$ direction. The Hall fields arising from the transverse displacement of the electrons and holes are thus opposed. Figure 6.12 is constructed in the manner of Fig. 6.11 except that for clarity most of the geometrical construction is omitted. Since both electron and hole fluxes are simultaneously present, Fig. 6.12 is actually two superimposed constructions of the type shown in Fig. 6.11 (for holes, the template has to be reversed).

In Figure 6.12 the vectors $\mathbf{J}_1$ and $\mathbf{J}_2$ are proportional to the hole and electron current densities, respectively. The Hall fields arising from these flows are labeled H_1 and H_2, $\mathbf{E}$ is the total electric field in the sample, and $E_\perp$ is the net Hall field. The left-hand figure shows the case for an un-compensated metal. As the magnetic flux density $\mathbf{B}$ is increased, the minor-ity band (which in this case contains the holes) suffers most from the im-balance between the Lorentz force and the Hall force. When $\mathbf{B}$ is increased sufficiently, $\mathbf{E}_1$ and $\mathbf{E}_2$ both move away from the x axis so that both J_1 and J_2 are diminished, but J_1 ultimately becomes negligible compared with J_2. In this situation the two-band conduction has been eclipsed: $\mathbf{E}$ tends to lie very close to $\mathbf{E}_2$, and the whole problem resembles closely that of the single-band situation considered in connection with Fig. 6.11. Therefore, by analogy with preceding arguments that showed that a single-band metal

---

* Whether a metal is compensated depends upon the product of its valency and the number of atoms per unit cell of the crystal lattice. If this product is odd, it turns out that the metal can never be compensated, but if it is even the metal can be compensated and invariably is so when nonmagnetic. This subject is discussed in more detail in Sect. 6.4.1.

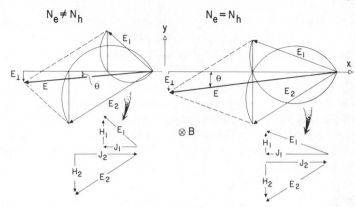

Fig. 6.12 A vector diagram showing the effect of a transverse magnetic flux **B** applied to a metal carrying separate electron and hole fluxes (*Left*) Showing the case of an uncompensated metal. (*Right*) Showing the case of an compensated metal for which the mobilities of the electron and hole quasiparticles are unequal. The diagrams are exactly similar to Fig. 6.11 except that for clarity most of the geometrical construction has been omitted. J_1 and J_2 are proportional to the hole and electron current densities, respectively, which are assumed to be confined to the x axis. The Hall fields arising from these separate fluxes are labeled H_1 and H_2, **E** is the total electric field in the sample, and $E_\perp$ is the net Hall field. The text explains how the behavior of the transverse magnetoresistance observed when **B** is increased can be predicted from these diagrams.

has no transverse magnetoresistance, the same parameter in the two-band case reaches a field-independent value for sufficiently large values of **B**.

When the metal is compensated the situation is different. The right-hand side of Fig. 6.12 shows the case when the mobilities of the two groups of particles are unequal ($\mu_e \neq \mu_h$). Then the Hall fields arising from the separate bands H_1 and H_2 almost cancel—they would exactly cancel if $\mu_e = \mu_h$. Now when **B** is increased, since J_1 and J_2 have to be approximately equal, **E** does not move close to either $\mathbf{E}_1$ or $\mathbf{E}_2$ as in the uncompensated case; it stays close to the x axis as in the figure. Neither group of particles experiences appreciably better balancing between the Lorentz and Hall forces. The net Hall field ($\mathbf{E}_\perp$) remains small (and is zero when $\mu_e = \mu_h$) and the transverse magnetoresistance increases indefinitely as **B** is increased.

When open orbits are possible in any cross section of the Fermi surface normal to **B**, their influence becomes pronounced in the following way. As the high-field limit is approached, two contributions to the electron motion in the cross section can be distinguished: That from electrons on closed cyclotron orbits, which we have seen decreases to zero in the high-field limit, and that from open orbits in which the electron is trapped between reflecting lattice planes with a motion that becomes independent of **B** in the high-field limit. As this limit is approached, the open orbits, therefore, completely dominate electron motion in the plane under consideration. Suppose a band of such orbits conducts along the direction shown in Fig. 6.13 (that is, inclined at some angle ϕ to the axis of the sample). For large enough **B**, when the closed orbit contribution is negligible and the magneto-resistance contribution from this source is saturated, all electrons on closed orbits for which the Hall and Lorentz forces are unbalanced are deflected along trajectories like that of Fig. 6.10*b*. Any further increase in **B** makes no further contribution to the magnetoresistance: It just tightens the loops of the trajectories of the already deflected electrons. However, this is not the case for electrons flowing in the open orbit. For them an increase of **B** means an increase in the current flowing along the open orbit's direction (at the expense of the previous equilibrium current density along the sample's axis) which in a finite sample continues only until a new trans-verse Hall field is built up to oppose further arrivals along the open orbits. Thus a new equilibrium current density is established along the sample's axis, having a smaller value than the first, which suffers a further reduction by the same mechanism if **B** is increased again. The magnetoresistance is, therefore, never saturated when open orbits are operative, except in the case when $\phi=0$ (Fig. 6.13) so that the entire current density is channeled along the open orbit. In that case, since the electrons are confined by re-peated reflections to follow the orbit's direction, they are unable to respond to the Lorentz force to produce a transverse current flow when **B** is varied. The net magnetoresistance is then determined in the high-field limit by the behavior of the closed orbit contribution, and so it saturates for large enough values of **B**.

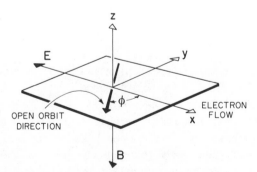

Fig. 6.13 Showing the arrangement of axes, forces, fluxes, and the direction of an open orbit for the case discussed in the text. The *x* direction is taken to be the axis of the sample along which the current density is constrained to flow.

These illustrations of the disruptive effect upon an itinerant electron's motion caused by a transverse magnetic field relate to the very idealized case of two entirely separate free-electron groups of quasiparticles. Nevertheless, it is an adequate model to describe the salient effects leading to the magnetoresistance and Hall effect in real metals. The former represents the inability of a net Hall field to balance exactly the Lorentz force on all the electrons, and the latter reflects the competitive electrostatic effects of the separate groups of electrons that are deflected from the mainstream and piled up against an inside face of the sample. In simple terms, the Hall effect samples the electronic motion in planes perpendicular to **B**. It is important to note that the electron's transverse displacement produced by the Lorentz force is not the only possible cause of magnetoresistance and Hall effects: The asymmetric scattering mechanisms described in Sect. 3.4.6, when coherently aligned by some externally applied magnetic flux, lead to a transverse displacement of the electrons from their general forward drift. This produces both resistive scattering and eventually a transverse concentration gradient of electrons. Consequently, both Hall and magnetoresistance effects can arise from these scattering mechanisms. Galvanomagnetic effects arising from such nonclassical sources (that is, from other than the classical Lorentz force) are frequently referred to as the anomalous effects; in general an anomalous effect is, of course, accompanied by a classical component, and together they form the total galvanomagnetic effect observed in the metal or alloy.

Galvanomagnetic effects in metals are a broad subject, and the preceding introductory description of some of its aspects in s/p bonded metals has necessarily been very selective. Hopefully it is sufficient to emphasize the objective of the chapter: The whole, that is the system of itinerant electrons together with the lattice of fixed ions, shows characteristic features that cannot be understood from a consideration simply of their separate components.

6.4 QUASIPARTICLE MOTION IN D BANDS

The description of electron motion given so far in this chapter relates principally to the simplest case drawn from the categories defined in Sect. 2.3.2: The s/p bonded class. In those metals all the itinerant electrons have energy states lying in a single energy band (the s/p band). Now the description turns to the more complicated cases of the metals from the d block of Fig. 2.1. The first increase in complexity arises from the fact (which is illustrated in Fig. 2.12 for the first transition series of the d

block) that the _d_ levels lie high enough in energy to be only partially occupied. This can be regarded as an extension of the picture described for Cu in connection with Fig. 6.2. The _s/p_ and _d_ bands of energy states overlap in space and energy so that there is hybridization between them to varying degrees along different crystallographic directions. But the _d_ levels in the transition metals are so positioned in energy that they can intersect the Fermi level in certain directions (unlike the case of Cu, for example, where they lie roughly in the middle of the _s/p_ band's range of energy for all directions). This means that not only is there distortion of the Fermi surface in reciprocal space produced by the asymmetric hybridization (as described for Cu in Sect. 6.2) but also that electrons in purely _d_-like states can be itinerant. Part of the metal's Fermi surface is thus made up of states having predominantly _s/p_ symmetry, and other parts are made up of of those with mainly _d_-like symmetry. These parts of different predominant symmetry may lie in different Brillouin zones, in which case they are generally referred to as separate _s/p_ and _d_ bands. The parts of the Fermi surface lying in these different zones are sometimes regarded as separate constructions appropriate to the band. Thus we see such references as "the Fermi surface of the _d_ band" when the description applies to the parts of the Fermi surface of predominantly _d_-like symmetry that happen to lie in a given Brillouin zone. (Just as the hole surface of Al shown in Fig. 6.3, which is that part of the Fermi surface that happens to lie in the second Brillouin zone, can be called the "second zone's Fermi surface.")

The second increase in complexity over the case of metals having simply _s/p_ energy bands arises from the sharp difference between the density of energy states (Sect. 4.5.1) appropriate to the _s/p_ and _d_ regions of the Fermi surface. Since the _d_ band's energy has generally a much narrower spread than the _s/p_ band's energy in the metal, its density of states is correspondingly higher—a feature that is accentuated by the fact that a _d_ shell in the atom can accommodate up to ten electrons. The metal's Fermi surface, therefore, consists of regions of quite different densities of representative states, and this influences the conduction electron's motion in the manner described in Sect. 6.4.1. The final complication arises from the fact that the electrons of up and down spin configurations in the narrow _d_ bands may be split in energy by an exchange interaction (Sect. 3.3) to the extent that separate _d_ bands (and hence, separate sections of the Fermi surface in reciprocal space) exist for the two spin orientations. Metals with this property have a net magnetism, since one spin orientation for their itinerant _d_ electrons is preferred over another, and they form the magnetic transition metals of the _d_ block. (This group comprises ferromagnetic Fe, Co, and Ni and antiferromagnetic Cr and Mn.) A consideration of this extra complication produced from the inextricable binding of the metal's magnetism with the transport of charge is postponed until Sect. 6.4.2.

6.4.1 Nonmagnetic d Transition Metals. A first noticeable difference between the metals of the transition series and those of the s/p block is that the former have appreciably larger electrical resistivities at all temperatures. A qualitative explanation of this observation was given by N. F. Mott over thirty years ago, and it has survived to the present day. It rests on the assumption that charge transport in these metals is provided largely by the electrons in states of s/p symmetry; those in states of d symmetry are assumed to be quite immobile by comparison because of their large effective mass. (We recall that a narrow energy band—the d band in this case—implies strong localization of the electrons upon their ions. This is manifested in the quasiparticle's effective mass, which is about ten times as large for the d states as for the s ones.) Consequently, if an electron is scattered from an s/p state to one of d symmetry (as a result of, for example, a scattering by a phonon, by another electron, or by some fixed point defect) it is removed from the conduction process and so is resistively scattered. States of d symmetry are, therefore, sinks in the sense introduced in Sect. 6.3.1.

The conductivity mechanism in the nonmagnetic transition metals is in principle no different from that in the purely s/p ones. As is pointed out above, in the present case the metal's Fermi surface consists of regions of different predominant electron symmetries that give the electrons distinctively different responses to externally applied fields. The regions on the surface made up of points representing states of predominantly d-like symmetry are the sinks in the present context. The effectiveness of these in determining the metal's overall electrical resistivity is much greater than those discussed in the context of s/p metals—as the higher resistivities of the transition metals indicate. This increased effectiveness arises from the high density electron states on the d regions of the Fermi surface. Since the probability of an electron scattering event terminating with an electron in a state of given symmetry depends upon the number of vacant states of that type that are available, it is clear that a resistive scattering from an s/p state is much more likely to terminate in a d state than in another of s/p symmetry. (These processes are frequently simply referred to as s-s and s-d scattering, respectively.) This immobilization of conduction electrons by scattering from s/p states to ones of d symmetry is known as Mott scattering. Subject to certain simplifying assumptions, theory shows that at low temperatures ($T \ll \theta_D$) it contributes a term in T^3 to the metal's electrical resistivity, and one linear in T at ordinary temperatures ($T > \theta_D$).

It is regarded as being responsible for the extra electrical resistance found in the nonmagnetic transition metals compared with the s/p class.*

Since in a transition metal the d band also makes a contribution to the itinerant class of electrons, being pushed in energy above the Fermi level, it is of interest to know what is its internal structure produced by the hybridization between the s/p and d states during the atoms' condensation to form the metal. This is precisely the subject discussed in Sect. 6.2 in connection with Cu (with the results expressed in Fig. 6.2) but in the present case we are interested in the effect of the hybridization upon the d band rather than upon the s/p one, as is the case in Sect. 6.2. It should be possible in principle to illustrate in a diagram analogous to Fig. 2.11 the formation of a narrow d band from the overlap of atomic d wave functions, but is rather more difficult to do this successfully in the case of d wave functions because of their spatial distributions. These are illustrated and described in Appendix A, but we recall from Sect. 3.4.4 that the five d wave functions of the free atom comprise three (the t_{2g} set) with symmetry of the type xy, zx, yz, and two (the e_g set) with symmetry of the type y^2-z^2, z^2-x^2. To give an extremely simplified example, suppose an energy band is formed from the overlap of the e_g functions of symmetry y^2-z^2. It has upper and lower energy limits (analogous to $s1$ and $s4$ of Fig. 2.11) corresponding, respectively, to the antibonding and bonding configurations shown in Fig. 6.14. (In the bonding case the d electron density is increased in the region between the ions, and it is reduced in the antibonding one.) Energy states lying between these limits correspond to similar two-dimensional arrays, each having a different symmetry combination of the d wave functions. (Exactly analogous to the way intermediate states arise in the s and p bands of Fig. 2.11. Thus the first energy state above the bonding ground state is obtained by the rotation of the orbital on one ion through $\pi/2$.) But this example is simplified to the point of being inadequate. First, there is the fact that d wave functions of all five symmetries are involved in the formation of the band, and, second, there is mutual hybridization between them (which has to be accounted for) as well as hybridization with the s/p levels in the metal. It is impractical to try to describe all of this in a sketch like Fig. 6.14. It is more convenient to go back instead to the variation of a d electron's energy expected in reciprocal space, after the fashion of Figs. 5.12 or 6.2.

---

* It would not serve the present purpose to repeat detailed discussion of the evidence supporting this view. It was summarized nearly forty years ago in the book by N. F. Mott and H. Jones cited at the end of Chapter 2 and was brought up to date with more modern data in 1964 by N. F. Mott on p. 399 of his review that is also cited there.

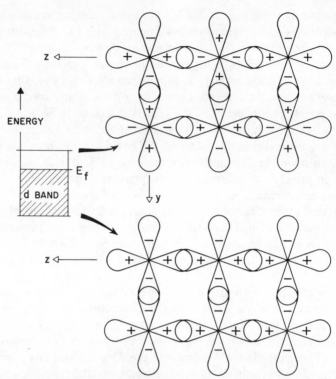

Fig. 6.14 A schematic view, in the style of Fig. 2.11, showing the formation of a narrow d energy band from overlapping e_g wave functions of $d_{y^2-z^2}$ symmetry. Since it is not practical to show the arrangements of orbitals for all the energy levels that make up the band, only those appropriate to the band's extreme energy limits are shown. (*Upper*) The antibonding configuration gives the top limit to the band, and electrons occupying this level are excluded from the regions of low electrostatic potential between the ions (as for the $s4$ and $p4$ levels in the bands of Fig. 2.11). (*Lower*) The bonding configuration gives the d band's bottom limit, and here the electrons are concentrated in the interionic regions. (After J. Friedel.)

Figure 6.15 shows schematically, in what can be regarded as a more detailed view of circumstances such as those described for Cu in Fig. 6.2, the expected form of the d bands before and after their hybridization with states of the s/p band. The sketch represents the energy variation of the different d wave functions as **k** is varied along some general crystallographic direction in either a f.c.c. or b.c.c. structure. Calculation shows that the d band (when unhybridized with any s or p components—as in the extreme left of Fig. 6.15) consists of five overlapping subbands that are degenerate in energy at certain points. These arise from the five atomic d

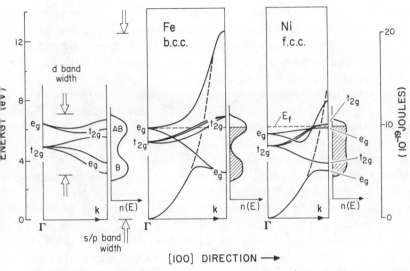

Fig. 6.15 Schematic illustration of the expected form of the *d* energy band before and after hybridization with the *s/p* levels in a metal. The extreme left shows for the five atomic functions that make up the *d* band the expected variation of $E(\mathbf{k})$ as $\mathbf{k}$ is varied along an arbitrary direction in either an f.c.c. or b.c.c. lattice. The mutual hybridization of these atomic *d* wave functions leads to the band structure shown and described in detail in the text. To illustrate the effect of an admixture of *s/p* energy states, the calculated $E(\mathbf{k})$ variations for a given direction in the hypothetical nonmagnetic states of the Fe and Ni are shown as examples. In both cases there is some hybridization between the *d* bands and suitably disposed levels of the metal's *s/p* band, but in the b.c.c. case the *d* band's salient features survive: A double-humped density of states is found with bonding and antibonding lobes. In the f.c.c. case the *s/p–d* hybridization produces a more thorough mixing of the symmetries so that they are spread rather more uniformly over the *d* band's density of states curve. The latter is thought to show a much less pronounced minimum between its bonding and antibonding regions than in the b.c.c. case.

wave functions (Appendix A) consisting of three of equivalent shape with symmetry of the type xy, yz, zx (the t_{2g} set) and two of equivalent shape with the symmetry type y^2-z^2, z^2-x^2 (the e_g set). At the center of the Brillouin zone (Γ) they are arranged as shown in Fig. 6.15: The t_{2g} set are triply degenerate and bonding wave functions (with electron density concentrated between the ions) and the e_g set are doubly degenerate and arranged in an antibonding configuration like that in Fig. 6.14. Moving out towards the edge of the Brillouin zone, the variation of their energy in any direction is as shown (when the mutual *d–d* hybridization is included). The degeneracy of each symmetry set is lifted and they both spread out somewhat and cross over to give a band structure at the Brillouin zone having upper and lower energy limits corresponding to antibonding configurations of purely xy symmetry and bonding ones of purely y^2-z^2 symmetry, respectively. Thus the lower part of the band is predominantly of e_g character, and the upper part of t_{2g} character. In the middle of the band the energy states having these two symmetries are distributed fairly uniformly, but their overall density [$n(E)$ of Sect. 4.5.1] is low. The density of states for the whole d band has, therefore a double-humped shape with a deep minimum in the center of the band. The hump (B) at lower energy corresponds to the bonding states of predominantly e_g character, and the higher energy hump (AB) is made up of antibonding states of predominantly t_{2g} character. AB has room for six electrons, and B can take the remaining four. (In some cubic nonmetallic compounds of transition metals, oxides, sulphides, selenides, etc., these e_g and t_{2g} bands are separated by an energy gap so that the density of states between the humps goes to zero. Some older theories also tried to apply such a picture to the metallic situation, but it is now believed to be an incorrect approach.)

When such d bands hybridize with the energy levels of an s/p band in a metal, the result is a further mixing of the different symmetries. The calculated $E(\mathbf{k})$ dependences observed for a given crystallographic direction in the hypothetical nonmagnetic states of Fe and Ni are shown in Fig. 6.15, and they illustrate this for the b.c.c. and f.c.c. cases, respectively. In both cases there is an appreciable hybridization between the d levels and those suitably disposed from the s/p band, but in the b.c.c. structure the d band's salient features are seen to survive. The band retains its double-humped form for the density of states, and its energy range is again limited by states of predominant symmetry: t_{2g} at the top and e_g at the bottom. In the f.c.c. case, the d–s/p hybridization produces a rather more thorough mixing of the states of different predominant symmetry so that the e_g and t_{2g} components of the density of states are spread out in energy rather more uniformly than in the b.c.c. case. As a result, it is thought that the f.c.c. transition metals do not generally possess the marked minimum in their d band's density of states. This strong subdivision of the d band in the b.c.c. case, in which energy states are crowded together in energy towards the top and bottom of the band, has numerous important consequences particularly for the thermal, electron transport, and magnetic properties of the metal. Some of these will be encountered in following sections.

In figures like 6.2 and 6.15, which are known self-evidently as energy band structure diagrams, the position of the Fermi level E_f within the band structure is of great importance, particularly when the structure is complicated by many interacting energy levels of different symmetries. In such a case, a slight movement in energy of the Fermi level can sometimes completely change the character of the electrons that are itinerant. The position of E_f is determined by the metal's valency and crystal structure. (Its shift in energy across a transition series is illustrated in Fig. 2.12.) In the cases illustrated in Fig. 6.15, it clearly intersects branches of both the d and s/p bands, so that there are regions of d and s/p symmetry on the Fermi surfaces corresponding to itinerant electrons of these two symmetries. This contrasts with the case of Cu, for example, shown in Fig. 6.2, where the d levels are not intersected by the Fermi surface and so do not contribute any electrons to the itinerant class.

It is convenient at this point to consider briefly some consequences of the relative movement of the Fermi level among the bands of a transition metal. Figure 6.16 shows a hypothetical and very simplified example that resembles Ni in its nonmagnetic (Fig. 6.16b) and magnetic form (Fig. 6.16c). The metal has an s/p band of the simplest form together with a d band formed by a single branch of the atomic functions of Fig. 6.15. In Fig. 6.16a this branch is assumed to be entirely below the Fermi level, so that the d band is completely filled and makes no contribution to the pool of itinerant electrons; Fig. 6.16a thus resembles the case of Cu (Fig. 6.2). Suppose that some influence forces up the energy of this d band at the larger values of $\mathbf{k}$ in the vicinity of the face of the Brillouin zone. Let this be to the extent shown in Fig. 6.16b, where the d level lies above E_f at the zone's face. Electrons that find themselves in energy states lying higher than the Fermi level prefer to relocate in ones of lower energy so as to

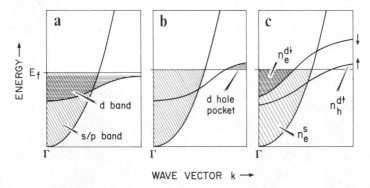

Fig. 6.16 A schematic view of the energy band structures along some arbitrary direction in reciprocal space for three characteristic cases. (a) A metal like Cu, in which the d band is entirely full. (b) One like the (hypothetical) nonmagnetic Ni, in which a single d band penetrates the Fermi level. (c) Magnetic Ni, in which exchange effects have split the d band into separate up- and down-spin components (but see the comment in the text).

keep the total energy of the electron fluid to a minimum. Since the only vacant energy states are in the s/p band, the d electrons from the states labeled as forming the d hole pocket in Fig. 6.16b spill over into the metal's s/p band and leave a number of vacancies in the d band. As we pointed out in Sect. 2.3.4, it is a matter of taste whether the response of the electrons in the d band is described in terms of the behavior of the quasiparticles in its occupied states or of the behavior of those that have a hole-like response and occupy the vacated levels. When the Fermi surface of the d band is constructed, it can be regarded either as an electron surface centered on Γ or as a separate hole surface centered about some point of the Brillouin zone's face. Surfaces of the latter kind, which are not centered about Γ and enclose a relatively small volume of reciprocal space, are known as pockets or needles depending upon their shape. Hence, the states vacated by the upward displacement of the d band considered here are said to give rise to a d hole pocket in the Brillouin zone.

Since whatever electrons are drained from one band must reappear elsewhere in another, a balance sheet can be drawn up to express the relationship between the occupied and unoccupied volumes of reciprocal space. We pointed out in Sect. 6.2.1 that the total number of occupied electron states allowed per energy band and within the Brillouin zone is $2N$, where N is the number of unit cells in the sample's crystallographic lattice. (It is convenient to work "per unit cell" throughout, so that the Brillouin zone's capacity becomes just two electron states per unit cell.) Suppose the metal has F full bands like the 3d bands of Cu in Fig. 2.9. The total number of occupied states contained in these filled Brillouin zones is thus $2F$. Incompletely filled zones can be looked upon in two ways, as outlined above. When the zone is almost full, it is convenient to emphasize its empty regions; conversely when it is almost empty, the occupied space is stressed. (As when water is splashed over a polished table, it is a matter of taste whether the result is regarded as islands of water or interconnected islands of table.) Concerning the case when the zone is almost full, suppose there is a total of N_h hole states per unit cell when summed over all the incompletely filled Brillouin zones regarded as hole zones. If there are J of them, the number of occupied electron states in these zones is clearly $2J - N_h$ per unit cell. Turning to the second case, when the zone is almost full, suppose there is a total of N_e of electron states per unit cell in the incompletely filled electron Brillouin zone. (N_e is obviously just the number of occupied states per unit cell in this zone.) All the occupied states in all zones have now been accounted for and since this number per unit cell is just xz (from Sect. 6.2.1) we can write the following:

$$xz = 2F + (2J - N_h) + N_e$$

or

$$N_e - N_h = xz - 2(F + J) \qquad (6.9)$$

When the left-hand side of Eq. (6.9) is zero, the metal is said to be compensated, and we described in Sect. 6.3.3 how some of its high-field galvanomagnetic effects depend upon this quality. Eq. (6.9) shows that a metal can never be compensated if xz is odd; when xz is even it turns out in practice that nonmagnetic metals are invariably compensated. For reasons outlined in the following, the situation is more complicated in the cases of magnetic metals.

The point of introducing this topic here is that when an energy band, such as the d band in Fig. 6.16a and Fig. 6.16b, is moved through the Fermi energy, the character of the corresponding sheet of the Fermi surface can change from holelike to electronlike. In the hypothetical case of Fig. 6.16b, for example, if the d band is forced upwards in energy a great deal more than is shown (say to the position of the down-spin level in Fig. 6.16c) a point is reached when it is no longer appropriate to describe its Fermi surface as a hole pocket centered on the Brillouin zone's face. The Fermi surface would eventually be more appropriately described as an electron surface centered on Γ. Eq. (6.9) expresses a balance which has to be maintained whatever view is taken, and its application is perhaps best illustrated in the more realistic case of Fig. 6.16c in which the d zone's Fermi surface is split by exchange effects when the metal becomes spontaneously magnetic. (This is strictly within the scope of the following paragraph, since here we are supposedly restricted to the nonmagnetic metals, but for our purpose it is convenient to anticipate this one aspect.)

Figure 6.16c shows a simplified view of the band structure along an arbitrary direction in ferromagnetic Ni. The single branch of the atomic d levels shown in Fig. 6.16b has been split by presently unspecified exchange effects (discussed in Sect. 6.4.2) that are assumed to be of negligible importance among electrons in the s/p band. As a result, the d electrons of down spin are displaced upwards in energy, and those of up spin are displaced a corresponding amount downwards. The end result is that separate d bands, Brillouin zones (spin zones) and Fermi surfaces (Sect. 6.2.1) exist for the spin-up and spin-down electrons.* (Of these, the bands are sketched in Fig. 6.16c.) The same effects are assumed to be of negligible

---

* In Ni the up-spin levels are believed to be pushed down in energy to such an extent that they lie entirely below the Fermi level. At absolute zero they are therefore, completely filled by their complement of five electrons. But for illustrative purposes we persist with the hypothetical case of Fig. 6.16c since the following description is in any case valid when $N_h^{d\uparrow}$ is zero. Fe has almost certainly the band structure of Fig. 6.16c at absolute zero, in so far as its d electrons can be regarded as itinerant.

proportions in the s/p band, and so its Fermi surface and band structure are taken to be unchanged from the nonmagnetic case. Our interest is to know whether or not such a model predicts Ni to be a compensated metal. According to the invariable empirical rule for nonmagnetic metals, when xz is even the metal is compensated. Since an isolated and neutral atom of Ni has two $4s$ and eight $3d$ electrons to populate the s/p and d bands in the metal ($z=10$) and since $x=1$ for the f.c.c. structure, xz is even so that the metal is expected to be compensated. Experimentally, however, it was shown to behave as an uncompensated metal,* and this lead to the following interpretation.

The objective is to draw up a balance sheet of occupied states in the different zones after the fashion of Eq. (6.9). The first point to note from Fig. 6.16c is that the down-spin d band has been pushed upwards in energy to the extent that its Fermi surface assumes an electron character. Suppose its spin zone (Sect. 6.2.1) contains a total of $n_e^{d\downarrow}$ occupied electron states. (As above, magnitudes per primitive unit cell of the crystal's lattice are implied throughout.) The corresponding number of unoccupied electron states in the zone is $1-n_e^{d\downarrow}$ since the spin zone has a capacity of one electron state per unit cell. (If the down-spin zone was to be regarded as a hole zone, then $1-n_e^{d\downarrow}$ would be written as $n_h^{d\downarrow}$.) The spin-up band, on the other hand, retains its holelike character (its Fermi surface is still best described as a $d\uparrow$ hole pocket centered on the face of the Brillouin zone). Let this up-spin zone contain a total of $n_h^{d\uparrow}$ unoccupied electron states. Finally, let the number of occupied states in the metal's s/p Brillouin zone be n_e^s.

Two equations can be written down to relate these quantities. The first expresses the fact that, since there are only ten electrons per atom available to populate both the s/p and d bands and since the d band itself can accommodate up to ten electrons per atom, whatever electrons exist in the s/p band must equal in number the vacancies in the d band. In other words, the number of electrons in the s/p band (n_e^s) must be equal to the sum of the numbers of holes left in each split d band by the electrons' relocation. Thus

$$n_e^s = (1 - n_e^{d\downarrow}) + n_h^{d\uparrow} \qquad (6.10)$$

The second equation expresses the fact that, since n_e^s is assumed to be unchanged by the spontaneous magnetization, the difference between the total numbers of occupied electron and hole sheets for the whole system

* E. Fawcett and W. A. Reed, *Phys. Rev.* **131**, 2463, 1963.

$(N_e - N_h)$ can be written just as the algebraic sum of their separate contributions

$$N_e - N_h = (n_e^s + n_e^{d\downarrow}) - n_h^{d\uparrow} \tag{6.11}$$

Adding these two equations gives $N_e - N_h = -1$. According to this
model, magnetic Ni is, therefore, an uncompensated metal, essentially because of the change in the hole- or electron-like character of some of its
itinerant electrons upon the metal's spontaneous magnetization. Thus when
an external magnetic field is applied with sufficient strength to render all
cyclotron orbits on all pieces of the Fermi surface into the high-field
condition, the metal's transverse magnetoresistance should, therefore, tend
to a field-independent (saturating) value (Sect. 6.3.3)—as is observed
experimentally.

6.4.2   Magnetic d Transition Metals.   Two questions dominate the description of the quasiparticle's motion in a magnetic transition metal. First,
what is the nature of the correlating force between the electrons in the d
band that separates the energies of the spin-down and spin-up configurations in the manner of Fig. 6.16c, and, second, how is the experimental
fact to be accounted for that in some d transition metals the quasiparticles
show simultaneously pronounced localized and itinerant characteristics?*

In many respects the points at issue are the same as those relating to the
magnetism of a foreign ion in an alloy discussed in Sect. 3.4.4, Sect. 3.4.5
and Fig. 3.11. In that context the physical picture is of an ion surrounded
by a region of space in which previously itinerant electrons linger and adopt
temporarily to a marked degree the features of an atomic precursor state
before returning to the pool of genuinely itinerant electrons. Electrons
whose motion is temporarily restrained in this way experience correlating
forces with those already resident in the ion and the surrounding region;
there is the straightforward Coulomb repulsion U (of Eq. 3.22) acting
between electrons with antiparallel spins, and there is the direct intraatomic
exchange force (Type 1 of Fig. 3.9). In the Anderson model of the ion
it is the former influence that separates in energy the spin configurations
of the constrained electrons, but in the Friedel model it is the latter (Sect.
3.4.4). But in both models the fluctuating occupancy of the spin-polarized
states gives rise to the steady net magnetic moment of the ion that is observed by the typical time-averaging probes of an experimentalist.

The same elementary picture applies in the case of the ferromagnetic
metals, except that here the ions are, of course, not isolated foreign ones

---

* Notably Ni and Co. In Fe and Mn the features characteristic of itinerant d electrons are much less pronounced.

but typical members of the lattice. Their proximity is such that the d levels interact and give appreciable itinerancy to the resident electrons because the resulting energy band is only partially filled (Fig. 6.16) and it is these electrons, hopping from ion to ion, that produce the fluctuating ionic moments in this case. The point is that despite this degree of itinerancy these electrons experience the intraatomic-correlation effects when they are temporarily resident upon the ion, and so they also behave to an appreciable extent in the manner expected of localized ions. However, there is considerable uncertainty about the exact correlating force that produces ferromagnetism. Is it the intraatomic exchange coupling (Type 1 of Fig. 3.9) responsible for Hund's rule, or is it the straightforward Coulomb repulsion between electrons of unlike spin configurations? Perhaps the different points of view are best illustrated with a specific example, and for this we return to the case of Ni, first encountered at the end of Sect. 6.4.1.

The ground state of the isolated atom is . . . $3d^8\ 4s^2$ (Fig. 2.6) but we know that other states like . . . $3d^9\ 4s^1$ and . . . $3d^{10}$ lie within only 2 or 3×10^{-19}J of this because the $4s$ electrons penetrate into the d shell and lower the latter's energy through electrostatic screening (Sect. 2.3.1). The same kind of thing is expected to occur for the ion in the metal. As the d electrons hop from ion to ion, the ion's d shell has time between hops to settle down to an optimum electron configuration; thus metallic Ni is probably made up of ions that have instantaneously $3d^8$, $3d^9$ and $3d^{10}$ configurations. Since each unpaired d electron contributes its spin moment of $1\mu_B$, these ions can have maximum instantaneous spin moments of 2, 1, and $0\mu_B$, respectively. The metal is, therefore, viewed as a collection of ions having these different and constantly changing configurations. But for uncorrelated d electron motion, in which the hopping of a d electron into or out of a d orbital is just as likely to leave the orbital with a net spin up as with a net spin down, the time-averaged moment for each ion is zero. For ferromagnetism, which is the appearance of a nonzero spontaneous magnetic moment on each ion, there has thus to be some correlating influence that favors a given spin direction for each ion. It must favor parallel alignment for the two unpaired spins in the $3d^8$ configuration, or a persistent up-spin alignment for the unpaired spin in the $3d^9$ configuration. In the event that either or both of these circumstances exist, the time-averaged moments on the different ions can be equal to the above instantaneous values, and their mean value averaged over all the ions can be a nonintegral number of μ_B—as is observed in practice. [For Fe, Co, and Ni these values are 2.2, 1.7 and $0.6\mu_B$, respectively. The orbital angular momentum makes no significant contribution to the ion's magnetism in these metals (Sect. 3.2.5).]

However, we must remember that at all times the ion tries to remain electrostatically neutral. It does this by piling up about itself a cloud of itinerant electrons that are in dynamic equilibrium with those of the s/p conduction band (Sect. 3.4.1 and Sect. 3.4.2). Since the d electrons' speeds as they hop from ion to ion are small compared with those of typical electrons in the s/p band, the latter are able to respond quickly enough to changes in an ion's d shell occupancy. Thus, although a $3d^{10}$ ion in Ni does not require any localized s electrons for this screening, a $3d^9$ ion requires them to form a charge cloud equivalent to one electron per ion; a $3d^8$ ion likewise requires the equivalent of two electrons (of opposite spins) around it. The aim in all cases being to maintain a balanced electrostatic charge within the atomic cell about the ion (Fig. 2.8a).

The presence of these screening electrons from the s/p band and their ability to adjust and compensate for the charge polarity of the ion produced by the arrival or departure of a d electron is very important when considering qualitatively the feasibility of the suggested model. For example, when the role of the itinerant electrons from the s/p band is ignored, the cost in energy of making separate $3d^8$ and $3d^{10}$ ions from two of $3d^9$ configuration [that is, by the reaction $2(3d^9) \rightarrow 3d^8 + 3d^{10}$ is at least 19.7×10^{-19}J (12.3 eV)] from the known ionization energies of the free atom, and this is somewhat too high to make the event probable in the metal. But when we consider the more realistic circumstance in this case, in which the s/p electron distribution forming the screening charge is admitted, the total cost in energy* of the reaction $2(3d^9 4s^1) \rightarrow 3d^8 4s^2 + 3d^{10}$ is only 2.9×10^{-19}J. The latter screened reaction costs much less energy than the former primarily because energy is gained when the negative screening charge is brought up around the uncompensated $3d^8$ ion, which, of course, has initially a net positive charge. In addition, important readjustments in the shapes of the d orbitals are brought about as a result of their penetration by the screening s/p electrons. Since this reduces the attractive influence of the ion's nucleus upon the outer d orbitals, they become less compact. The electrons in them are, therefore, able to spread out more, and so the total energy U arising from their mutual repulsion is reduced compared with the case when no s/p electrons are admitted into the picture. Looked at from the point of view of the d orbitals' occupancy, if an extra d electron hops onto an ion, its negative charge repels the itinerant electrons in the surrounding fluid to produce a correlation hole (Sect. 4.5.2) in the fluid about the ion. This hole's presence reduces the electrostatic potential at

* C. Herring, "Exchange Interactions among Itinerant Electrons," *Magnetism*, Vol IV, Ed. G. T. Rado and H. Suhl, *Academic Press*, New York, 1966, p. 187 and following, and Fig. 29.

the ion and so effectively reduces U for the d orbitals. Thus U for the d orbitals is believed to decrease continuously across the series $3d^8$, $3d^9$, $3d^{10}$.

Since Ni shows a net spontaneous moment per atom of $0.6\mu_B$, it is obvious that ions having instantaneously the configurations $3d^9$ and $3d^{10}$ must exist in it; and since Co has correspondingly a net moment per atom of $1.7\mu_B$, it is equally certain that ions having the configurations $3d^8$ and $3d^9$ must exist in it. The question is whether or not ions having the configuration $3d^8$ can also exist in Ni, and those with $3d^7$ in Co. Supporters of one interpretation of ferromagnetism maintain that they can, but others of a different viewpoint maintain that they cannot—or at least if they can exist, it is a very unlikely event. The first group argue that although U is largest for the $3d^8$ configuration in Ni and the $3d^7$ one in Co (as described above) it is still reduced sufficiently by the effects of screening for the configurations to exist in the respective cases. They argue that consequently two unpaired electrons can be simultaneously present or absent* in an ion's d orbital at any given instant. In their view the origin of ferromagnetism is then the intraatomic exchange coupling J of Eq. (3.32)—namely Type 1 of Fig. 3.9, the coupling responsible for Hund's rule—acting between the two d electrons (or d holes) in the orbital. The configuration with spins parallel always has the lowest energy, and so states of opposite spin can only be occupied at a certain cost in energy (Sect. 3.3.1). Hence, the tendency towards ferromagnetic alignment within the orbital, and hence the origin of the energy separation (the exchange splitting) of the two spin bands. [We recall that this separation is the energy required to reverse the spin of a quasiparticle in the d band without changing its wave vector **k**. (Fig. 6.16c).] In the case of Ni the existence of ions with the $3d^8$ configuration is thus crucial to this viewpoint, since the $3d^9$ and $3d^{10}$ configurations contribute nothing to the ferromagnetism because with only one or zero unpaired spins there can be no correlation from the Hund's rule coupling. In Co, where it is certain that the $3d^8$ configuration exists, it can be argued that the Hund's rule coupling gives rise to ferromagnetism in these ions. Thus it is not crucially important to establish the existence of the $3d^7$ configuration in this case.

Once the spontaneous moments on the individual ions are established by the intraatomic exchange in this picture, the tendency to ferromagnetism can be expressed in the elementary way of Sect. 4.7. Although that description is not concerned with the problem of the localization of the spins, as upon the ions in this case, its results can be taken over to the present context to give a condition for ferromagnetism. U is the mean energy necessary

---

* This can be referred to alternately as the hopping of two d holes onto the ion.

o reverse the spin of a *d* electron of fixed **k**, and this favors the parallel alignment of spins with lowest energy—providing the inevitable increase in the Fermi energy can be overcome (Sect. 4.7). All of this is expressed in Eq. (4.18), which becomes infinite when $U n(E) = 1$. Such a divergence is usually taken to imply ferromagnetism. Hence, the condition for ferromagnetism (sometimes known as the instability criterion) is put as

$$U n(E) > 1 \tag{6.12}$$

This suggests that ferromagnetism occurs only in metals for which the density of states at the Fermi energy is large, as in the transition metals, and where *U* is also relatively large because of the intraatomic Hund's rule coupling.

Then why are metals like Pt and Pd, which have a high $n(E)$ due to the presence of the *d* band close to the Fermi level (as in Fig. 6.16*a*) not ferromagnetic? This pointed question leads into controversial areas, but in terms of the preceding view of ferromagnetism the reason is thought to be that, unlike the case of Ni and Co, ions having instantaneously two unpaired *d* electrons cannot exist in these metals.

Supporters of the second view of ferromagnetism maintain that, although *U* for the ion's *d* orbitals is undoubtedly reduced by the screening effects described above, it remains large enough to prevent two quasiparticles from the *d* band (that is, either electrons or holes) from hopping onto the same ion and being present simultaneously. They maintain, therefore, that the repulsive energy *U* is large enough to exclude the configurations $3d^8$ and $3d^7$ from Ni and Co, respectively. In Ni at least, the intraatomic Hund's rule coupling cannot in this view be the origin of ferromagnetism as outlined above. Ferromagnetism in this second view arises instead from a favored and persistent spin direction produced for the unpaired electron in the $3d^9$ ions by the repulsion *U* (of Eq. 3.22) between electrons having antiparallel spins. (The $3d^8$ configuration in Co is not excluded, but, when it occurs, there is also certainly some effect from the intraatomic Hund's rule coupling that forms the basis of the first viewpoint.) This repulsion establishes the favored spin direction in the following way.* Suppose a *d* electron wishes to hop into an incompletely filled *d* orbital on a nearby ion. The intraatomic correlations arising from *J*, and responsible for Hund's rule, have aligned all the uncompensated spins in the *d* orbital. Suppose that at the instant under consideration the ion has its total spin in the up direction. If the incoming *d* electron is of down-spin, it is repelled by these intraatomic interactions since it is unable to enter the ion. An up-spin electron,

* J. Hubbard, *Proc. Roy. Soc. Ser. A* **276**, 238, 1963.

on the other hand, is attracted. Consequently, the ion tends to attract electrons having a spin orientation like its own and to repel those having one opposed. This has two effects. First, the ion's spin state tends to be self-perpetuating since these correlations keep a preponderance of parallel spin electrons in the ion's neighborhood. Second, since the repulsed electrons having spins antiparallel to the ion's are thus constrained and have less volume to move around in, their kinetic energy is increased (as in the description relating to Fig. 2.11, where the number of nodes in the wave function is proportional to the electron's kinetic energy). Hence the splitting in energy of the spin-up and spin-down bands. We should not imagine that the ion harbors the same group of d electrons of like spin in its neighborhood as a result of the above correlation, for the process is again a dynamic one—rather like those discussed in Sect. 3.4—in which the actual d electrons are constantly changing as a result of their itinerancy. It is just that the correlation effects keep in the ion's vicinity a surplus population of parallel-spin electrons that is in dynamic equilibrium with the rest of the d band's occupants.

These different views of ferromagnetism are compared schematically in Fig. 6.17. Both lead to an expected energy splitting of the d band in Ni— which is the crucial case, for reasons described above—of about 8×10^{-19} J (5 eV). This is in reasonable agreement with expectations from experiment, and so does not favor one view over the other. Fe and Co have not received as much attention as Ni, perhaps because they are not such crucial cases, but, as already pointed out, it is certain from the electronic configurations that must exist in them that the intraatomic Hund's rule coupling plays some part in their ferromagnetism.

Against this background of the possible microscopic origins of ferromagnetism, certain distinctive features resulting from the quasiparticles' motion in the d transition metals can be understood qualitatively as follows. First, the temperature dependence of the electrical resistivity of Ni above its Curie point is found to be not qualitatively different from that of other nonmagnetic transition metals at the ends of their series (notably nonmagnetic Pd). It is, therefore, thought that the same Mott scattering mechanism described in Sect. 6.4.1 dominates the resistivity mechanisms in paramagnetic Ni as in Pd. As the temperature is reduced in these cases, and with it the probability of this s-d scattering, so the resistivities of these metals decrease monotonically in qualitatively the same fashion. But for temperatures below the onset of ferromagnetism in Ni (630°K), the resistivity of this metal falls at a much greater rate with reducing temperature than is expected in the nonmagnetic phase. This is due largely to the fact that once the temperature falls below the Curie temperature (so that ferromagnetism

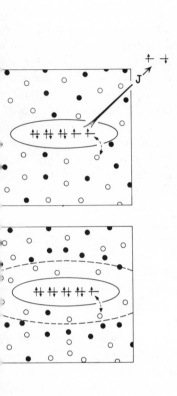

Fig. 6.17 Illustrating schematically the two views of ferromagnetism described in the text. The figures show the spin configuration of a single ion's *d* orbitals set against the background of the itinerant *d* electrons as they hop from one such ion to the next. (*Upper*) Ferromagnetism arises from ions for which the *d* orbitals contain at least two unpaired electrons (like the $3d^8$ configuration shown). There is time between the constant arrivals and departures in the orbital for the intraatomic coupling *U* (Type 1 of Fig. 3.9) to produce the parallel alignment of uncompensated spins required by Hund's rule. Hence the ion's nonzero magnetic moment that is required for ferromagnetism. (*Lower*) Here such a configuration cannot arise; the energy required to produce two vacancies in a *d* orbital is said to be too great. Configurations having only a single uncompensated spin are the source of ferromagnetism. Their persistent magnetic moment is assured because an ion having an instantaneous net spin surrounds itself with a correlation hole (shown dotted) from which most electrons of like spin (here solid circles) are excluded by the interatomic repulsion *U*. Consequently, electrons jumping into and out of the ion's orbital are drawn from a pool with a preferred spin orientation. Hence the ion's moment tends to be self-perpetuating.

ccurs and the Fermi surface splits in the manner suggested by Fig. 6.16*c*) and is continuously reduced, the spin-up levels approach closer to their full complement of five electrons. Consequently, there are fewer vacancies in the *d* band into which *s*/*p* electrons can be resistively scattered. Ultimately, when the spin-up band has no vacancies that persist for times comparable with the *s*/*p* lifetime τ, only half of the *s*/*p* electrons (namely, those with spin-down) are exposed to the risk of Mott scattering because at these lower temperatures the chance of an *s*/*p* electron changing its spin orientation in a collision with a phonon is very slight indeed.

This blocking of the Mott scattering for half the electrons in the *s*/*p* band cannot be the sole source of resistance in paramagnetic Ni, for each ion has a net localized spin due to the uncompensated electrons in its *d* orbitals and their disordered arrangement in the paramagnetic state produces electron scattering. (This source is known as spin-disorder resistivity; a subject encountered again in Sect. 6.5.) But since Ni and its nonmagnetic homologue show qualitatively the same behavior, as described above, it is presumed that the spin disorder contribution is dominated by the Mott contribution in this case. In the case of Fe exactly the converse is probably true. First, since the band structure of Fe is believed to have qualitatively the form of Fig. 6.16c, with vacancies in both spin-up and spin-down bands at all temperatures, it is clear that the blocking of Mott scattering for half the s/p conduction electrons in Fe cannot exist. Second, the temperature variation of the electrical resistivity of Fe above its Curie temperature is linear and, therefore, qualitatively different from the behavior typical of Ni and Pd. For this and other* reasons it is believed that the scattering from the spins' disorder is the dominant mechanism in the metal's paramagnetic range.

Turning finally to antiferromagnetic Cr and Mn, we come to two of the most complicated and least understood common metals in the periodic table. Consider first b.c.c. Cr. Its magnetic properties are unique among pure metals in the following respects. We recall that in a normal magnetic metal each ion has a net magnetic moment localized upon it. Depending upon the context, this can be regarded as the result of either a permanent vacancy in the ion's d (or f) shell (as in Sect 6.5) or an effective vacancy produced there by the constant hopping of itinerant electrons into and out of the ion, as described in preceding paragraphs. In the magnetically ordered state—whether ferromagnetic or antiferromagnetic—the directions of these individual moments are aligned cooperatively over long distances in the lattice in a strictly prescribed manner. A ferromagnet in its ground state has them all exactly parallel, and the classical antiferromagnet has essentially two interpenetrating lattices, each with a ferromagnetic alignment but opposed to one another. Whatever the case, a point is reached as the temperature is raised (known as the Curie temperature T_c for the ferromagnet and the Néel temperature T_N for the antiferromagnet) above which the thermal energies of the ions are able to overcome the exchange forces coupling their individual spins. The ordered arrangement then breaks down so that in the simplest view the direction of each ion's spin becomes randomized and no longer is related to that of its neighbors. But in Cr above its Néel temperature (311°K) experiment shows no evidence of any localized magnetic moment on the ions; either such moments do not exist in the magnetically ordered state, or they somehow disappear when the antiferromagnetism is destroyed. A second piece of experimental evidence points to the most likely of these alternatives: It is found that the period

* N. F. Mott, *Adv. Phys.* **13**, 325 1964, Sect. 14.6.

of the spin polarization* in the antiferromagnetic state is not commensurate with the period of the lattice measured along the same direction, as it must inevitably be in the classical antiferromagnet. A spin polarization that is out of step with the interatomic spacing in this way—the polarization repeats itself approximately every twenty-eight unit cells in the lattice—cannot arise from the ionic lattice; it must, therefore, originate in the metal's other structural unit: The itinerant electron system.

The contemporary view is that Cr does not possess localized moments in manner of classical ferromagnets or antiferromagnets, but rather that the effectively ordered moments are formed by the spin polarization of the fluid of itinerant electrons. This is the one convincing example currently demonstrated in a pure metal of the formation of a spin density wave (SDW) in the manner described at the end of Sect. 4.7. We pointed out there that this rather unusual instability possessed by the electron fluid, which leads to its spontaneous polarization in two sinusoidally varying components (Fig. 4.18) requires special conditions. Briefly, the explanation is that, when a spin density wave having wave vector $\mathbf{Q}$ is set up, so that the polarization of each spin component in the electron fluid varies sinusoidally with a period $2\pi/\mathbf{Q}$ in reciprocal space, an electron in a state with wave vector $\mathbf{k}$ can lower its energy by resonantly associating with that in the state $\mathbf{k} + \mathbf{Q}$. As in any resonance, the effect is most pronounced when the energies of the two electrons are nearly equal ($E_\mathbf{k} \simeq E_{\mathbf{k}+\mathbf{Q}}$) and in addition their common energy has here to be close to the Fermi energy. The whole effect is analogous to condensation in superconductivity (Sect. 4.6; indeed, the simpler mathematical models of the two effects are identical): Electrons in states $\mathbf{k}$ and $\mathbf{k}+\mathbf{Q}$ are mutually attracted and pair off in the triplet configuration by correlating their motion (pace the Cooper pairs) and the spectrum of the allowed energies of these quasiparticle pairs (which are actually excitons in Cr since the states at $\mathbf{k}$ and $\mathbf{k}+\mathbf{Q}$ are in that case electronlike and holelike, respectively) is split by an energy gap. T_N is thus the temperature at which this condensed state is broken up by the ion's thermal energy. This analogy between superconductivity and the formation of a spin density wave is only a mathematical one, however; physically, the two states are very different, for the quasiparticles in the magnetic case show none of the superfluid properties characteristic of those in the superconductive one.

To make the effect of a spin density wave most pronounced requires, therefore, the existence of two parallel pieces of Fermi surface separated by $2\pi/\mathbf{Q}$ and extending as long as possible in reciprocal space. Such an

* This is the distance in the lattice measured between positions where the direction of the net magnetic moment exactly repeats itself.

arrangement produces the maximum number of itinerant electrons at the same energy, and in states separated by a wave vector $\mathbf{Q}$, which can bene fit from the spin density wave's condensation. Such requirements are rarely met in practice; indeed, as we have already pointed out, it is only in b.c.c Cr (and perhaps b.c.c. Mn) where this favorable situation arises. A more de tailed description would go beyond our present scope, but we can note that for Cr there exists a hole sheet of its Fermi surface (having an octohedra shape and rather straight sides) that is separated from an oblate electron sheet by a wave vector commensurable with that deduced from the perio dicity measured for the metal's antiferromagnetism. It is, therefore, believed that a spin density wave formed by the correlation between the charge carriers on these two surfaces is responsible for the metal's magnetism. Its magnetic polarization $\mathbf{P(Q)}$ is found to be perpendicular to $\mathbf{Q}$ in the range $T_c > T > 115°K$, but this switches as the temperature is reduced through $115°K$ (which is known as the spin-flip temperature) to become parallel to $\mathbf{Q}$ (Fig. 4.18) at lower temperatures.

6.5   THE ROLE OF F BANDS

The least interactive and thus the least itinerant group of electrons dis tinguished in Sect. 2.3.3 and Sect. 2.3.4 is typified by the $4f$ electrons of the rare-earth series (Fig. 2.1 and Fig. 2.6). As is pointed out there, the interaction between such electrons on contiguous ions is so slight that they are generally assumed to have no appreciable itinerancy. Although only partially occupied, the $4f$ shells lie deep within the ion's interior and beneath the far more interactive $4p$, $5d$, and $6s$ shells. Consequently, there is no question of a direct contribution of such electrons to the metal's cur rent carrying capacity—as there is for electrons in the s/p or d bands of preceding sections—and there is correspondingly no question of the Mot mechanism (Sect. 6.4.1) contributing to the total scattering suffered by the metal's itinerant electrons. But the f electrons do, nevertheless, play an important role in this total scattering. Because of its partially occupied $4f$ shell, in which the intraatomic exchange (Type 1 of Fig. 3.9) aligns the residents' spins in accordance with Hund's rule (Sect. 3.2.5) each ion in the series possesses in general a net and very localized magnetic moment arising from the spins of the uncompensated electrons (Fig. 3.4). If in the solid each of these individual ionic moments can be exactly aligned along some arbitrary direction—a feature that can only be produced in a per fectly pure sample at absolute zero—there is no scattering from this per fectly ordered system of magnetic moments. (Just as the perfectly periodic potential of the ideal lattice produces no scattering of the itinerant electrons; Sect. 2.2.) But at any nonzero temperature, where each ion possesses a certain thermal energy that is able to partially randomize the orientation of its net moment, departures from this ideally ordered state are inevitable. The result is a contribution to the metal's electrical resistivity that is known as the spin-disorder resistivity (already encountered in Sect. 6.4.2).

As already remarked in Sect. 3.3.3, the rare earth metals show various forms of magnetism, including ferromagnetic, antiferromagnetic, and heliomagnetic ordering of their ionic spins. Often these states have relatively high Curie or Néel temperatures, indicating that a strong interaction exists between their ionic moments. Since the direct interaction between them (Type 1 of Fig. 3.9) is known to be extremely weak, however, it must be an indirect exchange interaction that is responsible, and this is generally believed to be the oscillatory RKKY interaction described in Sect. 3.3.3 and Sect. 3.4.3. The exchange coupling energy between the spin **s** of a conduction electron and the moment of an ion having a total angular momentum **J** is expressed in Eq. (3.41) as Γ ($\beta-1$) **s·J**, and a general expression for the energy of the coupling between two ionic moments $\mathbf{J}_1$ and $\mathbf{J}_2$ at $\mathbf{r}_1$ and $\mathbf{r}_2$ is given by the form

$$(\text{constant}) \ \Gamma^2 \ (\beta - 1)^2 \ F(\mathbf{r}_{12}) \ \mathbf{J}_1 \cdot \mathbf{J}_2 \tag{6.13}$$

where F ($\mathbf{r}_{12}$) is the oscillatory component in the coupling arising from the halos of spin polarization about each ion (Sect. 3.4.3). Subject to numer-

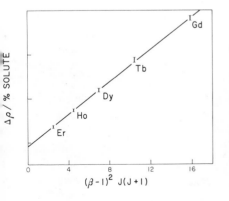

Fig. 6.18 Showing the dependence of the spin disorder component of an alloy's electrical resistivity upon the factor $(\beta-1)^2 \ J(J+1)$ predicted by Eq. (6.14). The ordinate is the change in the residual resistivity produced in pure Lu (which is nonmagnetic) by the addition of 1 atm % of each of the (magnetic) solutes indicated. The satisfying agreement with the predictions of Eq. (6.14) supports the view that spin-disorder scattering is correctly described in terms of the RKKY interaction discussed in Sect. 3.4.3. (After A. R. Mackintosh and F. A. Smidt.)

resistivity at temperatures well above the Curie temperature; it is of the ous simplifying assumptions that would not serve our present purpose to discuss in detail, an expression can be obtained* for the spin-disorder form

$$\text{(constant)} \frac{m^*}{E_f} \Gamma^2 (\beta - 1)^2 J(J + 1) \qquad (6.14)$$

where m^* is the dynamic effective mass (Sect. 6.2) of an itinerant electron. The spin-disorder component of the metal's resistivity is, therefore, predicted to be independent of temperature at high temperature $(T \gg T_c)$ and to vary as $(\beta-1)^2 J(J+1)$. This dependence has been confirmed (Fig. 6.18) from the extra resistivity produced when a small amount of one of the magnetic elements in the series Gd→Er is dissolved in a nonmagnetic matrix of pure Lu.*

6.6 CHAPTER SUMMARY

The salient features of the motion of the itinerant quasiparticles in the system consisting of the set of itinerant electrons and the lattice of fixed ions are best considered in terms of the classification first introduced in Sect. 2.3.4. Three degrees of itinerancy are distinguished that are exemplified by the electrons originating from s/p, d, or f energy levels in the atom. In the simplest metals only electrons originating from s/p energy levels are itinerant; typical examples being the monovalent alkali metals or Al. But when the possibility exists of itinerancy among d electrons, as in the transition series of d block of Fig. 2.1, then the situation is very much more complicated because the itinerant electrons are involved not only in the metal's current carrying properties but also in its magnetic ones. In spite of much effort in this field since the earliest times, an adequate description of the magnetic members of the d block metals still remains elusive. Finally, in the extreme case where the electrons' itinerancy is essentially zero even though the ion contains them in an incomplete shell, the f electrons of the rare earth series exemplify the influence on the metal's conductivity of highly localized magnetic moments.

* T. Kasuya, *Progr. Theoret. Phys.* (*Kyoto*) **16**, 58, 1956; P. G. de Gennes and J. Friedel, *J. Phys. Chem. Solids* **4**, 71, 1958.
* A. R. Mackintosh and F. A. Smidt, *Phys. Letters* **2**, 107, 1962.

RELATED READING

Specific research papers are referenced where cited in the text.

THE FERMI SURFACE

A. P. Cracknell and K. C. Wong, 1973, *The Fermi Surface* (Clarendon Press, Oxford). A generally advanced level description of this concept, although it is aimed at readers not necessarily specialized in the physics of metals.

W. Mercouroff, 1967, *La Surface de Fermi des Metaux* (Masson et Cie: Paris). An intermediate level text.

J. M. Ziman, 1962, *Electrons in Metals: A Short Guide to the Fermi Surface* (Taylor and Francis: London). A lucid and elementary introduction but less comprehensive in scope than Mercouroff's description.

D. Shoenberg, "Metallic Electrons in Magnetic Fields," *Contemporary Physics*, **13**, 321 (1972). A generally nontechnical introduction to Fermiology.

A. R. Mackintosh, "The Fermi Surface," in *Sci. Amer.*, July (1963) p. 110. An elementary account of how electrons in a metal behave.

ELECTRONIC MOTION

A. B. Pippard, 1965, *The Dynamics of Conduction Electrons* (Blackie and Son: London). An authoritative but rather difficult book that is not aimed at beginners.

F. Garcia–Moliner, 1968, "Electron Dynamics and Transport in Metals and Semiconductors," in *Theory of Condensed Matter* (International Atomic Agency: Vienna), p. 229. One of several lectures presented by different specialists at an international course that dealt at an advanced level with the movement of itinerant electrons in a magnetic field.

H. Ehrenreich, "*The Electrical Properties of Materials,*" *Sci. Amer.*, September (1967) p. 195. One of several relevant articles appearing in this special issue devoted to "Materials."

RIVAL VIEWS OF FERROMAGNETISM. Among the supporters of the view that Hund's rule coupling is primarily responsible are:

J. C. Slater, *Phys. Rev.* **49**, 537 (1936).

J. Friedel, *J. Phys. Radium,* **16**, 829 (1955).

J. C. Phillips, *Phys. Rev.* **133**, A1020 (1964).

N. F. Mott, *Adv. Phys.* **13**, 325 (1964).

C. Herring: *Magnetism*, Vol. IV, Ed. G. T. Rado and H. Suhl (Academic Press: New York) 1966.

In the opposite camp are:

J. Kanamori, *Progr. Theoret. Phys.* (Kyoto) **30**, 275 (1963).

J. Hubbard, *Proc. Roy. Soc. Ser.* **A276**, 238 (1963); *ibid*, **A277**, 237 (1964).

The Shapes of Atomic Orbitals and Their Electron Density Distributions

It is pointed out in Sect. 2.3.3 that the various s, p, d, and so forth orbitals under discussion are solutions of the Schrödinger equation for the one-electron or hydrogenic system. Such a wave function, say Ψ_{n,l,m_l}, has an important property: It can be written as the product of two separate functions, the first being a function only of the radial distance of the electron from the nucleus and the second being a function only of the angular coordinates of the electron with respect to the nucleus. Thus

$$\Psi_{n,l,m_l} = R_{n,l}(\mathbf{r}) \; Y_{l,m_l} (\theta,\phi) \tag{A.1}$$

where $R_{n,l}$ and Y_{l,m_l} are known as the radial and angular functions, respectively. ($\mathbf{r}$, θ, and ϕ are the usual spherical coordinates.) Although some idea of the physical meaning of Ψ follows from the fact, already emphasized in Sect. 2.3.3, that $\Psi^2 dv$ is the probability of finding an electron in a volume element dv, it is not very easy to give a simple graphical representation of Ψ because it is a function of θ, ϕ and $\mathbf{r}$. In fact, to do so requires a four-dimensional drawing showing simultaneously the variation of Ψ with $\mathbf{r}$, θ, and ϕ.

Consequently, it is the common practice to show the radial and angular functions in separate drawings. In each case there is a choice between alternate presentations. The radial dependence of the wave function can be illustrated by plotting as a function of distance from the nucleus the radial wave function $R_{n,l}$, its square (which is known as the radial probability density), or the corresponding quantity $4 \, \mathbf{r}^2 R^2_{n,l}$ (which is known as the radial distribution function and is the probability of finding an electron in a spherical shell of thickness dr and radius r). To illustrate the angular dependence, on the other hand, is even more of a challenge because some of the functions Y_{l,m_l} —which are known as spherical harmonics—are complex; in other words, they have imaginary and real parts and cannot be represented in simple real space. But in the absence of any magnetic field applied to the atom, the Y_{l,m_l} solutions of Schrödinger's equation obtained for a fixed value of l are degenerate in energy (Sect. 2.3.1). Furthermore, quantum mechanics shows that it is a general property of a set of such functions that any linear combination of their members is an equally good solution of the same Schrödinger equation.

This property has some important consequences. First, it means that real expressions can be constructed in this way from the complex Y_{l,m_l} to give a convenient representation in real space of the angular variation of Ψ. Second, it means that a chosen representation will not be unique: There will be a choice between alternatives that amount to the same thing from a mathematical point of view but that may vary in convenience in the particular context.* Frequently the choice reduces to the convenience of having either equivalence or independence among the degenerate angular functions corresponding to a given l value. The first refers to geometrical equivalence. In other words, in addition to having (of necessity) equal energies, equivalent orbitals have the same shapes in space; the only distinction between them is the directions in which they point. The second refers to independence in the mathematical sense: An independent orbital cannot be expressed as a function of the others in the set.

Let us turn now to the representation of the angular functions for specific values of l (Fig. A.1 and Fig. A.2). The s, p, and d orbitals (corresponding to $l = 0$, 1 and 2, respectively) are illustrated in numerous chemistry textbooks and are familiar to students of the subject. The f orbitals, however, occur less prominently, probably because they are believed to play little significant part in chemical bonding. But they are important in the present context because they determine the magnetic properties of a significant series in the periodic table. As Fig. A.1 shows, an s orbital, which has just one angular distribution corresponding to $m_l = 0$, is found to have spherical symmetry; the variation of the charge distribution resulting from the electrons in these orbitals is independent of the chosen direction in space. For a p orbital there is a set of three angular distributions corre-

---

* Even when a particular option has been selected, there remains the choice of depicting either the angular function or its square—just as in the radial case. Apart from slight differences between the shapes of the two, the former representation has regions of differing sign, whereas the latter is evidently positive everywhere. The sign of the orbital has no physical significance, of course, but it is crucial in predicting the effects of hybridization of orbitals of different symmetry (as Sect. 2.3 describes). Consequently, for present purposes it is more appropriate to illustrate the straightforward angular function rather than its square.

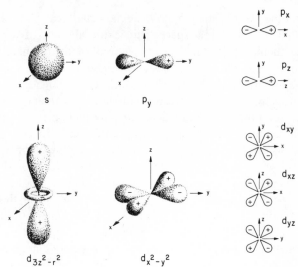

Fig. A.1 Representation of the angular parts of the *s*, *p*, and *d* hydrogenic orbital functions.

sponding to the $2l + 1$ possible values of m_l, and it turns out that in this case a set can be constructed whose members are both independent and equivalent. They are a natural choice for representation, and these orbitals, known as the p_x, p_y, and p_z, are illustrated in the figure.

For *d* orbitals the choice is less straightforward. The problem is that it is not possible to obtain a set of five orbitals that are both independent and equivalent. An equivalent set consists of the designations* d_{xy}, d_{yz}, d_{zx}, $d_{y^2-z^2}$, and $d_{z^2-x^2}$ (where the shapes of the two last named, which are not independent functions, are obvious by analogy with that of the $d_{x^2-y^2}$ shown in the figure). A linearly independent set can be constructed from these by taking the sum and differences of the two dependent orbitals to give instead orbitals of $d_{x^2-y^2}$ and $d_{2z^2-x^2-y^2}$ (or $d_{3z^2-r^2}$) symmetry (the latter is conventionally shortened† to the designation d_{z^2}). However, this is at the cost of equivalence since, as Fig. A.1 shows, the $d_{3z^2-r^2}$ orbital has a different shape from the other four. Finally, for *f* orbitals there are two real sets in use and each has its particular convenience. The first, known

---

* This is a standard notation where the polynominal subscript refers to the form of the angular variation of the orbital expressed in Cartesian coordinates.

† The more complicated polynominals in the *d* and *f* orbitals are frequently simplified by dropping terms containing only r^2 ($= x^2 + y^2 + z^2$).

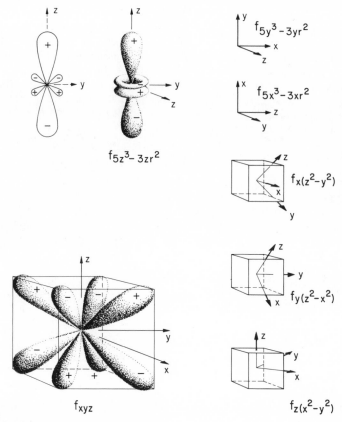

Fig. A.2   Representations of the angular parts of the f hydrogenic orbital functions.

as the general set, consists of the designations $f_{5z^3-3zr^2}$, $f_{5xz^2-xr^2}$, $f_{5yz^2-yr^2}$, $f_{z(x^2-y^2)}$, f_{xyz}, $f_{x^3-3xy^2}$, and $f_{3yx^2-y^3}$. This set turns out to be particularly useful when considering an atomic environment that does not have cubic symmetry. The second set, which is derived from the above, is known as the cubic set since it is particularly applicable in cubic systems. The cubic set retains the first, fourth, and fifth of the above orbitals, but the rest are combined to produce four new orbitals designated $f_{xz^2-xy^2}$, $f_{yz^2-yx^2}$, $f_{5y^3-3yr^2}$, and $f_{5z^3-3zr^2}$. The angular dependence of this cubic set is shown in Fig. A.2. We see that among the seven orbitals two subsets can be distinguished: The first contains three equivalent orbitals of the type $f_{5z^3-3zr^2}$, and the second has four equivalent orbitals of the type $f_{xz^2-xy^2}$.

RELATED READING

H. G. Friedman, G. R. Choppin, and D. G. Feuerbacher, *J. Chem. Educ.* **41**, 354 (1964); C. Becker, *Ibid.* **41**, 358 (1964). These publications describe pictures and models of the angular portions of the hydrogenic wave functions.

A. Streitwieser and P. H. Owens, 1973, *Orbital and Electron Density Diagrams* (Macmillan: New York). Beautiful examples of three-dimensional drawings of the electron probability distributions in hydrogenic orbitals and their hybridized resultants in simple molecules.

AUTHOR INDEX

319

SUBJECT INDEX